Library of
Davidson College

The Cybernetic Society
PUES-15

Pergamon Unified
Engineering Series

Pergamon
Unified Engineering
Series

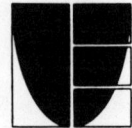

GENERAL EDITORS

Thomas F. Irvine, Jr.
State University of New York at Stony Brook

James P. Hartnett
University of Illinois at Chicago Circle

EDITORS

William F. Hughes
Carnegie-Mellon University

Arthur T. Murphy
Widener College

William H. Davenport
Harvey Mudd College

Daniel Rosenthal
University of California, Los Angeles

SECTIONS

Continuous Media Section
Engineering Design Section
Engineering Systems Section
Humanities and Social Sciences Section
Information Dynamics Section
Materials Engineering Section
Engineering Laboratory Section

The Cybernetic Society

Ralph Parkman
California State University
at San Jose

Pergamon Press Inc.
New York · Toronto · Oxford · Sydney · Braunschweig

PERGAMON PRESS INC.
Maxwell House, Fairview Park, Elmsford, N.Y. 10523

PERGAMON OF CANADA LTD.
207 Queen's Quay West, Toronto 117, Ontario

PERGAMON PRESS LTD.
Headington Hill Hall, Oxford

PERGAMON PRESS (AUST.) PTY. LTD.
Rushcutters Bay, Sydney, N.S.W.

VIEWEG & SOHN GmbH
Burgplatz 1, Braunschweig

Copyright © 1972, Pergamon Press Inc.
Library of Congress Catalog Card No. 78-185338

All Rights Reserved. No part of this publication may be reproduced, stored in a retrieval system or transmitted in any form, or by any means, electronic, mechanical, photocopying, recording or otherwise, without prior permission of Pergamon Press Inc.

Printed in the United States of America
08 016949 X (H)
08 017185 0 (S)

Contents

Preface		ix
Chapter 1	**Introduction**	1
	References	9
Chapter 2	**Technology in the Industrial Age**	10
	PART I THE INDUSTRIAL REVOLUTION IN ENGLAND	10
	Mechanical Technology – Textiles	11
	Mechanical Technology – The Steam Engine	17
	Metallurgical Technology	23
	High-pressure Engines	25
	New Energy Sources	27
	PART II THE ADVENT OF MASS PRODUCTION	29
	The "American System"	33
	PART III THE DEVELOPMENT OF THE COMPUTER	40
	Charles Babbage	46
	The Automatic Computer	52
	Computers and Logic	56
	The Age of Communications	58
	References	59
Chapter 3	**Industrialization, Cybernation and the Working Man**	61
	Post-World War II Employment Problems	62
	The Knowledge Industries	63

	The Worker in the United States: 1950–1975	66
	Cybernation and Industry	66
	The White-collar Worker	68
	The Blue-collar Worker	70
	The Technological Society, Unemployment and the Poor	77
	England – Seeds of Social Discord	79
	Economic Theories and Industrial Change	83
	The Laboring Class in Victorian England	85
	Seeds of Environmental Destruction	88
	Early American Patterns of Industrialization	89
	Contemporary Economic Concepts and Unemployment	90
	Structural versus Aggregate-demand Unemployment	95
	Work and Leisure	98
	The Guaranteed Income	100
	References	102
Chapter 4	**Technology and Government**	**105**
	The Winning of Elections	108
	The Maintaining of Government	112
	Urban Transportation	114
	Economic and Social Indicators	135
	Technology Assessment	139
	Data Centers and the Issue of Privacy	145
	The Technological Elite	147
	References	155
Chapter 5	**The Systems Planners**	**158**
	Definition of a System	159
	Classes of Systems	159
	Systems Engineering and Operations Research	162
	Systems Problems and Techniques	167
	Systems Design for Social Systems – Beginnings	194
	A Critical Look at the Systems Approach	197
	References	199
Chapter 6	**Cybernetics – Control and Communication**	**201**
	Feedback Control	203
	Communication	205
	Telecommunication and the Two-State Code	208

	Information and Choice	209
	Unlike Alternatives	211
	Information and Entropy	212
	The Relevance of Cybernetics	214
	References	218
Chapter 7	**Man's Mind and the Computer**	**220**
	The Creation of Intelligent Beings — Early Concerns	221
	Mental Prodigies and Idiot Savants	223
	What is Intelligence?	228
	The Brain and Neural Networks	230
	Perceptrons	233
	Game-playing machines	236
	Simulation of Human Thought	243
	The Language Machines	244
	Machine Translation	245
	Self-reproducing Machines	251
	Are the Robots Coming?	253
	Mind and Matter	255
	References	259
Chapter 8	**Technology, Science and the Arts**	**262**
	The "Two Cultures"	263
	What is Art?	266
	The Relationship of Science to the Arts	271
	Technology and the Arts	283
	Early Interactions Between Technology and Art in the United States and Europe	285
	Technology and Art in the 20th Century	292
	The Bauhaus	293
	Dadaism and Surrealism	295
	Art in the United States	297
	Electronics and Music	301
	Computers and Music	303
	Computers and the Visual Arts	307
	Computers and Literature	309
	Outlook for the Arts	311
	References	313
Chapter 9	**Technology and Education**	**315**
	Early Developments in Theory and Methods of Learning and Instruction	318

	Educational Theories and Methods in the 19th Century	321
	The Monitorial System	322
	The Psychologizing Influence in 19th Century Education	323
	Theories and Practice of Learning and Instruction in the 20th Century	326
	Educational Technology in the 20th Century	333
	References	356
Chapter 10	**Intimations of the Future**	**358**
	Early Views of the Future	359
	Methods of Technological Forecasting	362
	The Cybernetic Society	374
	References	381

Index 383

Preface

The aim of this book has been to bring together, in some coherent way, facts and ideas which help give perspective to a study of man's role in an increasingly technological environment.

In 1964 I began, with several colleagues at San Jose State College, to develop an interdisciplinary course examining various aspects of the interactions between technology and man. This was a relatively early attempt by a school of engineering to ally itself with other disciplines and thereby broaden channels of communication connecting engineering, the humanities and arts, and the social and physical sciences. Until very recently curricula in none of these major disciplines have dealt adequately with the many ways in which technology and science affect the organizing and functioning of human societies.

Our initial attention was primarily directed toward the implications of computers and automated processes because at that time it was rather generally felt that the new information technologies would radically transform the lives of everyone in the Western nations. So they yet may, although it has become increasingly apparent in the intervening years that many of the more optimistic (or pessimistic) projections of highly evolved decision making systems, at least, are proving far more difficult to realize than was at first anticipated. Social change is indeed proceeding at a precipitous rate and different forms of social and political institutions are emerging; but it is not at all clear to what extent these forms are due to new technologies or how much they still relate to unresolved problems and anxieties carried over from the beginnings of the industrial age. At the present stage of understanding one can only attempt to organize some of the information by which the fact of change is determined, thereby making it more useful to students from all disciplines. That is what I have tried to do, although the selection of subjects from the vast literature of social and technological change, as well as the depth to which each could be explored, has had to be restricted because of my own limitations in addition to those of space. I should like to think that the general student will be encouraged to turn to the more specialized studies relating to technology and society with a heightened appreciation for the significance of the ties linking them together.

I have learned much from guest teachers such as Alice Mary Hilton, R. Buckminster Fuller, R. Wade Cole, W. H. Ferry, David Kean, Louis Fein, Louis Kelso, Michael Novak and many others who, over the years, have visited my classroom to interact with the students. To avoid egregious error in preparing the manuscript I have sought the help of various colleagues and friends when I felt the need. Gordon Chapman, Peter M. Dean, Edward A. Dionne, William D. Donnelly, Louis Fein, Joseph Gans, Donald Garnel, William George, Duke Hatakeda, Albert H. Jacobson, Jr., Rajinder P. Loomba, Gene Mahoney, Ron J. McBeath, David C. Miller and Robert Newick have all made useful suggestions. The errors that still remain are my own responsibility.

Special acknowledgment is owed to former Dean of Engineering, Norman O. Gunderson, without whose foresight and steadfast support the idea of such an interdisciplinary course, originating in engineering, would never have succeeded, and to S. P. R. Charter who, as co-teacher and confrere, has supplied a virtual treasure of seminal insights.

Additional acknowledgment is due to Mrs. Phyllis Ashe for typing a part of the manuscript.

Finally, I would like to thank my wife, Yetty, for her unfailing help and encouragement, and my children, Phil, Judy and Donny for accepting, with a minimum of reproaches, the need for deferring good times together over the past three years.

Los Gatos, California RALPH PARKMAN
August, 1971

1
Introduction

Revolutions are not all alike. The sudden, violent outburst against long-festering conditions, typified by the French Revolution, is an example of one kind. The active participants and the would-be unengaged were acutely, and often tragically, aware of the wrenching dislocations they had to endure. The period of chaos and bloodshed which is identified as the revolution itself, was relatively short, although there were many years of causal events leading up to it, and other years of reaction following.

The Industrial Revolution, of which the French Revolution was historically a part, took much longer to unfold. A slow building up of insistent social and economic pressures occurred over many generations prior to the latter half of the 18th Century – sufficient to distort the old structures to their limits of toleration. A cascade of mechanical inventiveness, concentrating centuries of technological and scientific preparation, breached the boundaries of traditional organizations and new forms began to take their place. Most of these new forms were not to reach an equilibrium, but, notwithstanding, the general effect was to erase previously existing limitations to economic growth and to increase the material wealth of the majority of people in industrialized nations.

The history of technology provides many instances of the transforming power of technological innovation. The plow, the cannon, the printing press and the steam engine were all potent forces for change. Human events surely would have progressed on a different timetable and very likely in other directions had these devices not been invented; or (since the social milieu may have made some new technologies inevitable) had the functions they served been satisfied by other means. But the earliest technologies, products of cut and try, flowered sufficiently slowly through decades or centuries for the affected cultures to make a reasonable adjustment to them. This has become less true in recent times. Perhaps man's ability to adapt bears a direct relation to the immediacy of his involvement in the functioning of his technologies. In any case, as technologies proliferate and become more complex, the pace of change in the structures and values of society lags tangibly behind the need for new approaches.

There are those who say that new technologies have now precipitated a second Industrial Revolution leading to a post-industrial age whose emerging characteristics appear fundamentally different from anything society has faced before. Others dispute this. It is never easy to determine to what extent social changes represent logical, even (perhaps only in hindsight) inevitable, extensions of past developments. Generally clear to students of the industrial society, however, is the fact that in the middle years of the 1940s technological advances of surpassing importance took place in the United States, and as a result, many of our institutions have been indelibly influenced.

The first of these developments was the successful fissioning of the atom in 1945, brought about by an overwhelming concentration of scientific and engineering resources. The primary effects thus far have been twofold. Nuclear fission, after two decades, is now promising to prove economically competitive with older methods as a source of industrial power in a world of shrinking natural energy reserves. It has also become apparent that every potential confrontation involving great nations is monitored by the specter of nuclear devastation. National policy makers are trying to adjust to this reality as they bring physical and social scientists into their councils.

If diplomacy between nuclear powers should fail, then any speculation about a post-industrial age becomes irrelevant. The nuclear stalemate is the minimum prerequisite for a sensible discourse on the nature of the machine society. Accepting this, one can then turn to a second great technological landmark of the 1940s, the development of the electronic computer.

In 1946, when Eckert and Mauchly built the ENIAC (Electronic Numerical Integrator and Calculator) there were not many who were perceptive enough to foresee the range of consequences, many of which even now are only dimly apparent. And yet, some 100 years before, Charles Babbage, who devised many of the original concepts of digital computers, had written, "As soon as an analytical engine exists, it will necessarily guide the future course of science." He might have added that the manner of life in a science/technology dominated society would be changed as well. The computer, as it controls automatic production methods, processes information, and in general assists in the systematization of our institutions, may indeed be introducing momentous transformations into the life of Western man, although some of the more euphoric, early projections of computer usage are still a long way from becoming reality.

Actually, of course, the computer is one major link in a huge complex of public and private communications media that include television, (discovered before World War II, but not utilized commercially on a large scale until the immediate post-war years), and other electronic devices for replicating visual symbols or sound. It is this communications complex which is central to the late 20th Century dynamic of change.

A primary objective of the following chapters will be to try to describe the nature and significance of important technological innovations (with some emphasis on the role of the computer), relating them to the spectrum of human concerns. Many of these concerns are strongly rooted in the history of technology and science. For our purposes we shall distinguish between these two words, which are often confused, by defining technology as man's use of devices or systematic patterns of thought and activity to control physical phenomena in order to serve his desires with a minimum of effort and a maximum of efficiency. Contemporary science is a way of thinking based on experimentation, observation and theoretical speculation to help man understand himself and his total environment.

Chapter 2 traces the evolving utility of machines from the time that man discovered potent new ways of converting natural forms of energy to useful mechanical work, and it leads up to developments in computer technology which make it possible to eliminate the need for humans to perform certain kinds of routine mental labor. The decisive elements of modern technology with which we shall be concerned are energy and its transformation, organization or synchronization, and information. The separate parts of this chapter have been arranged to reflect this division.

The unfolding picture takes on sharp relevance to the current scene with the beginnings of the modern factory system in the late 18th and 19th Centuries. From the first, the working classes felt a pervading fear that the proliferation of machines would, in time, cause widespread, permanent unemployment. The coupling of the computer to mass production systems has revived and intensified this fear, not only among workers in the mechanical industries, but also among many "white-collar" employees, as it becomes involved in information processing situations of increasing intricacy. Chapter 3 will examine the problems of industrial job displacement, unemployment (or underemployment), and poverty from the time of the first Industrial Revolution to the present cybernated era, and will consider some of the economic and political solutions which have been proposed. It will also focus attention on the large increase in the numbers of people who are being educated in the intellectual professions, and explore some of the questions this raises.

Unemployment was only one of the vexing social dilemmas attending the creation of the technological society. The results of the initial attempts to systematize the organization of men and mechanized equipment were complex and convulsive. It is true that the daily stint of hard manual effort was reduced for many workers, and their chances of gaining material rewards were enhanced. On the other hand, there was an appalling abuse of common humanity in the economically enforced labor of women and children; a growing sense of trauma induced by the tyrannous monotony of machine-tending; and a polluting of the environment brought about by concentrations of industries that is still part of our heritage. Then, as now, bitter controversy fed upon the conflicting claims of various strata of society to the bounty of industrial progress. Now, as then, thoughtful men of good will seek to find the root causes of maldistribution and discord.

A critical political and philosophical question is: to what extent can the resources of government be employed to redress the ills of the technological society? The broad-ranging social problems with which the government leaders at all levels—local, state and federal—must grapple may seem to defy rationalization, but the most useful approaches need the aid of technology at the very least for gathering and winnowing information. Maintaining open communication channels and coping with multiplying chores of record-keeping require data processing machines to maintain even minimal coherence among governmental functions. Chapter 4 takes up some of these points by examining ways in which technology contributes to patterns of social change, and discussing how political leaders may try to employ current technologies, and the insights of scientists and technologists, to keep the processes within manageable bounds.

It is not only the devices themselves which have been proposed as tools to help find rational solutions to the problems of the times, but also some of the methodologies which have proven so helpful in the design and management of large technical systems. Chapter 5 looks at some aspects of these methods and contrasts the possibility of applying the "systems approach" to contemporary social structures with much earlier attempts to find a core of scientifically describable order in the interactions between men in society.

Man is a system and human societies are also systems; but one of the causes of difficulty and of paradox, as we try to trace relationships between technology and the life of man, is that discoveries made about man-in-the-group may not be very useful for describing man-the-individual. Although each human mind has developed as a social organism, it is also one of a kind, and thus, in the most important sense, unmeasurable. Collectively, men are measurable; there is a valid presumption that they can be studied to a degree by the same kinds of statistical methods used successfully to describe other physical systems. When human organizations are likened to machines, the parallels are often legitimate. If the individual human is thought of as a machine, then it must be a very strange and special kind.

The technologist, by nature, tends to feel that control is the end purpose of understanding. Without controls a system may descend quickly into disorder. There are readily understood principles which permit control of relatively simple inorganic machines, but the more complex systems of which man is a part have, as yet, no fully developed control theory. However, computers, as they have begun to replace certain human functions in man-machine systems, show some similarities to the human nervous system. Norbert Wiener was one of the first scientist-philosophers to study the apparently unifying principles connecting nominally disparate systems. From his reflections, and the contributions of his associates, came "cybernetics," which he described as the science of control and communication in the animal and the machine. He has stated in his book, *The Human Use of Human Beings*[1], "... society can only be understood through a study of the messages and the communication facilities that belong to it; and ... in the future development of these messages and communication facilities, messages between man and machines, between machines and man, and between machine and machine are destined to play an ever-increasing part." In the attempt to understand relationships between organic and inorganic systems, cybernetics has the potential for providing useful insights. Some of the possibilities are examined in Chapter 6.

The notion that there are machine processes that bear ever closer resemblance to human thought disturbs some people and exhilarates others.

Computer specialists take divergent attitudes not only about whether machines will be able to exhibit intelligent behavior, but also about what intelligence really is. Answers to this kind of question will not, it seems, come soon but scientists and philosophers cannot help seeking for them at a time when new developments in engineering technology and the physical and natural sciences seem to have the potential for reshaping man's place in the universe and altering his idea of himself. The nature of some of the arguments about the uniqueness of the human mind *vis-à-vis* "intelligent machines" is taken up in Chapter 7, along with descriptions of certain of the advances made and problems encountered in devising machines which simulate specific human characteristics.

No matter how convinced they may be about the potential of the computer to perform certain mental chores, few people are willing in the end to relinquish the notion of man's having inimitable gifts, of which the creative imagination is perhaps the most impressive sign. Artists, scientists and technologists all have the impulse to create, and the results of their efforts have been among the monuments of human progress. However, to growing numbers of the young, progress in modern technology and science has been predominantly of negative effect, and the outcomes inhumane. Among the current school and college generations, an unmistakable trend away from studies in the engineering and physical sciences has occurred. Many other reasons have been suggested for this phenomenon, but whatever they are, there is no mistaking the serious rejection of technology or physical science as the basis for a life's work at the same time as enrollments in the social sciences and the arts have soared. Young people who elect the social sciences presumably hope to find there the clues for dealing with social disorganization and individual alienation in a technological world, while many of those moving into the arts, besides assuming that only thus can their creative dispositions be satisfied, seem impelled to look for meaning outside of the so-called rational patterns of organization. Those in the latter group are apt to encounter certain ambivalent relationships between art, technology and science, as Chapter 8 discusses.

The art of a society has often been among the most responsive indicators of mutations in its cultural processes. Science has been, too, although it is quite likely that science causes, more than it mirrors, social change. Art is a sensitive reflector of humanness—humanness capable of error and the creation of beauty and endless surprise. Since technologies, as embodied in the machine and its functions, are designed (even when closely engaged with valid human purpose) for efficiency and error-free performance, few conjunctions would seem less likely than that between art and technology. Before this century, at least, the fine arts exemplified man's self-awareness and sense of beauty, and would surely have been thought antithetical to the mechanized image. Today many young artists find it helpful to turn to the current technologies and the newer materials as they experiment to find the most satisfying means of fulfilling themselves. The experiments include using rational technologies irrationally as one way of rejecting the ordinary uses to which they may be put. As it happens these kinds of experiments are not fundamentally different in kind from those which some of their predecessors tried fifty years and more ago. Other uses of technology in art, however, are more purposefully directed toward creating new visual and aural experiences which would never otherwise be possible. The bringing of new experiences to light is one of the many definitions of art. Modern devices of communication and the commercial forces that support them also have a telling influence on the world of the artist. He must often work in a tense commercial milieu, and here, paradoxically, the very richness of opportunity sometimes reduces the time he can find to explore the range of his creative potential and thus retain his individuality.

Issues of individual freedom versus group order, surface again and again in studies of technology and man — never more obviously than when one examines the problems of educating citizens to meet the needs of the contemporary world. The average educator would probably express the basic aim of the educational system as that of preparing young people for living in a democratic society — assuming that the technological society is also democratic. In some ways it is; in more ways it could be; but it is never ideally so. Still, no matter how the aims of education are viewed, the task of the educator is overwhelming. In the United States, the affluence fostered by technology theoretically offers each young person an opportunity to make the most of his abilities through education. Instead the opportunities are unequal and poorly distributed. Yet even when they are taken, the programs which are designed to prepare students for jobs with a high content of technical knowledge quickly obsolesce because of the rapid changes in the technological world. The educator is continually being forced to re-evaluate his task. He sometimes comes to believe that the basic structures of education are inadequate; and this conclusion is often shared, even influenced, by a sizable proportion of the young who question the aims, and even the benefits, of the technological culture.

Should, or can, the structure of education be remade, given the existing or potential resources? Can the computer and other devices and methods of instructional technology, applied according to accepted precepts of learning theory, help restyle educational systems to meet the demands of a post-industrial society? Are there really effective, laboratory-tested solutions to basic educational problems waiting to be put to work on a large scale? These are questions which are addressed in Chapter 9. The root need seems to be for a system which confers the maximum number and kinds of quality opportunities directed toward valid aims, along with the maximum freedom for the qualified student to pursue an individual study track with the resources and at the pace he finds the most congenial. Proponents of educational technology suggest that a systematic utilization of imaginatively programmed devices and methods can go a long way toward meeting those requirements. Some other educators are pessimistic that the massive education establishment can ever overcome the inertia and acquire the mobility to meet rapid change, and there are many more who take intermediate positions between these two.

Whether the topic is education or something different, one is struck, when sampling the literature on technological and social change, by the strong orientation toward the future encountered in so much of the writing. This is not surprising. Scientists and technologists are notably future-directed, and the average person now assumes that the course of each of the most fundamental patterns of his life is being determined by how they think and what they do. The most instinctive reactions are of distrust or hope. The distrust springs from the feeling that the hubris of technical man has set in motion new and destructive forces which are essentially out of control. Technological and scientific advances have always aroused such fears and, although the most vivid of them have not come to pass, post-World War II developments seem to give them a more ominous validity. Still, it would be a poor world if men lost all faith in the future. Many of those who have any involvement with science, especially, incline to the optimistic belief that seemingly intractable human problems for which partial theoretical solutions, at present, yield only tentative answers will, in the future, succumb in fact to those kinds of solutions finally perfected. They would not accept the irony of La Rochefoucauld's maxim: "Philosophy triumphs easily over past evils and future evils; but present evils triumph over it."

The complexity of the new society surely makes it difficult to anticipate what the future holds. Nevertheless, the urge to predict the course of events is compelling, and it has produced a class of scholars whose primary occupation is the forecasting of the future, or, more correctly, finding better ways to do it. The "futurists," whose activities are described in Chapter 10, seek to devise methods which essentially remove emotion and bias from forecasting. Their intention is constructive. By and large they recognize that, while the future has already begun, it may take alternative forms. A knowledge of the alternative probabilities can permit recommendations having a good likelihood of leading to the most desirable outcomes or of avoiding the most menacing.

The most pressing questions to be considered in the following pages, then, are, at the end, speculations about man's potential for bringing himself into harmony with his technological environment in the foreseeable future. If there is cause for hope that he can, it may very well come from recognizing the beginnings of a new revolution of understanding which is in some ways a return to older wisdoms. Man will never find ideal solutions for an ideal world, but he is capable of smaller victories. His most significant victory of all would be to recognize that he can become master of his technological environment or, indeed, his total environment, to the extent that he becomes master of himself, for he and his environments are inescapably linked.

REFERENCES

1. Wiener, Norbert, *The Human Use of Human Beings*, Doubleday-Anchor, Garden City, N.Y., 1954, p. 16.

2

Technology in the Industrial Age

"It is the Age of Machinery, in every outward and inward sense of that word."
THOMAS CARLYLE, *Signs of the Times*

PART I   THE INDUSTRIAL REVOLUTION IN ENGLAND

Current events live off happenings from the past and nourish those yet to come. So if we say that the twenty-five years between 1765 and 1790 are a crucial jumping-off point into the era of fundamental change in Western life which has been labelled the Industrial Revolution, we do so realizing there is a continuum; that the developments in those years were integrally related to other important ones that occurred earlier, and some still later in the following century.

The reasons for considering this period as a critical time of change are, however, impressive. The American Declaration of Independence in 1776 and the French Revolution in 1789 foretold new concepts of man's self-governance. In 1776, too, Adam Smith published his *Wealth of Nations* and thereby provided a resounding sanction for the essential right of industrial and commercial entrepreneurs to pursue their quest for growth and profits unfettered by government interference.

From the standpoint of technology, notable progress was made in many fields. Some of the advances had identifiable beginnings earlier in the century; but the spinning jenny (1768), the water frame (1769), and the mule (1779), the boring mill (1774), cotton printing machinery (1783) the puddling process for wrought iron (1774), the first useful threshing machine (1786), Watt's steam engine (1776), and the method of manufacturing of interchangeable parts (1785), were all devised or perfected at this time. Represented in this incomplete list are inventions significant to the growth of the textile industry, agriculture, metallurgy and also to the basic manufacturing processes of the new factory system.

MECHANICAL TECHNOLOGY — TEXTILES

Perhaps the single industry upon which England's growth in the 18th Century most depended was the production of textiles, particularly cotton goods. Besides being basically a cheaper material than wool, linen or silk, cotton is easy to print and wash, and its natural elasticity lends itself to fabrication by mechanical methods. At the beginning of the century, however, textile making was still in a comparatively primitive state. The arts of spinning and weaving were tasks for human arms and fingers much as they had been since long before the birth of Christ. We have no sure way of determining how soon man began making fabrics, but twisting strands of natural materials into lines, and working lines into nets for fishing and trapping, baskets for containing, and fabrics for wearing were very early crafts. There are, in the writings of Homer, Confucius and the Book of Genesis, references describing spinning and weaving as already venerable arts. Herodotus (445 B.C.) speaks of cotton in India: "They possess, likewise, a kind of plant which instead of fruit produces wool of a firmer and better quality than that of sheep: of this the Indians make their clothes." There had been some minor processing inventions in the intervening centuries, such as the 15th Century flyer which, when added to the spindle of a spinning wheel made it possible to spin and wind at the same time. Still the methods in use, though capable of producing beautiful fabrics, would never have sufficed to supply the demands of a rapidly expanding market.

We have stated that many, though not all, of the mechanical inventions which spurred the Industrial Revolution came toward the end of the century. One of the notable exceptions was the fly-shuttle of John Kay, which helped transform the weaving trade. Weaving consists of crossing and interlacing a series of strands of some flexible material into an expanded whole. Plain cloth, for example, is a woven product in which the warp, or longitudinal thread, and the weft, or transverse thread, are the same, or approximately the same thickness. Many different kinds of woven materials can be made by varying the thickness of thread or the intervals of intersection. Kay, the son of the owner of a woolen mill, worked for his father and made a number of improvements in existing processes before his most famous invention which he patented in 1733.

His shuttle, running on four wheels, carried the weft through the separated warp threads so that the weaver with a slight pull at a cord could move the shuttle from side to side for a distance greater than his arm span. (*See* Fig. 2-1.) Whereas, before, weavers had to stand on each side of the wider

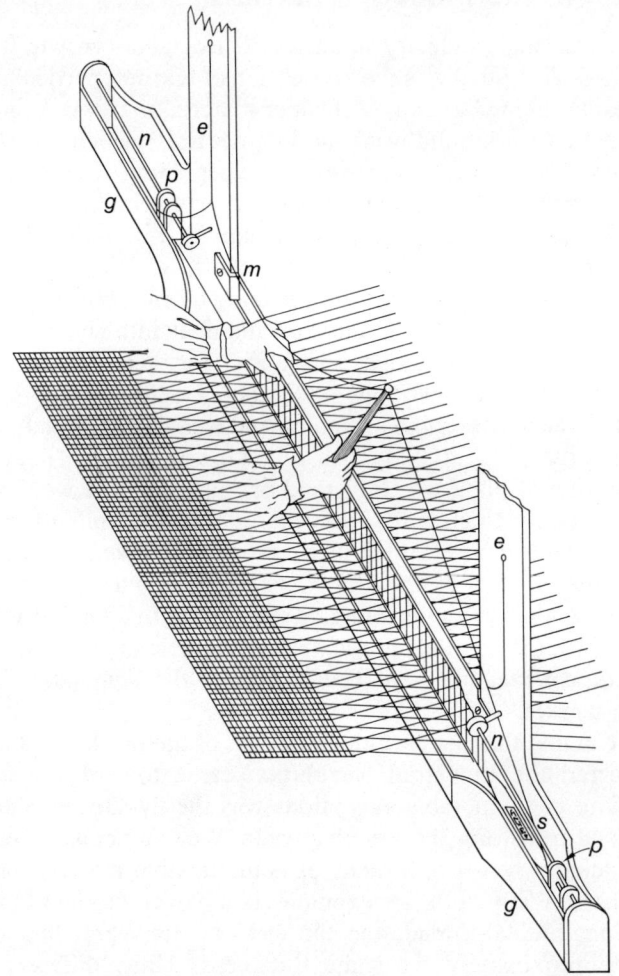

Fig. 2-1. The fly-shuttle of John Kay (as the peg is jerked with the right hand the shuttle flies through the open shed to the shuttle box at the other end, trailing a line of the weft behind. With a hand shuttle the left hand must help, but with the fly-shuttle it can beat up the weft with the batten).

looms, now one man could do the work of two, and still later improvements permitted several men to be replaced by a single operator. The work force in the textile industry was at once out of balance, for the spinners were incapable of providing enough yarn to keep all the weavers busy. It would not be correct to say that Kay's invention was the sole impetus for the subsequent mechanical improvements in spinning because men such as Lewis Paul and John Wyatt were working on spinning machines at the same time Kay was perfecting his shuttle. However, the primary improvements did not come until more than thirty years had gone by.

The object of spinning is to rearrange loose fibers into a somewhat parallel order, then pull them lengthwise (attenuation), twist them to provide greater strength to resist strain, and finally to wind the finished yarn. Probably the most primitive method was to rub the threads between the palms of the hands; or between the hand and the thigh, which left one hand free. A further conservation of human energy was effected by the early discovery of the first spinning device, the hand spindle, which developed as a slender implement (generally of wood) a foot or so long, tapered at either end. Later a disk, called a whorl, was attached part way down the shaft to steady its rotation. One of the methods of hand spinning with a spindle was to draw a partially twisted thread as the raw material, from a distaff held under the arm. The distaff was a larger rod around which the material to be spun was wound. The thread was fed by hand to a spindle being twirled by the fingers of the other hand. The spindle twisted the thread which at the outset was caught in a notch near the upper end, and by its own weight stretched it.

To improve production it was necessary to develop mechanical devices for rotating the spindle and drawing out the fibers; and also mechanical means of coordinating the rapid rotation of a large number of spindles working together. The hand wheel, brought to Europe from the East about the 14th Century, and the previously mentioned flyer, represented basic improvements in the spinning process, but there was little else for several hundred years.

The first of three major inventions which promised to readjust the labor imbalance caused by Kay's shuttle was made by James Hargreaves. Hargreaves was an impoverished textile worker who apparently also doubled as a carpenter. Not enough is known about his early history to say when his considerable mechanical ingenuity began to show itself, but by 1760 at least he was working on improvements to a carding engine Lewis Paul had designed. It is the spinning jenny, however, for which

Hargreaves is remembered. The story of the intuitive conception which came to him one day when his daughter accidentally overturned a one-thread spinning wheel is often told. As the wheel continued to revolve when the spindle was thrown upright, he recognized that it should be possible to spin many threads at the same time if a number of spindles were placed upright side by side. He worked on this idea for several years, and when the spinning jenny was patented in 1770, his daughter's name was immortalized. (*See* Fig. 2-2.) Many smaller mills and cottage

Fig. 2-2. Hargreaves' spinning jenny (the horizontal moving carriage is moved back and forth with the left hand. The drive wheel, turned by the right hand, rotates the spindles by means of endless bands and a cylinder).

industries used the jenny, but since its action did not produce a thread strong enough for the warp (which must withstand greater tension than the weft) it was not the final answer to the needs of the spinning trade. A second primary requirement in spinning machinery was for some practical means of providing a fast drawing out of the slivers of cotton to a desired fineness and strength. The answer was found in roller spinning. Again evidence shows that Paul had an idea for using multiple rollers rotating at different speeds, so that each pair would turn faster than the one preceding, thus drawing the thread finer for twisting by a flyer. But it was Richard Arkwright who in 1769 obtained patents for a machine which became commercially successful.

Many accounts of the history of textile manufacturing make Arkwright out to be less an inventor than a manipulator of men and money. He, like Hargreaves, began in very humble circumstances, but from the time he apprenticed to a barber he showed a forceful business acumen. When on his own, he warred with other barbers by undercutting their prices. Later, as a hair and wig merchant, he obtained a secret process for dying hair, and was able to sell the hair he dyed to wig makers for premium prices. Traveling about as he did in the textile-making country, sharp-eared and keenly alert to the gossip about business possibilities, he would not have remained long unaware of interest in higher-production spinning processes. He joined forces with a clock-maker, John Kay (not the inventor of the fly-shuttle) and together they began to experiment secretly toward the perfecting of a spinning machine. The result was the "water-frame" (*see* Fig. 2-3). It was a more powerful machine than the spinning

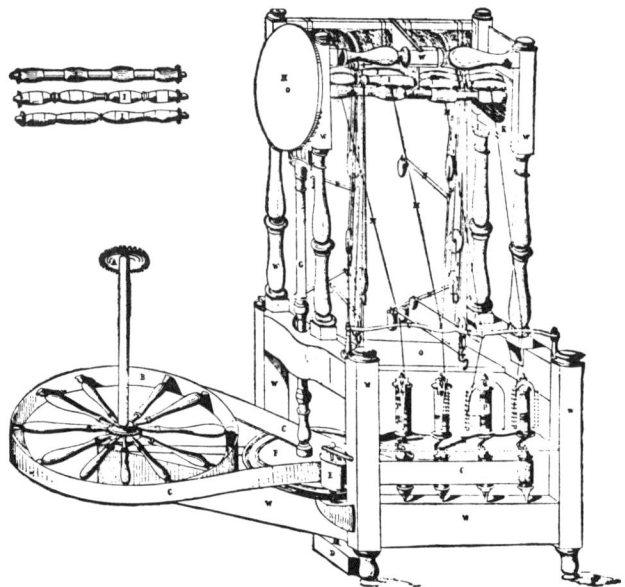

Fig. 2-3. Arkwright's water-frame (the wheel at the left drives the entire mechanism. The draft rollers are at the top of the water-frame and the flyer spindles are below).

jenny, and was capable of producing the warp yarns for weaving. The water-frame was originally designed to be driven by a horse, but the first large mill erected on the banks of the Derwent River in Derbyshire in 1771 utilized water power. The development of this machine was accompanied by the same sort of difficulties that bedevilled the other inventors of this time—periodic shortages of capital, hostile workmen, and, later, widespread patent infringements—but Arkwright was a man of special stripe, and before many years was the master of a growing industrial empire. Thomas Carlyle was to describe him as "...a plain, almost gross, bag-cheeked, pot-bellied Lancashire man..." the very image of the insensitive, hard-driving entrepreneur. Working often from five in the morning until late in the evening, he would race from one of his mills to another in a coach and four, continually making modifications on his devices and methods of manufacturing, training work corps, searching out new markets, and building one of the first great fortunes of the new mechanical age.

The third major innovation in spinning was in the form of a machine designed and built by Samuel Crompton. Crompton as a young boy had to help his widowed mother with her spinning and weaving. She was a competent but impatient woman and little inclined to be tolerant of delays when her son took too long joining together the soft yarn which was continually breaking on the jenny. He in turn was a studious lad, mechanically and artistically gifted; so it seems natural that he would consider how to overcome this difficulty, perhaps with an idea to finding more time for practicing on the violin he had made.

For about seven years, from 1772 to 1779, he worked secretly in the attic of the large old house for which his mother was caretaker. At last he succeeded in producing a machine which could spin a yarn of exceptional fineness, quite comparable to the famous hand-spun muslins of India. It combined the most useful features of Hargreave's jenny and Arkwright's sometimes horse-powered water-frame (the name "mule" which was later applied reflected its diverse parentage) but it also had an independently adjustable spindle carriage to prevent the thread being overstrained before it was completed. Crompton did not acknowledge any copying from Arkwright's or, more precisely Paul's, idea of roller spinning, but, clearly, obtaining a patent for his mule would have been a difficult proposition. He chose instead to make his device known to local manufacturers, relying on their generosity in fulfilling a public subscription to provide a just compensation. His trust was ill-rewarded. A relative pittance was subscribed, and even less eventually returned to him. He lived until 1827, but little else that happened in his dealings with those in the textile trade would dispose him to a warmer view of his fellows, although a later subscription was somewhat more beneficial. He continued to produce his fine yarns on a small scale, but he was often undersold, his designs for fabrics appropriated by others, and his trained workers hired away. Crompton never really understood the art of business management, and he died in relative poverty and obscurity.

Mules and water-frames continued to proliferate in the many new cotton mills being built in England, and it was evident now that the weaver in his turn would have difficulty keeping up with all the cotton being spun. Some of the most useful ideas for increasing the mechanization of weaving were conceived by the Rev. Edmund Cartwright, a novice inventor, but they were never really commercially successful. Developing a true power loom proved to be an obstreperous problem which was not satisfactorily solved in the period we have been discussing.

MECHANICAL TECHNOLOGY – THE STEAM ENGINE

In 1785, in Nottinghamshire, a steam engine operated a spinning machine for the first time. This was a noteworthy event, but steam as a new source of power had been employed in other practical ways and to an increasing degree since the beginning of the century. In fact a French experimenter, Salomon de Caus[1], was considering the effects of its elastic force at least as early as 1615. He described a means of using steam pressure to jet heated water out of a pipe extending near to the bottom of an enclosed metal vessel. One of the first Englishmen to discourse in a similar vein was the Marquis of Worcester, who in 1663, in his book *Century of Inventions* wrote about "An admirable and most forcible way to drive up water by fire...."

More scientifically distinguished than either of these worthies was Denys Papin, a native Frenchman, born at Blois in 1647, who had fled to England to escape religious persecution and had, in time, become a Fellow of the Royal Society. He was a skilled experimentalist. One of his models was the first piston-and-cylinder steam device. The expansion of steam under the piston forced it to rise in the cylinder, whereas condensing the steam back to water caused the piston to be pressed down again in the cylinder by atmospheric pressure. Papin felt that his "engine" could have many uses, from moving ships to pumping water from the mines. His model was never applied in any such practical way, however.

The British patent records of this period attest to the attempts of numerous inventors to perfect a means of pumping water from metal mines and collieries many of which extended several hundred feet down into the earth. In the 17th Century the most common kinds of pumps were endless chains of buckets, or chains of plates passing up like pistons through a column of pipes and carrying the water before them. In either case the chain revolved about an axletree at the mouth of the pit, and the water emptied into a trough. Rushing water was used to drive water wheels which rotated the axletree, or, where no streams were nearby, horses walking in a circular path provided the power. These limited methods of pumping water were ages old at that time and are found even today in some of the less-developed regions of the world.

The problem of water seepage was especially vexatious in the deepening tin mines of Cornwall, whose operators were often individuals with no reserves of capital to provide extensive pumping facilities. Thomas Savery, a contemporary of Papin, was a military engineer from Cornwall who set about to invent a way of pumping to solve the problem. In 1698 he was able to demonstrate to King William III the results of his work and to secure a patent, and in 1702 he described the device in a small book entitled "The Miner's Friend, or an Engine to Raise Water by Fire described, and the manner of fixing it in mines, with an account of the several other uses it is applicable unto; and an answer to the objections made against it." Savery's pump consisted of a boiler and a separate vessel or vessels into which steam from the boiler was conducted. When the outside of the boiler was quenched with cold water, water from a lower level was forced upward into the vacuum thus formed.

The main objection to it for the job it was designed to do was simply its inherent inability to raise the water to a height anywhere near that required for the average mine. Staging the pumps at successive levels in the mines left too many practical difficulties to be solved, and so Savery's "fire engine" found limited use outside of pumping water on farms or estates.

Thomas Newcomen, an ironmonger, succeeded in inventing the first steam engine with sufficient power to drain the mines. He, too, secured a vacuum by condensing steam in a vessel, but his idea differed from Savery's in fundamental ways. (*See* Fig. 2-4.) For one, his vessel (like Papin's) was an open-topped cylinder with a piston. The steam produced was condensed by jetting cold water inside the lower section of the cylinder. Atmospheric pressure forced the piston down, and it rose again largely because of the weight of the pump rod pulling down the opposite end of the rocking beam. Savery's patent, however, was so broadly drawn that Newcomen had to construct his engine under Savery's license, and for this reason the name of Newcomen never received the honors his work merited. The first of the successful Newcomen engines began operating in 1712, and for the next fifty years they were widely used in England, on the Continent, and even in America. Ironically they found

Mechanical Technology—The Steam Engine   19

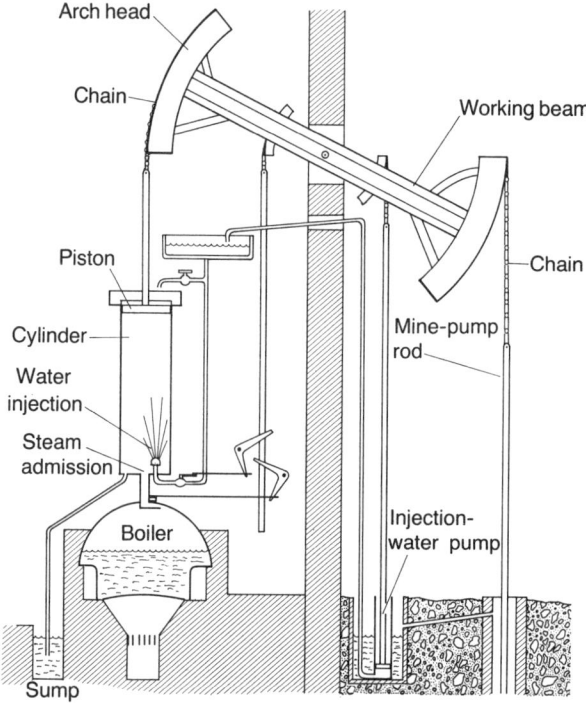

Fig. 2-4. Newcomen's steam engine (1712). (Courtesy, Scientific American, Inc.)

less favor with the Cornish miners, whose needs Newcomen wished to satisfy, than with the colliers farther north. The great pistons, two feet and sometimes much larger in diameter, fit badly in the roughly machined cylinders, and the continual quenching to condense the steam was inherently inefficient, so that the cost of fuel was too high for the regions where coal had to be imported. In the years after Newcomen's death in 1729 other workers made minor structural improvements in his engine, but it remained basically inadequate to meet the rising demand for steam power at the last third of the century.

Ingenuity was matched to need when James Watt, working as a mathematical instrument maker at the University of Glasgow, was asked to try his hand at repairing a demonstration model of a Newcomen engine which had never worked very well, besides being quite expensive to run. Watt pondered on the problem for many months before a solution came to him suddenly one fine Sunday afternoon in May of 1765. If excess fuel had to be burned because of the necessity of bringing the cylinder back up to temperature after each quenching, why not provide instead a connecting, evacuated vessel into which the steam would rush and condense? The cylinder would then remain hot throughout the cycle.

Though simple, in retrospect, (*see* Fig. 2-5) the concept of the separate condenser was not the kind of notion that would likely have occurred to

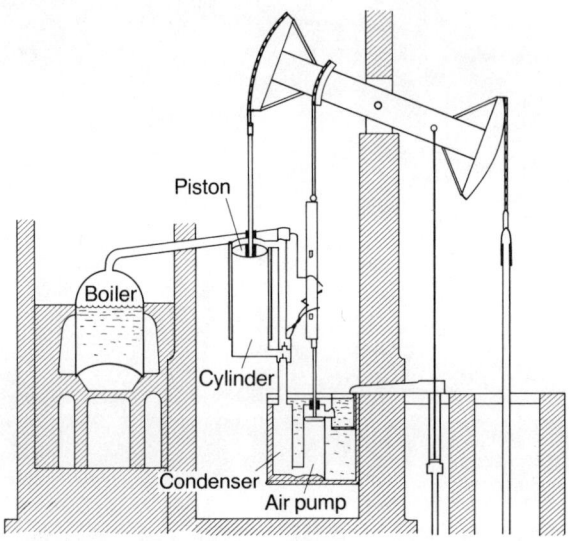

Fig. 2-5. James Watt's steam engine (1775). (Courtesy, Scientific American, Inc.)

an untutored mechanic of this era. Watt, however, was more than just a laboratory technician. His grandfather had taught mathematics; his father had his own instrument shops in which young James had been able to pick up the fundamental mechanical skills. He was an untiring student, gifted in mathematics, well-read in philosophy and the arts. If there were promise that the clue to a difficult problem might be found in a work written in another language, he was fully capable of learning that language sufficiently well to seek the answer. Professors at the University recognized his quick perceptive mind and he associated closely with some of them, including Dr. Joseph Black, the discoverer of the theory of latent heat.

Talented as he was, Watt hadn't but a novice's notion of how best to make his invention a commercial reality. His close-fisted, cautious ways were not those of the entrepreneur. For several frustrating years he labored on his engine, beset by mechanical and monetary difficulties.

Mechanical Technology—The Steam Engine

Financial assistance was obtained from Dr. John Roebuck, the founder of a substantial iron works at Carron, in return for a part interest in the engine; but Roebuck seriously overextended himself in a coal mining venture and went bankrupt in 1773. Fortunately for Watt, Mathew Boulton, a prominent manufacturer who had been keenly interested in the engine for some time, was prepared to step in and buy out Roebuck's share.

Thus began the famous company of Boulton and Watt, a partnership of rare felicity. Boulton's water-powered Soho works in Birmingham employed a large corps of able workmen producing a variety of articles such as inlaid buttons, toys, and metal ornaments for domestic trade and export. Here were the kinds of resources Watt required to perfect his engine, and when the firm was able to get his patent of 1769 extended to 1800, he also found the needed time. Success came slowly. Watt's gloomy, doubting disposition had often to be countered by Boulton's natural sanguinity. The first two engines were ready at last in 1776—one to drain a colliery and the other to provide the blast for the furnaces of the iron master, John Wilkinson.

Watt continued to experiment to find ways of simplifying his engine for ease and efficiency of construction and operation. In some twenty years of brilliant innovation, he changed the conception of the steam engine fundamentally. The separate condenser has already been described. Besides this, he incorporated the double-acting principle in which steam and vacuum were applied alternately to each side of the piston. Not only was this more efficient than the single-acting cylinder (especially when he cut off the steam admission early in the stroke) but it led directly to another significant development. With the power stroke now in both directions, the chain, formerly used to pull the rocking beam on the downstroke, no longer sufficed, nor would a simple rod connection do, since the end of the beam moved in an arc. Watt's solution was his "parallel motion" device, a linkage arrangement which guided the rod in a nearly straight line. His adaptation of a fly-ball governor (Fig. 2-6) to control steam inflow was another imaginative idea that set the stage for present day self-regulating feedback control systems. His work was aided, too, by the contributions of others, at least once reciprocally. When John Wilkinson took out a patent in 1774 for a new machine for boring cannon it was almost immediately apparent that it would be ideal for boring engine cylinders too. He was asked to cast and bore one for a Watt engine, and it worked beautifully.

In spite of its obvious advantages for generating power, the steam

22   Technology in the Industrial Age

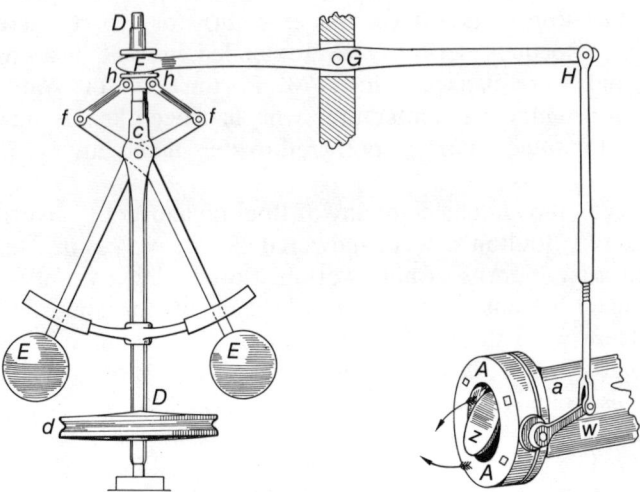

Fig. 2-6. The fly-ball governor. An early automatic control device. (Courtesy, Scientific American, Inc.)

engine would have been a limited machine had it remained restricted to the simple up-and-down motion used for pumping water. Watt realized the desirability of securing a rotary action, and was also aware that the familiar crank and connecting rod which had long been used on the treadle of a foot lathe could do the job. Considering it an old idea, he started to work on the application without bothering to seek a patent. When James Pickard, another inventor, did patent the notion, Watt was affronted and determined to find another way rather than be forced to meet Pickard's terms. The eventual method settled on was the sun-and-planet gear suggested by William Murdock, a valued worker in the organization. Before the basic patents expired in 1800, close to five hundred Boulton and Watt steam engines were turned out, over 60% of which were designed for rotary motion. With all of its improvements, this versatile prime mover was ideally suited to serve as a nexus for the flourishing enterprises of the new industrial age.

The mines and the mills, iron works and water works, all had need for a reliable, efficient source of power, and when it was available a whole array of fresh technological possibilities was spread out before alert innovators. A look at the growth of metallurgy can suggest how the steam engine served, and was served by, other technologies.

METALLURGICAL TECHNOLOGY

At the beginning of the 18th Century, Germany and Sweden were well ahead of England in ability to produce iron. In fact, England had had a more impressive ferrous metallurgical output in the 16th and early 17th Centuries, especially in certain heavily forested regions, but the devastation of the forests for wood to make charcoal had aroused concern that there would be insufficient timber for naval construction. Even in Queen Elizabeth's time the rampant destruction had caused the passing of a law prohibiting the cutting of trees of a certain size in many areas, but deforestation continued. By the end of the 17th Century it had become necessary to import high quality iron.

To realize the importance of charcoal in early iron making it is necessary to understand that iron as it is mined in the form of ore is chemically associated with oxygen and many other elements. The purifying process is designed to remove the oxygen by causing it to combine with carbon, for which it has a higher affinity at elevated temperatures than it does for iron. In Europe, until about 1300 A.D., iron to be wrought into useful forms was produced directly from ore. Lumps of ore mixed with charcoal were placed on a charcoal bed in a crude hearth, (in former times sometimes just a hole in the ground), and supplied with air to burn the charcoal. At first, natural draft provided the necessary air, but various hand- or water-powered bellows or other devices were used quite early. The charcoal acted as a generator of heat, and indirectly as a reducing agent. The reduced iron did not get hot enough to liquefy, but instead collected in a lumpy, pasty mass with slag impurities in its pores. Hammering this mass while it was hot would squeeze out most of the slag, and leave a small amount — at most a hundred pounds or so — of comparatively soft, malleable iron.

The forerunner of the blast furnace, or shaft furnace as it was sometimes termed, evolved from the hearths. Around the 14th Century it took the form of a shaft not more than fifteen feet high into which ore, fuel and limestone were charged from the top, and relatively low-pressure air was blown in through "tuyeres" near the bottom. By the end of the 17th Century, shafts twice this height were being used. With this kind of arrangement the hot iron could be in contact with the fuel away from the tuyeres long enough to absorb considerable carbon. The effect was to lower the melting temperature of the iron to the point where it liquefied. The result was pig iron, a high carbon material containing other elemental impurities from the charge. It was often so brittle when solid that it could not be used in any application requiring a reasonable measure of ductility without additional processing to remove some of the embrittling impurities. This second purifying stage most often involved some such method as reheating the pig iron with charcoal and limestone by a blast of air in a smaller furnace. The remelted pig iron, mixed with slag, became more pasty as purification proceeded and was then hammered into wrought iron at the forge.

Aside from the important problem of timber scarcity, charcoal had an inherent limitation as a blast furnace fuel. Being mechanically weak it could not sustain a heavy weight of charge without collapsing and packing down so as to inhibit the passage of the hot reducing gases. The height of the furnace, its capacity and efficiency, were thus restricted. Coal was an obvious possibility as a substitute and from the middle of the 17th Century on, a number of patents were granted covering the use of coal in iron making. None of the processes had much success at first, partly for the reason that sulfur and phosphorous in the fuel were injurious to the quality of the iron.

Early in the 1700s Abraham Darby and his son Abraham II, Quakers of Coalbrookdale, began to try making coke from coal and using it in the production of iron. Fortunately the coal in their area was low in phosphorous and sulfur, and the Darbys were tenacious iron makers. After many years of experimenting (most successfully by the younger Darby after his father had died) the process was achieving consistently favorable results. Other iron makers whose coal or ore supplies were of poorer quality were slow to adopt similar practices and many who did, found a serious limitation to improved efficiency in the fact that coke was harder to burn than charcoal. The problem of insuring a strong, constant blast was effectively solved by using the steam engine. John Wilkinson used a Boulton and Watt steam engine to provide the air blast for a furnace at his Willey Ironworks. The number of coke-fired furnaces and the output of pig iron jumped sharply. The making of cast iron (a high-carbon form of iron cast directly into molds from the liquid state) by remelting pig iron also increased although for many years after this period the main tonnage of finished metal in England was still to be in the form of wrought iron. The increased output of wrought iron was made possible by the perfecting of the "puddling" process in 1784 by Henry Cort and, almost simultaneously, Peter Onions. Cort used a shallow-hearthed "reverberatory" furnace in which the decarburizing gases from burning coal radiated from the furnace roof, so that no coal need come in contact with the molten metal. Stirring the metal hastened carbon removal to the point where the iron became malleable and it was then rolled to shape between grooved rollers.

A kind of technological inevitability seems to connect the events of this fruitful period. It is not easy to distinguish cause from effect. The use of coke in iron making encouraged the development of coal reserves. The need to save the coal and ore in the mines from inundation by underground water called for capacious pumps, and the steam engine was a consequence. For its part, the steam engine, its cylinder machined to fine accuracy by John Wilkinson's boring mill, freed the iron maker from his uncertain dependence upon water power, and the output of iron products could now be measured in tonnages that would have been inconceivable a half-century before. "Iron mad" they called Wilkinson. He was to make iron bridges, iron pipes for water supplies, even an iron pulpit for a church[2]; and when he decided to build a boat of iron there were few who felt that it would float—but it did. Each time another use for iron was found, new production was spurred that much more. Coal and iron were taking command in England.

HIGH-PRESSURE ENGINES

The following century—the 19th—began with the throbbing steam engines pumping energy through the arteries of the growing industrial centers. Yet for all of their popularity, the Boulton and Watt engines were limited in the amount of power each could generate. Working at atmospheric pressure, or only slightly above it, they were rated to deliver anywhere from 10 to 50 horsepower as a rule. Many water wheels of the early 19th Century were fully capable of matching this kind of power output. The primary Boulton and Watt patents expired in 1800 and by this time others had been thinking of new approaches for developing steam power, most particularly by designing for much higher pressures than Watt had considered either practical or prudent. Richard Trevithick, a talented Cornishman, was one of the foremost of these. In the history of steam power there are few stories more romantic than his. Indifferent to his formal school training, but possessed of great native ingenuity, he had begun working and modifying steam engines before he was twenty. At twenty-five he was embroiled with the law for collaborating with Edward Bull on an engine design which was adjudged to be an infringement on Watt's patent. This was in 1796. Within a year he was attempting to build a high-pressure engine which expelled steam directly into the atmosphere, reasoning that the increased force on the piston would permit a significant reduction in engine size. This was in fact, the case, although it later became apparent that the condenser made a fundamental contribution to engine efficiency.

An especial motive for Trevithick's seeking a more compact engine rested on his conviction that only thus would a self-propelled carriage be practicable. Although his first engine was used in the mines, one suspects that he found greater pleasure in his experiments with the steam carriage. He was a restless, impulsive man to whom the imminent danger of a high-pressure explosion was more challenging than daunting. His steam carriage had its first run on Christmas Eve, 1801. On another trial a few days later, it overturned and the wooden frame, ignited by the hot iron, was destroyed. Perhaps this was an omen. Trevithick obtained a patent in 1802 and other carriages followed, but ultimate success was precluded by explosions, abominable roads and the protests of startled citizens. In February 1804, at Penydaren Iron Works, Wales, his was the first locomotive to run on rails. Five wagons with seventy men and ten tons of iron were pulled successfully[3], but then heavier loads fractured the cast iron rails and the experiments stopped.

For many years, then, he worked at one time or another as a designer of steam barges and dredges and of the famous Cornish boiler, as an engineer on the Thames tunnel project, and as a maker of cast iron tanks for raising sunken ships. His irrepressible optimism and native talents never deserted him, but his luck often did. The few financial successes he achieved seemed destined to be followed by episodes of extravagant generosity, ill-timed investing, and even bankruptcy.

In 1816 Trevithick left his wife and children to pursue his star to South America[4]. He was to bring steam power to the silver mines of Peru, but in the war for independence most of his machinery was destroyed. He prospected for minerals, and at one time devised a diving bell to salvage a sunken ship in Chile. He also served with Simon Bolivar into whose army he was impressed as a kind of military engineer. Still another time he survived the capsizing of his boat in alligator-infested waters only through the fortunate appearance of a fellow Englishman who was hunting nearby. When he returned to England and his faithful wife after eleven years wandering, he was penniless. Five years later he was laid in a pauper's grave, but a window in Westminister Abbey now serves as a reminder of his contributions to the age of steam.

There were other inventors and entrepreneurs who made important contributions to the development of high-pressure steam engines, but in this abbreviated history only the name of Oliver Evans need be mentioned. Evans built, and in 1804 patented, a high-pressure engine with several original features. This is especially noteworthy because Evans was American, and it hints that his young country would have its own part to play in the spread of mechanical invention. He, too, constructed a steam carriage—an amphibious one at that—but it was unsuccessful. The railway and the steamship were far more promising means of providing large-scale transport than any conveyance that would have to travel on the inadequate roads of that time. Only three years later, in 1807, Robert Fulton's paddle-wheeled Clermont, propelled by a Boulton and Watt engine, began sailing regularly between New York and Albany on the Hudson River. Another eighteen years would pass before a commercially successful railway could be found. This was the Stockton and Darlington Railway in England, featuring the locomotives of George Stephenson.

NEW ENERGY SOURCES

As the 19th Century matured, technology and science stimulated a remarkable expansion in productivity. In the early decades industry was powered by steam in vaulting annual increments, by rushing water far more effectively than it ever had been, and by humans or beasts in decreasing proportions. In the later decades new energy sources would become available; often through the efforts of inspired amateur inventors, but more frequently now with the aid of a thriving science. These discoveries not only increased the power to be used by mushrooming industry, they also hastened the processes of qualitative change taking place in Western society. For the purposes of this short history, however, it should suffice to sketch their evolution in brief outline, because the seminal event had already occurred. Once the yoke of oppressive toil had been lifted from human shoulders by the steam engine—by the realization that man could multiply his capabilities through the mechanical manipulation and control of certain natural forces—the headlong rush into the Industrial Age was irresistible.

At the core of the new developments was the unfolding understanding of the nature of energy transformation. Energy, which has the sun as its ultimate source, is locked up in all forms of matter. The heat of burning natural fuels had served man for ages, but the steam engine was the first device which converted it into the useful mechanical work associated with turning a shaft or lifting a bar. Obviously other fuels than coal or wood could furnish energy, too. Gas, issuing from the earth, had been lighting lamps in China for centuries, but the production of combustible gas from the controlled burning of coal and its subsequent use for commercial illumination resulted to a considerable extent from pioneering work by two associates of James Watt—William Murdock and Samuel Clegg. With such a suitable fuel available, it was natural to think of employing it directly in a cylinder, thereby effecting a considerable simplification in the power producing cycle characteristic of the steam engine.

The one cylinder Lenoir engine, patented in France in 1860, was the first gas internal-combustion engine which had any kind of commercial success. A more powerful and efficient engine was perfected by Otto sixteen years later. Shortly before the Civil War large underground pools of petroleum were tapped by Drake in Pennsylvania, and not long after, experimenters began to try its many fractions as internal combustion engine fuel. Two Germans, Gottlieb Daimler and Karl Benz, separately produced original versions of the gasoline engine in 1885.

A parallel development of incalculable importance in the 19th Century grew out of the discovery that the energy in matter could be unleashed by means other than the burning of fuels. The peculiar forces which caused attraction and repulsion between objects of various materials were first systematically studied by William Gilbert, a court physician to Elizabeth I. In the 17th Century, Von Guericke produced an electric charge by continuously rotating a ball of sulfur against a piece of cloth. The "Leyden jar," an instrument for storing an electric charge on two conducting plates separated by an insulator, was discovered by Van Musschenbroek in 1745, and intense interest was shown in the electric shocks which were sustained by touching the charged jar. The Methodist, John Wesley, stated that electricity was "The soul of the Universe." But these experiments had limited scientific value because the transient electricity produced could not be used for more probing experiments. A continuous source of electric current was needed. It was found in 1800, when Allesandro Volta stacked successive layers of silver, brine-soaked paper and zinc on top of each other and then connected the bottom zinc plate with the top silver by a conductor.

The furthest reaching of the subsequent researches on electrical effects was carried out by Michael Faraday, one of the most brilliant intuitive experimenters of the 19th Century. His work on electromagnetic induction, following earlier observations by Oersted and Ampère about the magnet-like behavior of current-carrying wires, was fundamental. Faraday knew that a magnet would induce magnetism in a piece of soft iron, and wondered if, by changing a current in one coil of wire, he could, in like fashion, induce a current in another. In 1831 he coiled two spirals of wire around a wooden cylinder, interspersing the windings and insulating them from each other. One wire was attached to a battery, and the other to a galvanometer. Faraday hoped to see a galvanometer deflection indicating a current in the second wire, when electricity flowed through the first. No deflection could be noticed, to his disappointment, but then he observed a very slight movement at the moment he turned the primary current on. Immediately it was obvious that it took a change in current in one wire to produce electricity in an adjacent wire. The motion of a magnet had the same effect. This meant that mechanical energy — the moving of a magnet by a steam engine or the flow of water — could be converted into electrical energy. The self-exciting generator (called the dynamo), perfected by Siemens and others in 1866, turned this knowledge into practical reality. In 1873, it was learned that the dynamo could, with proper design, also serve as a motor.

Many refinements had yet to be made in the means of generating, transmitting, and transforming electricity. As another energy source, however, few would doubt that because of its flexibility (for example the ability to provide power at any distance from the generator) it would be even more effective in reducing physical drudgery than the steam engine had been.

In concluding this section we need to mention that as the 19th Century drew to a close, Marie Curie isolated radium. The potential, for good or ill, of this discovery for mankind could hardly be surmised then, but Newton's orderly world would never be the same. There were intuitive expressions of foreboding, (author Henry Adams wrote that radium "denied its God"), but the interest at that time was mainly scientific, rather than technological or sociological. Nearly two more generations would pass before the energy within the atom would be unleashed, first as an uncontrolled force of awesome dimensions, and later as a primary stage in the production of useful industrial power, promising another alternative to traditional sources.

PART II THE ADVENT OF MASS PRODUCTION

As each new machine was added to the inventory of the Industrial Age, as each new mill began to discharge its goods into the stream of production, the need to impose a discipline of synchronization became apparent. Manufacturing — from *manus* (hand) and *facere* (to make) — no longer meant making things by hands alone. The machines were ubiquitous, and had to be if the growing demands for their products or functions were to be satisfied. They were also complex, and very often, so were the things they made. The ways of the solitary worker could no longer prevail in this regime of interdependency.

In the earliest days of the individual craftsman, making an article had been a demanding but relatively uncomplicated task. Wood, horn and stone were the principal solid materials and their shape could be changed, albeit laboriously, to suit a need. Usually the finished articles were in one piece, or, if not, the parts were united by binding, hafting or joining in some simple way.

By the time of the 14th or 15th Century, skilled workers were fashioning products both complex and ornate. Many articles were still chiefly of wood, but metal tools, clocks, ornamental pieces and instruments of war were made in fair quantity. A smith of that period could produce superb plate armor shaped to the human body. Twenty or thirty separate parts or more—greaves, gauntlets, pauldrons, visors and the like—were hammered, fitted and riveted, and then for full protection the pieces were carefully overlapped. This painstaking work took as long as was needed, and it was done when it was done. If something fit poorly it might have to be reshaped or remade until it served. Time was not normally an oppressive taskmaster, notwithstanding an occasional impatient patron.

Now by the 18th Century the division of labor, which many social theorists consider one of the hallmarks of a technologically advanced society, was well established. Certain vital manufacturing processes consisted of numerous separate but related stages, and the worker's responsibility was restricted to one simple repetitive task in that stage. The time he took was more important than his skill. A famous passage from Adam Smith's *Wealth of Nations*, written in 1776, describes the manufacturing of pins:

> ...the way in which this business is now carried on, not only the whole work is a peculiar trade, but it is divided into a number of branches, of which the greater part are likewise peculiar trades. One man draws out the wire, another straightens it, a third cuts it, a fourth points it, a fifth grinds it at the top for receiving the head; to make the head requires two or three distinct operations; to put it on is a peculiar business; to whiten the pins is another; it is even a trade by itself to put them into the paper; and the important business of making a pin is, in this manner, divided into about eighteen distinct operations, which in some manufactories are all performed by distinct hands, though in others the same man will some times perform two or three of them. I have seen a small manufactory of this kind where only ten men were employed, and where some of them consequently performed two or three distinct operations. But though they

were very poor, and therefore but indifferently accommodated with the necessary machinery, they could, when they exerted themselves make among them about twelve pounds of pins in a day. There are in a pound upwards of four thousand pins of a middling size. Those ten persons, therefore could make among them upwards of forty-eight thousand pins in a day. Each person, therefore, making a tenth part of forty-eight thousand pins might be considered as making four thousand eight hundred pins in a day. But if they had all wrought separately and independently, and without any of them having been educated to this peculiar business, they certainly could not each of them have made twenty, perhaps not one pin in a day ...

Pin-making is thus seen as an involved manufacturing sequence. The production of muskets or locks for doors would surely require an even greater degree of technical integration. Machines would have to be thought of and designed according to their sequential relationship to other machines. Machine operations rather than human would become the basis of manufacturing processes, not only because the new energy sources promised to amplify man's physical strength and efficiency; but especially because machines, now made more rigid through metal construction, could impose new standards of accuracy on their output. Process accuracy was of less interest than speed or continuity for some products—textiles for instance—but where there was a question of fitting together components of an assembly, dimensional tolerance became a vital consideration.

High accuracy mass production was made possible through the use of machine tools—power-driven machines for cutting, shearing or in some way forming metal into different shapes. Norbert Wiener once said, "Each tool has a genealogy, and is the result of tools that served to make it." Certainly the machine tools and processes of the Industrial Revolution had their forerunners. The lathe, in which the work is held between rigid supports and rotated while being cut or shaped, was probably the most important basic machine tool, and its principle was known in antiquity. The notebooks of Leonardo da Vinci clearly show he was familiar with several forms of the lathe. At that time, the power was applied to the work piece itself through a cord wound around it. This limited the shape of work which could be turned in the lathe, but the development in the 16th Century of a method of applying power directly to a spindle or mandrel as the work was held between centers significantly extended its usefulness. Seventeenth and eighteenth Century clockmakers had much to do with designing small lathes which worked with greater precision.

Many figures played roles in the establishment of the British machine tool industry, none more important than Henry Maudslay. Maudslay, born in 1771, orphaned at nine, had his first job filling cartridges at Woolwich Arsenal, but at the age of fourteen he went to work as a blacksmith. Metal working was his metier. At eighteen he was approached by Joseph Bramah, who had heard of his workmanship from another smith. Bramah, a few years before, had invented an "unpickable" lock of ingenious design. A standing money prize for anyone who could open the lock without the proper key remained unclaimed for sixty-seven years until a well-known lock expert succeeded in doing so after fifty-one hours of effort[5]. It was a complicated mechanism, however, and its grooves and notches had to be machined with extreme precision to guarantee security. The demand for these locks could not be met with the equipment at hand, so Bramah wished to hire someone with sufficient skill to help build the proper machines. Maudslay was the right man. He worked eight years for Bramah and each must have learned much from the other, but then a dispute over wages caused him to go on his own.

His shop was successful from the start, and before long, larger quarters were needed. An advertisement for the firm from a few years later reads in part: "...Millwork, Water Works, and Machinery of every kind executed in their usual style of workmanship." Their usual style was one which reflected a passion for accuracy and exquisite attention to every detail. Maudslay insisted on his workmen's testing the precision of their work against an absolutely true plane surface as a standard.

About 1800 he constructed a screw-cutting lathe with a slide rest holding a traversing tool, which was one of the significant early developments in the machine tool industry. Screws for transmitting accurate movement for scientific instruments had long been the products of craftsmen working with painstaking skill. Ramsden had produced a screw-cutting lathe in 1770, following earlier developments by the Frenchmen, Thiout and Vaucanson, which had resulted in lathes with a tool-bearing carriage driven longitudinally by a lead-screw, so Maudslay's design was not entirely original. It was his contribution, however, to recognize the necessity for achieving machine rigidity through all-metal construction, and for training a corps of workmen totally committed to checking all measurements against accurate standards. His lathe was capable of cutting long screws of uncompromising uniformity, which could then themselves be used as lead screws for other precision machines. Thus his machines could be made to reproduce their vital mechanisms with no loss of fidelity.

One of the early important commissions for the Maudslay firm involved a collaboration with Marc Isambard Brunel and Sir Samuel Bentham, who between them had proposed to the British Admiralty a plan for manufacturing pulley blocks for the Royal Navy. These men had conceived an ensemble of special purpose machines operating in a prescribed sequence to produce a minimum of 100,000 finished parts a year at greatly reduced cost. They had heard of Maudslay from a friend of Brunel, and a visit to his shop was sufficient to convince Brunel that Maudslay, if anyone, could design and build the machines. So he did, over a period of five-and-a-half years and when they were finished, forty-four in all, their output was greater than 130,000 pulley blocks a year with only ten unskilled men required as the work force, where 110 had been used before. Maudslay and Brunel had introduced to England the very model of a mass production system.

THE "AMERICAN SYSTEM"

The story of mass production might appear from the foregoing paragraphs to be an essentially British phenomenon although on the Continent, the French, in particular, made an appreciable contribution in the form of novel ideas for machine tools; and in Sweden an isolated genius named Christopher Polhem (1661–1751) had anticipated many later developments. He left over 20,000 manuscripts recording a life-time of incredible activity, and it is apparent that as an inventor of mechanical devices, particularly for the mining and metal-working industries, he had a clear grasp of the requirements for large-scale integration of manufacturing processes, and for the mass production of replaceable parts. However, it was not in England, or on the Continent either, that mass production became the over-riding pattern of industrial growth, but rather in young America.

Americans in the last quarter of the 18th Century were filled with the hope that since they had won the right to make their own way in peace, abundance from their well-favored land would follow naturally. Unfortunately, abundance did not come quite that easily. England, as the colonial master, had provided a kind of security, even if an oppressive one; but was not now inclined to encourage any competition for its own growing industrial establishment. At the same time that it was dumping goods on the American market, it imposed strict controls on the emigration of skilled mechanics to the former colonies. Neither designs nor specifications for machines were allowed to be taken out of England, but money prizes offered for the services of machine makers often tempted men to evade the restrictions. Samuel Slater was one of these.

He had been an apprentice in the Belper mill of Arkwright and Strutt, and at twenty-one was helping to manage. This did not provide enough challenge for his venturesome spirit so he booked passage in 1789 and sailed for America disguised as a farmer, the secrets of Arkwright's textile machinery engraved in his memory. He found his way, after a time, to Pawtucket, Rhode Island, where Moses Brown, a Quaker, had been trying without much success to build reliable water-powered machines for his cotton-spinning mill. Slater contracted to build a mill like the one at Belper and, with the help of an expert mechanic, Sylvanus Brown, he did in about two years. So efficient was the Slater mill that before long it was unable to get enough raw cotton to keep it busy.

Establishing a strong base of manufacturing in the United States was the dream of many influential men but the difficulties were formidable. Apart from the obstructive policies of England, there was a lack of capital, the sparsely populated land had little skilled manpower, transportation and communication were primitive, and farms, seriously neglected during the war, were often impoverished and under-productive. The shortage of cotton for Slater's mill was an example of one of the problems. There were those, Alexander Hamilton and Tench Coxe among them, who hoped that the large-scale growing of cotton in the South might provide one answer.

The long-fibered, black-seeded, "Sea Island" cotton which flourished in the West Indies was planted in the mainland South, but it could be made to grow only on a narrow strip of the coast where the soil and climate were favorable. There was also a short-staple, green-seed variety which grew well in most parts of the South. The seeds, unlike the black seeds of the Sea Island cotton, clung tenaciously to the fibers so that cleaning cotton, a job done by slaves, was a slow, tedious process.

The prospect changed dramatically as the result of a simple invention in 1793 by a Northerner temporarily living in the South. This was Eli Whitney, whose story is so well known that the mere outlines will do here. Whitney, upon graduating from Yale in 1792, had accepted a post, tutoring the children of a Major Dupont in South Carolina. It was a job he never began. He had traveled South by ship with Mr. Phineas Miller, who was acting for Major Dupont although his actual job was as manager of the estate of the late Revolutionary War hero, General Nathaniel Greene. Mrs. Greene was also on the voyage, and this charming woman invited Whitney to visit her plantation, Mulberry Grove, near Savannah before he went to the Duponts. He stayed on for months, enjoying the gracious social life, but also actively engaged in applying his considerable mechanical ability to a project which he hoped might benefit the debt-ridden plantation and the whole region, not to mention himself. He proposed, following a suggestion of Mrs. Greene, to build a machine to clean the short-staple cotton.

His first simple model took only a few days to build, but he spent another several months perfecting a practical machine. The final product was the cotton "gin" (a corruption of engine). It consisted of a cylinder with bent wires projecting from it. As the cylinder rotated, the wires passed through slots just wide enough to admit the teeth pulling the cotton but not wide enough for the seeds. These dropped into a box, and the lint was brushed off by a second, smaller, bristle-covered cylinder rotating in the opposite direction.

The neighboring planters were delighted with the gin, which to them meant rejuvenation for their land and fortunes. It did indeed. In 1790, one-and-a-half million pounds of cotton were grown in the United States. By 1797, the figure was up to eleven million pounds and in 1825, when Whitney died, it was 225 million pounds. There was a comparable growth in the number of slaves imported into the South to work the plantations, with what pernicious results we are today only too well aware.

To Whitney and Miller, now partners, the visions of riches proved ephemeral. Their plan was for Whitney, after obtaining a patent, to manufacture gins back in New Haven, and then install them throughout the South in stations where planters would have their cotton cleaned. Unfortunately for them, the gin was too simple to make and the South's need too urgent for such a plan to succeed. The patent was violated almost with impunity. The partners had to spend years in maddening litigation, attempting to obtain just compensation.

Our concern has been with the development of methods for making interchangeable parts. The connection has often been made that, in planning to meet the great demand for his gins, Whitney seems to have been far-sighted enough to realize he would need to devise a system which would routinely fabricate the parts, using special machines. Because of the paucity of direct, recorded evidence, however, there are scholarly questions about the details of his total contribution to a labor-saving mass production process. By 1798, it appears, he had despaired to making his fortune through the gin, but his resourceful Yankee nature led him to conclude that his methods would be applicable to the production of a far more complicated mechanism, namely, the musket. The customer here would be the United States government, which seemed perilously close to war with France in 1798.

The French, in their continuing struggles with Great Britain, wished to forestall any trading with the enemy, and their privateers plundered many American ships on the high seas. When it was learned that French diplomats had offered to stop these acts for payment, an outraged citizenry took up the cry, "Millions for defense, but not one cent for tribute." Under these circumstances the Congress had passed a bill making $800,000 available for arms.

On May 1, Whitney wrote a letter to Oliver Wolcott, the Secretary of the Treasury under President John Adams, proposing to manufacture "ten or fifteen thousand stand of arms" with water-powered machinery. A contract was approved shortly afterwards in which Whitney agreed to deliver 4000 muskets by September 30, 1799 and 6000 by September 30, 1800 for a price of $13.40 apiece. This was an audacious gamble for at that time he had not even bought the land on which to build the factory. It is not surprising that by the summer of 1799 he was nowhere near ready to produce 4000 muskets. What he had been trying to do, once the factory was erected, was to "tool up" for mass production. He was designing and building his own forming and cutting machines along with special fixtures to hold work pieces tightly in place, and jigs to guide the movement of cutting tools so that the same operations would be repeated for each part with no deviation. He wrote to Wolcott for an extension of time, and the government had little choice but to accede to his request. No one else, including their own arsenals, was doing any better, and the inventor of the cotton gin had a formidable reputation.

He did not deliver his first five hundred muskets until September 1801, but the previous December he had journeyed to Washington to give a demonstration to high government officials. There he took apart the locks of ten of his carefully manufactured muskets and mixed them in a pile on a table. He then showed that the parts could be picked up at random and assembled into locks which worked perfectly each time. Before that time in America, musket parts had had to be made and assembled individually.

Thomas Jefferson was the new President-elect, and no one was better equipped than he to understand the significance of Whitney's achievement. A distinguished inventor in his own right, Jefferson had, some fifteen years before, served as Minister to France. He spent as much time as he was able, while abroad, absorbing manufacturing, scientific and agricultural knowledge, and later wrote of meeting a young Frenchman named Le Blanc who had developed an intriguing method of manufacturing interchangeable parts for guns. French gun-makers, proud of their art, apparently had little interest, and Le Blanc was not again heard from after the French Revolution. In the United States the conditions were right for acceptance. The government advanced Whitney the money he needed to finish the contract, and after that his factory continued to turn out superior arms in great number.

This was the time for the beginning of mass production of machines in America, although there was an earlier development which made an important contribution to the eventual integration of production processes. Oliver Evans was the central figure involved. Evans, who has already been described as one of the pioneers in high-pressure steam engineering, was the first competent inventor to recognize the advantages of the continuous conveyance of solid material through an industrial cycle.

Grain was a valuable commodity after the Revolutionary War, and the northern farmers grew corn, oats and wheat in abundance. Evans, after growing up on a farm, had turned to storekeeping probably to allow more time for his inventions. Seeing flour mills everywhere, his natural curiosity led him to question the efficiency of a system of milling which was tediously dependent on gravity and the strong backs of millers. The millers had to carry the heavy sacks of grain up narrow steps to an upper floor. There the sacks were emptied and the grain passed down through cleaning devices to the millstones, which were the only moving parts powered by water. The ground meal then again had to be carried to the top floor for raking and drying, and finally sifted downward before being barreled. By about 1787, Evans had built a mill in which water power was applied to an endless belt conveyor with attached buckets. All of the operations were automatic from the time the wheat was dumped in at one end until the flour passed into barrels at the other.

Evans wanted to install his automatic system in other mills but had to fight through years of indifference, suspicion, and patent infringements before there was any appreciable degree of adoption of his ideas.

As a result of the achievements of Whitney and Evans, essentially conceived before the 19th Century began, the ground work for the "American System" of manufacturing was firmly laid.

The term, "American System" may have been used first in 1853 in a report issued by the British Small Arms Commission[6]. It took at least that long for the changes to be incorporated into this country's industrial structure on a large scale. Why did the system eventually spread as it did in the United States rather than in Europe? The most compelling reasons are precisely those which have already been cited to explain the obstacles to establishing any kind of a manufacturing base after the Revolutionary War. The Napoleonic Wars had the effect of commercially isolating the young republic at a critical period when it had need for many industrial commodities but few hands to make them. The push was westward. Many of the most practical and able men had left the East to open up the land.

Who would take the places in shops and factories to produce all the goods needed by men too busy pioneering to make their own? England had a large body of rural unemployed ready to move to the cities to work in the factories, but there was no such group in America at first. Machines would have to serve, instead. So they did—first turning out guns and textiles, then axes and plows. Clocks and household utensils of many kinds were made by machines. In the 1850s a factory in Chicago used a production line to make the McCormick reaper which was sold on the installment plan, and helped make a vast grain basket of the western states.

It was a big country. If the food-growing and machine-building regions were to be interrelated, there would have to be transportation systems on the same scale. The building of the railways commenced about the same time as in England, and before many years the land was laced with networks of metal rails. Over 20,000 miles of track were laid in the 1850s, increasing to a peak of almost 70,000 miles in the 1880s. The use of steam boats throughout the extensive inland waterways was greater than in England and the rest of Europe from the start.

By the end of the 19th Century, with the Continent contained, many Americans had begun to turn from agriculture toward industrial and commercial occupations. The "American System" was still flourishing, to be sure. A large share of the manufacturing in many industries was performed by automatic machines, but these and other industries had expanded and diversified to such a degree that there were thousands of new jobs, including machine tending and operating, which required human workers. The difference was that now, unlike a century earlier, there was an ample additional supply of labor to be found among the hordes of immigrants who had arrived from Europe and Asia.

At the root of the unparalleled success of the industrial system was the fact that it was an essential facet of the nature of a typical American to feel that he was as good as the next man. Alexis de Toqueville after his journey to America had written in 1840, "I never met in America with any citizen so poor... whose imagination did not possess itself by anticipation of those good things which fate still obstinately withheld from him.... The love of well-being is now become the predominant taste of the nation." The primary objective of the system as it evolved was not to make things for the luxury trade, but to satisfy the demands of the average worker for material pleasures which would indeed be luxuries in other lands.

Nobody grasped this truth to greater effect than Henry Ford. The automobile had developed from basic experiments on the gasoline internal-combustion engine mainly conducted in France and Germany in the last quarter of the 19th Century. Early motor cars produced for sale in those countries in the 1890s were hand-tooled and expensive. Ford insisted that he could make cars cheap enough for the working man to afford. He was, of course, talking about a product far more complex than any that had been mass produced up to that time, but he had a clear idea of what would be required. It would take fabrication of interchangeable parts on a grand scale, a continuously moving assembly line, standardized procedures at every stage, and, in fact, complete synchronization of man and machine. None of these concepts were original with Ford, but he synthesized them and then formed the nucleus of a mass-consuming market by paying his workers in 1914 a minimum daily wage of five dollars, unheard of before that time. One kind of car he made, the plain, utilitarian Model-T. Over nearly two decades, some twelve million Model-T's were sold before the last one was assembled in 1927. Its day was over. The worker-consumer had become more affluent and now was able to demand more luxury and variety in his acquisitions.

Marvels of productivity were achieved in many other industries employing basic principles much as Ford had conceived them. It should, however, be re-emphasized that the accelerating use of machines did not eliminate the need for a large work force. There were many industrial jobs to be done, except of course, during bad times when both men and machines were idle. Men operated those machines which were not fully automatic, they performed certain manual assembly functions, they fed work to the equipment, transferred it between stations and removed it when fabrication and assembly were completed. The "automatic factory" anticipated by Oliver Evans was still nowhere a reality, in the sense of an industrial system completely without human workers.

Not until after World War II, was the term "automation" used in the United States. That was in 1947, when a group of manufacturing specialists in the Ford Motor Company was set up to design material-handling mechanisms. They were called "automation engineers." Automation has been variously defined since then by many people; but according to Bright[7], true automatic production must include (1) machines to perform the production operation, (2) machines to move materials from one work station to the next, including feeding the machines, and (3) control systems to regulate the performance of production and handling systems. The most significant post-war trend has been toward the development of more versatile and sensitive control systems. The use of electronic computers in these systems has begun to amplify this trend to the point where new types of industrial and social phenomena may be appearing. Donald N. Michael has coined the term "cybernation" to fit the whole process, and defined it as "the electromechanical manipulation of material things by automation, and the electromechanical manipulation of symbols by computers, and ... the simultaneous use of computers and automation for the organization and control of material and social processes."[8]

For the first time in the history of man's attempt to free himself from unrelenting labor, the prospect appears that the mental component of that labor may also be eliminated or at least many so believe. To see how this has come about, it will be useful to survey briefly the history of modern computers in the last section of this chapter.

PART III THE DEVELOPMENT OF THE COMPUTER

Labor, if it is properly defined, does not refer to physical effort alone, but in early times that is what it was. Before the 15th Century the reduction of mental toil could not be considered to have been a very pressing need. Drudgery in this form came, in large part, as an accompaniment to the accumulation of great wealth in the capitalist societies, and the application of mathematics to the unfolding problems of the new science. Banks began to extend their influence and responsibility in order to deal with large increases in the volume of money and the development of the credit system. Insurance companies were recognizing the need of statistics for figuring risks. A strong empirical foundation for astronomy and the arts of navigation and surveying was essential. All of this meant that huge arrays of numbers would have to be managed somehow.

Man had grasped the concept of numbers in prehistoric days. By the time he began to keep track of livestock, he required simple aids. This need could be met by tallying with knots tied in a cord, or with piles of pebbles.

The earliest mechanical counting device of any sophistication, the *abacus*, evolved from such methods. In its primitive form, the abacus was a board or tray, (*abax* is Greek for counting board) with grooves cut in the surface. Each pebble in the first groove represented a single unit, while in the second groove a pebble stood for ten units (if ten were used as the base of the number system), and so on. Many early civilizations throughout the Middle East and the Orient used the abacus in a variety of forms, and in some countries it is still a popular means of calculating. A typical abacus will have beads on wires in two groups separated by a bar. A bead in the upper group counts five, while a bead in the lower group counts one. Each bead will have a numerical value only when it is moved toward the separating bar. Figure 2-7 shows an abacus of the modern Japanese type,

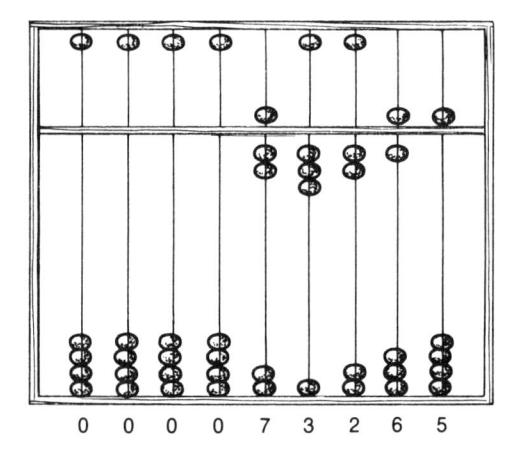

Fig. 2-7. A modern abacus.

representing the number 73265. A skillful operator of the abacus can perform arithmetic calculations with remarkable speed, but after lengthy periods of time, the physical and mental demands on him are considerable.

The abacus provided a physical means of representing large numbers according to the position of the digit symbols. Sometime around the 1st or 2nd Century A.D., mathematicians from India combined the positional representation of numbers with the decimal scale, including zero as one of the digits. Several centuries later, Arab scholars brought the concept to Europe through Spain, and thus the Western world came to call it the system of "Arabic" numerals.

In those countries which adopted the Arabic system, the abacus no longer remained an important aid to calculation. Pledge [9] points out that writers of commercial arithmetics from about the 14th Century on were able to employ this system to develop rules for calculation which would have been out of the question with the cumbersome Roman numerals. The availability of cheaper writing paper facilitated extensive calculations, and provided a means of preserving the record of those calculations, an advantage denied to the simpler calculating devices.

The 15th, 16th and 17th Centuries saw the growth of a class of professional mathematicians whose job it was to produce mathematical tables for commerce and science. The addition and subtraction of lengthy columns of figures must have been monotonous enough, but multiplication and division were even more laborious and exacting, the chances of error being so much greater. That there was a correspondence between arithmetic and geometric series had been known since early times. As seen below

$$1.\ 2.\ 3.\ 4.\ 5.\ldots$$
$$3.\ 9.\ 27.\ 81.\ 243.\ldots$$

the successive additions of the first number, (1), to each number in the top row produces a series which corresponds to that produced by multiplying the first by each succeeding number in the second row. e.g.,

$$1+2=3,\quad 1+3=4,\text{ etc.}$$
$$3\times 9=27,\quad 3\times 27=81$$

It would naturally occur to men engaged in computing to wonder how this kind of relationship could be used to make their work easier.

John Napier discovered a way. Napier was not a professional calculator, but a Scottish nobleman who spent a great deal of his time studying mathematics. In 1614 he announced the invention of logarithms in a work entitled *Mirifici Logarithmorum Canonis Discriptio*. Logarithms are only exponents but in Napier's time the theory of exponents was not commonly understood, so that he had to spend years of patient work to arrive at his result. The reason can be seen if we note that while the exponents in the following two equations:

$$4=2^2\quad\text{and}\quad 8=2^3$$

are clearly successive integers, the unknown exponent in the equation $5=2^x$ must be something between 2 and 3. Again, looking at a row of numbers above a matching row of logarithms

$$\begin{array}{lcccccc}\text{Numbers} & 1 & 2 & 4 & 8 & 16 & 32\\ \text{Logarithms} & 0 & 1 & 2 & 3 & 4 & 5\end{array}$$

we see that multiplying two numbers in the first row, e.g., $2\times 8=16$ corresponds to adding logarithms in the second row, e.g., $1+3=4$. The logarithms are, of course, the exponents of the numbers expressed as powers of 2. To produce a continuous table of numbers with corresponding logarithms, e.g.

$$\begin{array}{lllllllll} \text{Numbers} & 1 & 2 & 3 & 4 & 5 & 6 & 7 & 8 \ldots \\ \text{Logarithms} & 0 & 1 & - & 2 & - & - & - & 3 \ldots \end{array}$$

values for the blank spaces could only be reached by tedious approximations. The value of x in the equation $5 = 2^x$ would be approximately 2.3222, for example.

Napier's work was not, in fact, concerned with the logarithms of natural numbers but of trigonometric ratios since his primary interest had to do with astronomical tables. In 1616, Henry Briggs, an English mathematician, proposed that Napier's idea be made more useful by altering the scale to a base of 10, rather than 2. In 1624 Briggs published a book of tables containing the logarithms of 30,000 natural numbers to 14 places of figures.

Practicing mathematicians soon realized that logarithms could be plotted on a straight line, and then multiplication and division could be carried out by adding and subtracting lengths, using dividers. The next step was to construct two such lines so that they could slide past each other. This was the forerunner of the slide rule, an instrument which is still employed to provide flexibility and speed in the solution of many problems. The slide rule is an example of an *analog* computer. It differs from the digital computer in that, while the latter *counts* discrete numbers, the analog computer *measures* continuously varying physical quantities representative of calculations with numbers.

Truly mechanized methods of calculation also had their beginnings in the 17th Century. The eminent mathematical philosophers, Blaise Pascal and Gottfried Leibniz, each designed and built machines which are historically interesting. Pascal's father was the tax assessor for Rouen, a job which required interminable calculations. He regularly worked until the early morning hours copying long columns of figures. Young Blaise, who worked as his father's assistant, felt that it should be possible to make a machine to handle the dreary chores. He was superbly gifted in mathematics, (at sixteen he had composed the "Essay on Conics," containing a widely praised theorem), and the machine he finally completed in 1645 after many experimental models, demonstrated a high order of mechanical inventiveness as well.

A fundamental requirement for mechanical addition is a means of "carrying" automatically. Pascal's machine accomplished this with an ingenious arrangement of wheels and ratchets. In Fig. 2-8 we see a lower

Fig. 2-8. Blaise Pascal's arithmetical machine.

row of wheels each of which can be rotated to a desired setting, and the rotation transmitted by gears to a result wheel. The relevant number on the result wheel appears at the little window. The result wheels are connected to each other by a ratchet device, and when one wheel passes from nine to zero the ratchet causes the wheel to the left (representing the next significant digital position) to move one unit. Pascal never made his fortune selling these machines as he had hoped to do. Perhaps the costs were too high, and maintenance too difficult. In any event it was the precursor of the present day mechanical calculators. Pascal's sister later wrote in her brother's biography, "He reduced to mechanism a science which is wholly in the human mind."

Mechanizing multiplication was the next step, and this Leibniz managed to do in 1671. His machine had adding wheels similar in operation to Pascal's, but also included a unique feature, the "stepped wheel." Referring to Fig. 2-9 one can see the nature of Leibniz's innovation. A series of these stepped wheels mounted on a single shaft represents the digits of the multiplicand, which can be set by pushing the smaller pinion wheels horizontally to a position where they engage the appropriate number of teeth. Each multiplier digit is simulated by the appropriate number of rotations and horizontal position of the stepped wheels. The stepped wheels then actuate addition wheels similar to Pascal's.

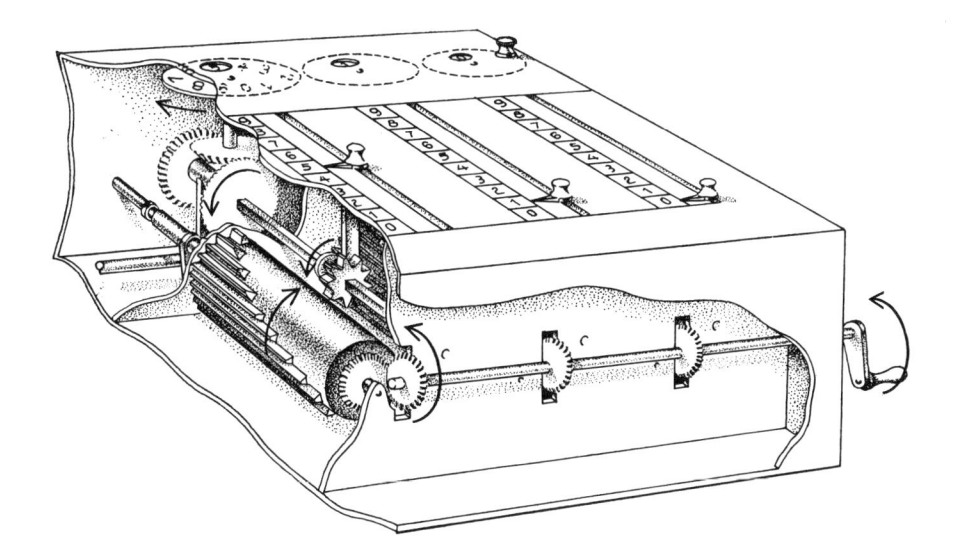

Fig. 2-9. Leibniz calculating machine. (© Deutsches Museum München.)

The Leibniz machine broke down too frequently to be of much practical use, but it provided a model for later mechanical calculators. Neither his machine nor Pascal's, however, was in any real sense automatic.

CHARLES BABBAGE

When, in the 19th Century, the dream of automatically powered machines had become a reality, it was reasonable to wonder how mathematical computations also might be performed without a human operator. The genesis of the solution came from the fertile mind of the English mathematician, Charles Babbage, and if anyone could be called pre-eminent among the pioneers of the modern computer it would be he. Babbage was born the son of a wealthy banker, in 1791, in the time of England's early technological ferment. Through adolescence and his years at Cambridge he was zealously absorbed in the study of mathematics, but not to the exclusion of other activities. He was quick-witted and gregarious, a popular host, and an indefatigable experimenter. Still one would not have expected him to spend a large share of his lifetime trying to build a machine with a seemingly preposterous conglomeration of more than 50,000 wheels, gears, cams and other parts.

In Babbage's own *Passages from the Life of a Philosopher*, he stated that while he was at Cambridge the possibility of calculating a table of logarithms by machinery occurred to him, but he did not at that time turn to the task. He took his degree in 1814, married, and fathered several children by 1820. That was the year he helped found the Royal Astronomical Society. Shortly after, he and his long-time friend John Herschel were asked to prepare certain tables for the Society. After developing the formulas they hired independent mathematicians to do the calculations, but were disturbed to find many errors in the results. The two had been spending wearisome hours verifying figures one day when Babbage cried "I wish to God these calculations had been executed by steam." Herschel replied, "It is quite possible."[10] From that time, and to the end of his long life, Babbage devoted his energies to building a "calculating engine."

At first things went promisingly. His conception of a machine capable of computing any table by means of differences was considered by a committee of the Royal Society which then reported a favorable opinion to the government, and as a result a decision was made to give funds to Babbage for construction. The government advanced substantial sums over a period of years (a total of seventeen thousand pounds was granted altogether) but it did not prove enough for what he wanted to do, even though he also spent large amounts from his personal fortune.

The difficulty was in the translating of his ideas to mechanical form. Babbage himself had, at the start, no experience with machining or manufacturing methods, and although he proved astonishingly capable of inventing new tools and techniques, he had to rely on others for the actual making of the parts. He engaged Joseph Clement (a former employee of Henry Maudslay) and other workmen of exceptional skill; but Maudslay himself would not have succeeded in overcoming the obstacles to Babbage's intentions. Not only was the metal-working technology of the time insufficiently advanced to produce all of the non-standard parts he designed, but also Babbage's dream was inconstant; he was continually seeing possibilities for improvement, and changing the designs after the workmen had begun.

His Difference Engine embodied a fairly simple, if novel, conception which promised to achieve the practical end of facilitating the preparation of valuable scientific tables using the "method of differences." Referring to Table 2-1, constructed for the function, $y = x^2 + 2x + 3$, we can see the

Table 2-1 Table of differences.

x	y	D_1	D_2
0	3		
		> 3	
1	6		> 2
		> 5	
2	11		> 2
		> 7	
3	18		> 2
		> 9	
4	27		> 2
		> 11	
5	38		

Function: $y = x^2 + 2x + 3$

principle involved. The second column gives values of y corresponding to unit increments of x in the first column. Column D_1, contains numbers representing differences between successive values for y, while column D_2 gives the differences between the successive numbers in D_1. Had the highest power term been a cubic there would be three columns of differences before a constant difference would be reached, and so on. Obviously, one could calculate as many additional values of y for unit step increases in x as he chooses by straightforward addition instead of substituting in the original equation each time. Tables of values of most functions can be computed in this way.

Mechanization of the basic process was entirely feasible; in fact much later George Schuetz, a Swede, after a number of years of work completed a machine based on a published description of the principle of Babbage's Engine. His machine could tabulate four orders of difference to fourteen decimal places. Had Babbage been satisfied with that kind of capability, rather than striving for six orders of difference and twenty places of decimals, he too, might have been more successful.

Instead, frustration piled on frustration, and in 1827 there was also profound tragedy. That year his devoted wife, aged thirty-five, his father, and two of his children died. This was almost too much for him to bear, and for the sake of his health he went traveling on the Continent for more than a year. A man of his irrepressible curiosity and energy could not, however, long allow his activities to languish. He spent a good deal of his time visiting foreign manufacturing establishments. Many of his observations are included in a book he published in 1832, entitled *Economy of Manufactures and Machinery*. This was an influential account of manufacturing methods of the times and, like most of his work, remarkably prescient. It anticipates many aspects of modern Operations Research, a form of management science.

The work on the Difference Engine had not proceeded very rapidly in his absence, nor would it when he returned. Moreover, some questions were beginning to be asked about the spending of more government money on such a doubtful enterprise. Some of the initial good will toward Babbage's idea had by this time evaporated, in no small measure because of his contentiousness in matters scientific, political and personal. His gift for making loyal friends was fully matched by his capacity for developing potent enemies.

The arrangements for the government to pay the bills had never been clearly defined, so the payments to Joseph Clement and his workmen were always irregular; increasingly so as time went on. Finally, in 1833, Clement left the project entirely; and, as he had legal sanction to do, took the special tools and some of the drawings with him. This was a serious blow to Babbage, but his restless mind had already been conceiving another machine that would do all the Difference Engine was designed to do, and much more. He again approached the officials in the government in 1834, to find out to what extent they were still interested in completing the Difference Engine; or if they would prefer that he try to develop the Analytical Engine, as he called it, instead. He was unable to get a forthright reply either then or for a number of years after. Finally in 1842 he received a letter stating that the government intended to abandon its interest in the machine in any form.

It is doubtful that any of the officials who concurred in this decision understood the true significance of the Analytical Engine. Far from being merely a more elaborate version of the Difference Engine (which could really do nothing but add) it was designed to carry out the operational requirements for *any* conceivable class of arithmetic calculation. The distinctive characteristic giving it this generality resulted from Babbage's incorporation of an idea for the automatic control of industrial looms which had been perfected in 1801 by the French inventor, Joseph Marie Jacquard. Babbage himself described the principle as follows: [11]†

> ... It is known as a fact that the Jacquard loom is capable of weaving any design that the imagination of man may conceive. It is also the constant practice for skilled artists to be employed by manufacturers in designing patterns. These patterns are then sent to a peculiar artist, who by means of a certain machine, punches holes in a set of pasteboard cards in such a manner that when those cards are placed in a Jacquard loom, it will then weave upon its produce, the exact pattern designed by the artist.
>
> Now the manufacturer may use, for the warp and weft of his work, threads which are all of the same colour; let us suppose them to be unbleached or white threads. In this case the cloth will be woven all of one colour; but there will be a damask pattern upon it such as the artist designed.
>
> But the manufacturer might use the same cards, and put into the warp threads of any other colour. Every thread might even be of a different colour, or of a different shade of colour; but in all these cases the *form* of the pattern will be precisely the same — the colours only will differ.
>
> The analogy of the Analytical Engine with this well-known process is nearly perfect.
>
> The Analytical Engine consists of two parts:
>
> 1. The store in which all the variables to be operated upon, as well as those quantities which have risen from the result of other operations, are placed.
>
> 2. The mill into which the quantities to be operated upon are always brought.
>
> Every formula which the Analytical Engine can be required to compute consists of certain algebraical operations to be performed upon given letters, and of certain other modifications depending upon the numerical values assigned to those letters.
>
> There are therefore two sets of cards, the first to direct the nature of the operations to be performed — these are called operation cards: the other to direct the particular variables on which the cards are required to operate — these latter are called variable cards. Now the symbol of each variable or constant, is placed at the top of a column capable of containing any required number of digits.
>
> Under this arrangement, when any formula is required to be computed, a set of

†From *Charles Babbage and His Calculating Engines* edited by Philip and Emily Morrison, Dover Publications, Inc., New York, 1961. Reprinted through permission of the publisher.

operation cards must be strung together, which contain the series of operations in the order in which they occur. Another set of cards must then be strung together, to call in the variables into the mill, the order in which they are required to be acted upon. Each operation card will require three other cards, two to represent the variables and constants and their numerical values upon which the previous operation card is to act, and one to indicate the variable on which the arithmetical result of this operation is to be placed.

But each variable has below it, on the same axis, a certain number of figure-wheels marked on their edges with the ten digits: upon these any number the machine is capable of holding can be placed. Whenever variables are ordered into the mill, these figures will be brought in, and the operation indicated by the preceding card will be performed upon them. The result of this operation will then be replaced in the store.

The Analytical Engine is therefore a machine of the most general nature, whatever formula it is required to develop, the law of its development must be communicated to it by two sets of cards. When these have been placed, the engine is special for that particular formula. The numerical value of its constants must then be put on the columns of wheels below them, and on setting the Engine in motion it will calculate and print the numerical results of that formula.

Every set of cards made for any formula will at any future time recalculate that formula with whatever constants may be required.

Thus the Analytical Engine will possess a library of its own. Every set of cards once made will at any future time reproduce the calculations for which it was first arranged. The numerical values of its constants may then be inserted....

What is fascinating about this description is that it contains important elements which can be recognized in modern automatic computers, including the memory and the arithmetic unit.

Some of the more specific details of the principles and operation of the Analytical Engine were never spelled out so clearly by Babbage as they might have been, but fortunately the machine did have, in L. F. Menabrea and Augusta Ada, Countess of Lovelace, two expositors of uncommon perceptiveness. Menabrea, an Italian mathematician, later to become a general with Garibaldi, wrote a lucid description in 1842, primarily referring to problems of programming the machine. This was translated into English by Lady Lovelace, who then appended an admirable series of notes of her own.

Lady Lovelace was the only child of Lord and Lady Byron. Byron had left England when she was an infant, never to return alive, and it is she to whom he referred in *Childe Harold's Pilgrimage*: [12]

"Is thy face like thy mother's, my fair child!
Ada! Sole daughter of my house and heart?"

She was a graceful young woman of extraordinary mathematical talents who became Babbage's cherished friend and confidante—his "lady fairy." About the same age as his daughter whom he had lost, and possessed of an ability beyond that of his surviving sons to comprehend his work, she was a constant encouragement to him as he tried to find ways to finance additional construction on his machine. These ways were often impractical. One plan was to build an automaton to play tick-tack-toe against human opponents, and to send it on tour. He was persuaded, however, that the possibility of a suitable monetary return for the time and effort he would have to expend on the project would be limited. Another scheme with a tragic aftermath, it turned out, involved an attempt to use probability theory to devise a winning system for betting on horse races. Lord and Lady Lovelace, both keen horse-fanciers, were collaborators. Lord Lovelace did not carry his interest very far. He assumed that his wife would also stop, but for her, betting (and proving the system) had become a fever, and she found herself deeply in debt to a group of bookmakers to the point of having to pawn the family jewels. Her time was running out. As early as 1850 her letters referred to her poor health. By 1852 it was apparent that she was in the terminal stages of an incurable malignancy. In her last agonized days she was forced to confess her situation to her husband, and to leave to him, her mother and her old friend Babbage the task of protecting her name. She was thirty-six when she died.

Babbage was to live on nearly to the age of eighty. He was often bitter toward what he considered the sycophancy of many of his fellow scientists, toward the organ-grinders and street-hawkers who disturbed his concentration and especially toward the fates which seemingly mocked his efforts to build the Analytical Engine, although his bitterness was sometimes tempered with a mordant wit. He is said to have remarked in his later years that he could not remember one completely happy day in his entire life[13]. One wonders if, most days of his life, Babbage had ever taken time to consider the state of his happiness. His mind had never been at rest. A man of as many parts as his machines, he had probed the natural and physical sciences, publishing papers in the fields of archeology, astronomy, on magnetism, and many in mathematics. He had suggested practical ideas for improving the safety of rail and ocean transportation, and it was his study of the postal system that had led to the establishment of the penny post in England.

Still, his monument was the unfinished Analytical Engine, a machine about which Menabrea had written "although it is not itself the being that reflects, it may yet be considered as the being which executes the conceptions of intelligence." He wrote this knowing that the engine, if completed, would be capable, as Lady Lovelace stated, of "feeling about to discover which of two or more possible contingencies has occurred, and of shaping its future action accordingly." In other words, it would have the power to choose the next stage of any computations based on the results of steps already completed. When, well into the 20th Century, computing machines with this kind of capability were finally built, Babbage's status as the pioneer of a powerful new technology was at last recognized.

THE AUTOMATIC COMPUTER

We note that approximately one hundred years would pass between Babbage's original conception of the Analytical Engine, and the development of the first fully automatic computer. This long interim was in part a measure of his genius, but, also, of the difference between 19th and 20th Century technologies and needs.

There was naturally a constant improvement in calculating devices of various kinds. In the later years of the 19th Century, technology had advanced to a stage where precision parts could be produced on an industrial scale. An inventor could hope to have a mechanism of moderate complexity manufactured in quantity. A number of devices which would qualify as forerunners of modern accounting machines appeared about this time. The earliest calculating machine with a keyboard was designed by Parmalee in 1850, but a key-driven machine acceptable for commercial purposes was not made until 1885. In 1892 Steiger's "Millionaire" became the first direct multiplication machine produced on a large scale.

By this period the use of electricity for industrial and commercial purposes was beginning to increase. In the 1880s, Herman Hollerith, a statistician with the U.S. Bureau of the Census, together with an associate, James Powers, conceived a way of combining the speed of electricity with the Jacquard punched-card control system to tabulate the 1890 census. To them this was an urgent need. It was obvious that the method which took seven years to complete the 1880 census, involving fifty million people, would be totally inadequate for subsequent decennial counts. In the process they succeeded in devising, the individual cards were coded by punching. The information denoted by the positions of the punched holes was recorded by means of wires passing through the holes and making electrical contact with a pool of mercury underneath the card. Each time contact was made, a mechanical tabulator was actuated. Most answers to census questions could be phrased to give a yes or no answer which would be indicated by the presence or absence of a current. The process worked, and Hollerith eventually established a company to exploit his patents. It merged with two others to form the Computing-Tabulating-Recording Company which years later became IBM Corporation.

Electromechanical punched-card machines which evolved directly from the work of Hollerith and Powers were essentially limited to single-purpose operations. Commercial data-processing installations often required a number of separate machines for purposes such as key-punching, sorting and tabulating and this is still the case. These are labor-saving devices of great importance, but they did not eliminate the need for data-processing machine operators, nor did they provide the flexibility needed for computing of a less routine kind. The capacity, for instance, for the machine itself to choose alternative operations, i.e., "branching," which Babbage had anticipated was basically lacking.

The modern era of automatic digital computers began during World War II. Three developments of this period stand out in any history of computers. The first was the construction of a machine called the Automatic Sequence Control Calculator, or Mark I, by Howard Aiken and his associates at Harvard University. Work on this machine was begun in 1939 and completed in 1944. Originally designed for the solution of ordinary differential equations, the Mark I followed a programmed sequence of instructions contained on punched tape. It was the first automatic digital computer and during its operational life of some fifteen years it carried on many useful functions. In its original form, however, it did not have an ability for branching and, furthermore, since its operation was strictly electromechanical the speed of computation was seriously limited.

A second development marked a decisive step into contemporary computer technology. This was the building of the ENIAC (Electronic Numerical Integrator and Calculator), the first electronic computer. The ENIAC was a collaborative project between the Moore School of Engineering at the University of Pennsylvania and the United States Army. In the war years there were many urgent military needs of an applied science nature. The calculation of tables describing the proper trajectories of shells for a great variety of weapons was such a need, but the magnitude of the job was defeating the efforts of a large staff of employees. To try to cope with the situation J. Presper Eckert, an electrical engineer, and John Mauchly, a physicist, at the Moore School, began to design a computer. Herman H. Goldstine, a mathematician on liaison duty with the Army Ordnance Corps, succeeded in interesting the military service in providing backing. The ENIAC when finally completed in 1946 was a massive thirty ton machine covering over 1500 square feet of floor space. It used 18,000 vacuum tubes in place of relays and other partially mechanical components. This caused wastage of 150 Kilowatts of power as heat and posed a continuous maintenance problem, since the tubes were relatively short-lived. When operative, it was an impressive performer—in one hour it could complete calculations which might have taken Mark I a week to do.

However, the continued evolution of computers would have been forestalled by the basic limitations of the vacuum tube, had it not been for the discovery of a new electronic amplifying device, the germanium transistor, by Shockley, Bardeen and Brattain in 1947. The transistor was a prime example of an important technological invention which resulted directly from a theoretical investigation. When developed, it made possible more reliable electronic equipment, designed on a much smaller scale.

In 1946 the war was over, and the work being provided for the ENIAC was more varied than the essentially single-purpose function for which it was first conceived. Each new kind of problem required a different program to be set up manually in the machine. This was time-consuming, and like the trouble with the vacuum tubes represented a fundamental disadvantage of the original ENIAC design. The answer to this difficulty came through the conception of the "stored program." This third basic advance in the design of computers was proposed in an influential report entitled "Preliminary Discussion of the Logical Design of an Electronic Computing Instrument." This report, published in 1946, was prepared by John von Neumann, in association with Herman Goldstine and Arthur W. Burks. Von Neumann, who died in 1957, is generally recognized as having had one of the great theoretical minds of this century. He had become interested in computer design through a chance meeting with Goldstine during the war years.

The idea of the stored program is to store, in the memory of the machine, codes which represent the basic instructional operations. With instructions and data both stored in the same way, the machine itself can automatically execute commands given to it in a coherent mode. The general progression would ordinarily be from one programmed instruction to the following instruction, but a stored program machine can also execute an instruction which is not in sequence by branching to another instruction. For example, if the machine has performed an arithmetic computation, the result can be tested to determine the execution of the next instruction; e.g., if the result is positive the normal sequential instruction will be executed, but if negative the machine will branch to another instruction. The machine can thus control the progress of the computation with great flexibility.

Von Neumann and his associates were aware that a positive-negative (or yes-no) dichotomy was crucial to the operation of the automatic digital computer. This was recognized in the design and construction of subsequent machines by various groups, including one supervised by von Neumann himself at Princeton. The basic function of the automatic control is to close or open circuits, or to detect circuits that are closed or open. It is understandable, therefore, that the instructions given the machine, as well as the calculations which take place in it, are best expressed as binary alternatives. That is to say, those instructions or the calculating processes are stated only by the symbols 0 or 1 in some ordered sequence. To comprehend the significance of this, one may find it helpful to recall that the mechanical calculators described earlier operated with wheels having ten teeth, each representing one of the ten digits of the decimal system. The numerical "weight" of each wheel depended on its position in a right-to-left sequence. The ENIAC also stored numbers in decimal form, but electronically. The intrinsic bi-stable character of the components of an electronic computer, (i.e., a pulse is present or not present, or a core is magnetized in a clockwise or counter-clockwise direction), makes it much more effective to use a binary number system. Briefly, the binary system employs the digits 1 and 0, only. A binary number is positional, and is made up of a series of additions, just like a decimal number, but with powers of two rather than ten. The table below shows how we would represent decimal 15 as a binary number.

Table 2-2.

Power of 2	3	2	1	0
Decimal number (15)	8	4	2	1
Binary number	1	1	1	1

Thus the binary code for 15 is 1111, meaning $2^3 + 2^2 + 2^1 + 2^0$.

The rules of multiplication and addition are a simple few in binary arithmetic. For example, for binary addition:

$$0 + 0 = 0$$
$$0 + 1 = 1$$
$$1 + 0 = 1$$
$$1 + 1 = 10$$

meaning: two different digits added together produce 1, while the addition of two similar digits, 1 and 1, requires us to carry 1.

For binary multiplication:

$$0 \times 0 = 0$$
$$0 \times 1 = 0$$
$$1 \times 0 = 0$$
$$1 \times 1 = 1$$

meaning: if either digit is 0 so is the result, but if neither digit is 0, the result is 1.

While the machine works on binary instructions, a programmer ordinarily finds it more convenient to use symbolic language, a kind of script which gives the computer directions for handling each message. The translation process which can readily be performed by the machine need not concern us here.

COMPUTERS AND LOGIC

If a computer were confined to performing ordinary arithmetic computations it would still be a most useful contrivance, but its capability is not so restricted. The binary concept also permits the characterization of the digital computer as a logical machine. This means that it can deal with information patterns of a qualitative, but systematic, kind; to the point in fact that we can begin to think of its operation in terms of simulating subtle human thought processes.

One simple definition of logic, appearing in *Webster's New Collegiate Dictionary*, is that it is: "the science of the formal principles of reasoning." Ever since the time of Aristotle, philosophers have sought to discover a method of correct thinking which would lead the way to truth. Aristotle contributed the *syllogism*—a set of three propositions of which the last follows from the accepted truth of the other two, e.g.,

> Major premise: man is a rational animal
> Minor premise: but Socrates is a man
> Conclusion: therefore Socrates is a rational animal

Statements of this kind have a recognizable resemblance to the proposition: if $A = B$ and $C = A$, then $C = B$. The syllogism as a logical device has demonstrable weaknesses, but it was a giant first step toward a system of notation representing ideas by numbers, letters or other symbols, which can then be formally manipulated to obtain correct conclusions, i.e., symbolic logic.

It is not necessary for our purposes to consider the historical progression of logical thought until we come to the 19th Century. Largely because of the work of Augustus De Morgan and George Boole, logic experienced a renewal at that time. Professor De Morgan, an eminent mathematician, was a friend of both Charles Babbage and Lady Lovelace. He recognized the importance of mathematical processes to the study of logic, and developed numerical logical systems. George Boole, an obscure English mathematician who taught at Queen's College in Ireland, published his seminal work "An Investigation of the Laws of Thought on which are found the Mathematical Theories of Logic and Probabilities" in 1854. He discussed the striking analogy existing between algebraic symbols and those indicative of logical forms and syllogisms. Boole's work was confined to the limbo of presumably impractical academic thought for many years, of interest only to some philosophers. In 1938, Claude Shannon, an M.I.T. student working for his M.Sc. degree, wrote a thesis entitled "A Symbolic Analysis of Relay and Switching Circuits" in which he demonstrated the applicability of Boolean logic to electrical circuitry.

In Boolean algebra, unlike ordinary algebra, the symbols representing variables are restricted to either one of two states. This corresponds to what we can call two-valued logic—it deals with propositions that are either true (truth value of 1) or false (truth value of 0). The two-state elements can be manipulated in a series of operative conjunctions (*and, or, or else*, and *except* are conjunctions Boole proposed) to arrive at logical conclusions. Shannon pointed out that if series connections in a network were to be represented by the operation, *and*, and parallel connections by the operation, *or*, the functioning of the networks could be expressed by this kind of algebra. An examination of computer logic is not at all within the scope of this book, but a very simple illustration may help show its significance to design problems.

Using the symbols $\wedge$ for *and*, and $\vee$ for *or*, one can construct a "truth table" which shows the truth of a statement or its falsity for all truth values of the elements of that statement. The elements are also propositions whose truth values were originally assumed.

Table 2-3.

Possible combinations for 2-element statement		Truth values for statements	
P	Q	$P \wedge Q$	$P \vee Q$
0	0	0	0
1	0	0	1
0	1	0	1
1	1	1	1

It can be seen, for example, that the *and* connective requires two trues for the statement to be true, but with the *or* connective one true will be sufficient to make the statement true. If we consider two switches in series in an electrical circuit, current will flow if, *and* only if, both switches are closed, i.e., P and Q are both true. This is an *and* circuit. Thus as shown in the sketch with only Q true, current will not flow. In the circuit shown below, with two switches in parallel, current will still flow if either one switch *or* the other is closed, *or* both are closed. This is an *or* circuit.

The examples are perhaps fairly trivial but extensions of these concepts make it possible to deal with very complex networks which would not yield readily to straightforward engineering reasoning. Moreover, the method can play a powerful role in program design. A computer programmed on logical principles can act as a decision maker, stopping a production line *if* a seam welder is turning out defective seams, *or* the temperature in a galvanizing bath is too hot, *and* a beta gauge shows finished sheet steel to be thicker than tolerance limits. In programming for decisions critical to the nation's defense, combinations of conditions numbering in the thousands may have to be checked almost instantaneously—an impossible task for unaided human capabilities. In a few short years, computer systems have become indispensable—although, to be sure, not infallible—in situations of this kind. They are not infallible, it must be added, not only because of the possibility of mechanical errors, but also because much remains to be learned about interactions among subcomponents of such complex systems.

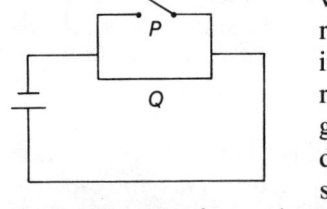

THE AGE OF COMMUNICATIONS

So it happened that by the middle 1950s, the abstract thought of mathematicians coupled with the development of ingenious electronic technologies created not only a new industry, but also a catalyst to help quicken the tempo and reshape the structure of industrial society.

In the United States, more than any other country, both the will and the way existed to allow breaking through to a condition of unexampled productivity and change. The way had been found much earlier. The systematic application of science, (often imported, to be sure), to technology inspired a commitment to research and innovation which promised an eventual technological answer to all material wants. The will was to some degree inherent in the American dedication to profits and consumption; but war was the spur that goaded the scientific, engineering and industrial communities to prodigious accomplishments. The fears which war perpetuates have since helped sustain an unprecedented level of support for research and development, resulting in a spiraling technical collectivism. Countless generations of technological achievement have been compressed into little more than a half century. Great jet airliners and space trips to the moon hardly seem to belong in the same millennium with the flight of the Wrights at Kitty Hawk. Multi-megaton weaponry and breeder reactors for nuclear power appear eons away from the tiny laboratories of Becquerel and the Curies. Many mature Americans fondly remember the days of their youth spent on isolated farms which have now nearly all disappeared as industrial methods of agriculture have taken over.

The changes have been so sweeping that some observers of the contemporary scene now proclaim the advent of a new kind of society in which the production of material goods through the expenditure of mechanical energy no longer serves as the basis for the technological system. Instead, they see the central functions required for human existence or amenities audited and controlled by information transmitted by energy in its electronic form. The importance of the media of communication in such a society is paramount and the computer as a tireless processor of energy is a vital link. One likely consequence of such a system would seem to be a changing relationship between man and his work. This, the following chapter will seek to explore.

REFERENCES

1. Stowers, A., "Pumps and Water-Raising Machinery," *Engineering Heritage*, Vol. I, Inst. of Mechanical Engineers, London, 1953, p. 42.
2. Fleming, A. P. M. and Brocklehurst, H. J., *A History of Engineering*, A. & C. Black, Ltd. London, 1925, p. 183.
3. Derry, T. K. and Williams, T. I., *A Short History of Technology*, Oxford University Press, London, 1960, p. 332.
4. Wailes, Rex, "Richard Trevithick—Engineer and Adventurer," *Engineering Heritage*, Vol. I, Institution of Mechanical Engineers, London, 1963, p. 70.
5. Hoare, R. A., "Joseph Bramah, Multiple Inventor," *Engineering Heritage*, op. cit., p. 97.
6. Burlingame, Roger, *Backgrounds of Power*, Charles Scribner's Sons, N.Y., 1949, p. 144.

7. Kranzberg, Melvin and Pursell, Carroll W., Jr. (eds.), *Technology in Western Civilization*, Vol. II, Oxford University Press, N.Y., 1967, p. 642.
8. Kranzberg and Pursell, ibid., p. 655.
9. Pledge, H. T., *Science Since 1500*, Harper & Bros., N.Y., 1959, p. 48.
10. Moseley, Maboth, *Irascible Genius, A Life of Charles Babbage*, Hutchinson & Co., London, 1964, p. 65.
11. Babbage, Charles, *Passages from the Life of a Philosopher*, Longman, Green, Longman, Roberts & Green, London, 1864, pp. 55–56. Reproduced in *Charles Babbage and His Calculating Engines*, Morrison Philip and Morrison, Emily Dover, N.Y., 1961.
12. Moseley, Maboth, op. cit., p. 156.
13. Bowden, B. V. (ed.), *Faster Than Thought*, Sir Isaac Pitman & Sons, Ltd., London, 1953, p. 18.

Recommended Bibliography

Ferguson, Eugene S., *Bibliography of the History of Technology*, M.I.T. Press, Cambridge, Mass., 1968.

3

Industrialization, Cybernation and the Working Man

"Work is no disgrace; it is idleness which is a disgrace."
HESIOD, *Works and Days*

The momentous industrial changes in the United States in the first three decades of the 20th Century brought a standard of living which would have been unimaginable for the laboring class in the preceding century. But the working man was soon again to know the sting of hard times and to be reminded throughout the 1930s that technology had not, after all, succeeded in eliminating poverty and hunger. Instead, it appeared to many that the industrial system had finally and irretrievably failed.

The coming of World War II brought new life to technology and industry, and this time it was as though the system had reached critical mass. The fuel was applied research, now institutionalized, and for the first time generously endowed by both government and business. Waves of innovation, most dramatically in the technologies of communication, have fostered new industries to challenge the old.

But there has persisted an unacceptably large group of citizens who have been left relatively unaffected by the general prosperity which has accompanied the technological transformation; and even for the working man and the other members of the society who have profited, the material benefits have had a cost. Change and adjustment have been demanded of everyone, but social adjustment occurs only with painful slowness. Many inadequate socio-political institutions and attitudes, some formed as long ago as the start of the Industrial Revolution, continue to serve as the basis for thought and action. The coupling of technological dynamism with these unsuitable means for dealing with change has produced serious torsions. This chapter will discuss the roots of these problems, and will examine remedies which have been proposed for the decade of the 1970s.

POST-WORLD WAR II EMPLOYMENT PROBLEMS

The end of World War II was a pivotal period in the United States for reasons besides the nascent developments in computer technology. For millions it was a time of joy and relief, knowing that the ugly travail of war was over at last; but mixed with this feeling was a sense of apprehension about the future. The commanding figure of Franklin Delano Roosevelt had passed from the scene and the new president was still a little known quantity. Through President Truman's order to drop the atomic bomb, men had recently become frighteningly aware of the secrets in the nucleus of the atom, finding there a malignancy which could infect their hopes for a new start.

For most men just out of uniform, however, the concern was more immediate and personal. What kinds of jobs would there be for them as civilians? Previous wars had been followed by rending depressions. Who would be bold enough to say that after this, the most devastating of wars, there would not be serious economic dislocations? After all, the memory of the Great Depression had not yet had time to fade.

In the face of this last concern, the Congress passed the Employment Act of 1946. Sometimes called the Full Employment Act, (although the "Full" was dropped during hearings), it was the first real attempt in the United States to bind the federal government to the permanent broad-scale commitment of providing jobs for its employable citizens. One could hardly have expected the representatives of the people to do less. The millions of veterans returning home from the war would not be disposed to stand quietly in breadlines as they, or their fathers, might have done a dozen years previously. They had by now seen lines enough to last their life-times. During the war there had been jobs for all who could work and to meet the voracious demands for more production automatic machinery and new techniques had been used as never before. The nation was geared for productivity, and no one knew if the pent-up demand for civilian consumer goods would take up the slack.

The Employment Act did not propose a firm answer to this question, nor did it suggest a mechanism for generating jobs, but rather it expressed more or less symbolically a recognition of Federal responsibility for maintaining an industrial and economic stability in the land. A jobless rate of above three percent, it was agreed, would be cause for concern and action. In the years since the passing of the Act the unemployment rate has almost always exceeded this figure and the response of the administration in power has varied, depending on world or domestic conditions as well as on the prevailing political and economic viewpoint.

Within this period too, a different dimension has been added to the problem. The focus of employment in this country has begun to shift away from the energy-consuming industries, such as steel-making and the manufacture of automobiles. This is not to say that these products are not being made in ever-increasing quantity; but instead that the numbers of workers required for their production have not grown in proportion to the output. The indexes of output in the steel industry in the United States, for example, increased from 100.0 (taken as a base) in 1957 to 113.4 in 1968 (with some negative yearly fluctuation in between). The indexes of output *per man hour* for the same period increased from 100.0 to 124.4.[1] The productivity per worker increase represented by these figures is characteristic of that found for many other manufacturing industries. Productivity has also increased in agriculture and is traceable to new methods of work organization and the introduction of more mechanical and chemical technologies. There has been a sharp reduction in the number of farms and the number of farm workers.

Steel-making and automobile manufacturing may never again need additional new workers to the degree that they have in the past. These are major industries (and there are others similarly affected) which provided sinews for America's economic growth in the maturing years by giving jobs to its millions. If the growth is to be sustained without the artificial stimulus of war, there must be other sources of employment for the country's expanding population — that is, unless the population can let itself be supported by the work of its machines.

The figures from Table 3-1 provide some feeling for the post World War II employment trends. For example, from the early 1950s until 1969, the relative increase, when it occurred, in the number of employees in "heavy" industries — mining, contract construction, manufacturing and transportation, and public utilities — was markedly less and more irregular than that for retail trade, finance, services, and state and local government.

THE KNOWLEDGE INDUSTRIES

The basis for this nation's economic growth is beginning to shift perceptibly toward what Fritz Machlup has named "the knowledge industries."[3] The term refers to the gathering, processing and

Table 3-1 Employees on nonagricultural payrolls, by industry division, 1947–1969 [2].

Year	TOTAL	Mining	Contract construction	Manufacturing	Transportation and public utilities	Wholesale and retail trade Total	Wholesale trade	Retail trade	Finance, insurance and real estate	Services	Government Total	Federal	State and local
1947	43,881	955	1,982	15,545	4,166	8,955	2,361	6,595	1,754	5,050	5,474	1,892	3,582
1948	44,891	994	2,169	15,582	4,189	9,272	2,489	6,783	1,829	5,206	5,650	1,863	3,787
1949	43,778	930	2,165	14,441	4,001	9,264	2,487	6,778	1,857	5,264	5,856	1,908	3,948
1950	45,222	901	2,333	15,241	4,034	9,386	2,518	6,868	1,919	5,382	6,026	1,928	4,098
1951	47,849	929	2,603	16,393	4,226	9,742	2,606	7,136	1,991	5,576	6,389	2,302	4,087
1952	48,825	898	2,634	16,632	4,248	10,004	2,687	7,317	2,069	5,730	6,609	2,420	4,188
1953	50,232	866	2,623	17,549	4,290	10,247	2,727	7,520	2,146	5,867	6,645	2,305	4,340
1954	49,022	791	2,612	16,314	4,084	10,235	2,739	7,496	2,234	6,002	6,751	2,188	4,563
1955	50,675	792	2,802	16,882	4,141	10,535	2,796	7,740	2,335	6,274	6,914	2,187	4,727
1956	52,408	822	2,999	17,243	4,244	10,858	2,884	7,974	2,429	6,536	7,277	2,209	5,069
1957	52,894	828	2,923	17,174	4,241	10,886	2,893	7,992	2,477	6,749	7,616	2,217	5,399
1958	51,363	751	2,778	15,945	3,976	10,750	2,848	7,902	2,519	6,806	7,839	2,191	5,648
1959	53,313	732	2,960	16,675	4,011	11,127	2,946	8,182	2,594	7,130	8,083	2,233	5,850
1960	54,234	712	2,885	16,796	4,004	11,391	3,004	8,388	2,669	7,423	8,353	2,270	6,083
1961	54,042	672	2,816	16,326	3,903	11,337	2,993	8,344	2,731	7,664	8,594	2,279	6,315
1962	55,596	650	2,902	16,853	3,906	11,566	3,056	8,511	2,800	8,028	8,890	2,340	6,550
1963	56,702	635	2,963	16,995	3,903	11,778	3,104	8,675	2,877	8,325	9,225	2,358	6,868
1964	58,331	634	3,050	17,274	3,951	12,160	3,189	8,971	2,957	8,709	9,596	2,348	7,248
1965	60,815	632	3,186	18,062	4,036	12,716	3,312	9,404	3,023	9,087	10,074	2,378	7,696
1966	63,955	627	3,275	19,214	4,151	13,245	3,437	9,808	3,100	9,551	10,792	2,564	8,227
1967	65,857	613	3,208	19,447	4,261	13,606	3,525	10,081	3,225	10,099	11,398	2,719	8,679
1968	67,915	606	3,285	19,781	4,310	14,084	3,611	10,473	3,382	10,623	11,845	2,737	9,109
1969	70,274	619	3,437	20,169	4,431	14,645	3,738	10,907	3,557	11,211	12,204	2,758	9,446

disseminating of information for some identifiable purpose. Peter Drucker calls information, "energy for mind work."[4] These are imperfect metaphors but the implications are clear enough. The "mind-workers" are those whose level of education and training surpasses that of the employees who formed the core of the work force a generation ago. Many of them are scientists, engineers and technologists, but their number includes formally trained scholars and professional people from a great variety of fields.

In the mechanical industries as well as in the information-handling enterprises there are highly educated professionals whose jobs have appeared to be inherently resistant to replacement by machines; just as there are workers in all fields whose skills are minimal but whose functions are non-routine enough that, unless their wages exceed a certain level, it would be uneconomic to devise machine substitutes for them. In those industries which process materials in a continuous flow, such as the oil refineries, fewer of the latter type jobs remain. The advantages of continuous flow are fully recognized in process design. As labor and production costs rise in other industries and if competition becomes more keen, there is a strong impetus to redesign and integrate processes to make them more amenable to machine control. The steel industry, for instance, has traditionally handled material in batches, but in recent years there have been many developments leading to continuous processing.

In the information-handling industries, while there are still subordinate jobs which do not require a high level of education, much of the information can and must be processed by devices — must be because its sheer volume would overwhelm any attempt to cope with it through a human work force. The classic example of this has been the telephone system, where the enormous increase in the use of telephones taxes even the most advanced electronic computer-controlled equipment. The day is long past that human operators, even if enough were available, could provide adequate service. It is almost a typical modern irony that the very use of computers magnifies the problem because computer networks at the present time are heavy users of telephone lines.

Clearly, computers in integrated systems can be used to perform many tasks other than the ones to which they have thus far been applied, and will be whenever this seems to be the most profitable course. It is equally obvious that these tasks need not be restricted to those at the least common denominator of intellectual content. Some redefinition and consolidation of jobs in lower and middle echelons of management has already begun.

Will, then, the future see a society in which critically large numbers of people are displaced from their jobs by machines? This is an enduring question which has troubled members of the working classes from the beginnings of the Industrial Revolution. It has also engaged the attention of many scholars concerned with the effects of technology and social change.

THE WORKER IN THE UNITED STATES: 1950-1975

Is it possible to decide whether the relationship between worker and machine has become primarily destructive or beneficent? Not simply, it seems; and not at all without tracing the work patterns of various groups and noting their links to changes in technology and social organization. Furthermore, distinctions between the effects of both energy-intensive and information-intensive technologies on the one hand, and the capitalistic culture on the other are important if often obscure.

The term "working class" has become an ambiguous way to describe those who work for a living at sub-managerial levels, and whose jobs might at first thought seem to be the most vulnerable to technological change. While there still may be some social distinction attached to the level of a person's education, his economic position, at least, is no longer strictly limited by a blue-collar status. Transcontinental truck drivers, plumbers, or rolling-mill operators in a steel plant will receive more pay than most teachers and many college-trained engineers, and their respective styles of consumption will often reflect this. At the same time, in spite of its status as an affluent country, the United States still has (in addition to a sizable body of unemployed), a "working poor," defined by an annual income below a certain level.

What are the basic effects of cybernation on job-holders in the many different categories? Opinions are mixed, but the fear that cybernation, like mechanization or automation before it, may cause displacement (however temporary) of jobs is one that is not easily laid to rest among working people. The computer revolution has indeed come too swiftly to allow us to capture its full scope and meaning, but we can try to broaden our understanding by examining certain trends in employment patterns.

CYBERNATION AND INDUSTRY

The use of computers in industry in the United States began in 1954 with the placing of a large-scale business computer in the Louisville plant of the General Electric Company. By 1961, at least 6000 computers had been installed by American manufacturers, and by the middle of 1968 it was estimated that they had leased or sold approximately 63,000 machines around the world of which some 45,000 or 50,000 were in the United States. Of these latter, one-third were in manufacturing establishments[5].

During a large part of this rapid growth period, from 1958 to 1964, the general unemployment rate was 5% or higher, and in 1961 it reached a level of 6.7%. This was the highest rate since the Depression, except for 1958 when it went slightly higher. The relation between the unemployment figures and the growth of the computer industry was not lost on some students of the national economy. They were also aware of some productivity figures for the total private economy which showed an average annual growth rate of 2% from 1909 to 1947, increasing to 3% from 1947 to 1962. Putting this information together, many found it logical to conclude that the accelerating use of computers and other new technologies was bound to have serious consequences for the employment market; and that within a decade or so it would be possible for, say, 10% of the work force to produce manufactured goods for the entire population. One of the most widely circulated public statements of this general position was made by the Ad Hoc Committee on "The Triple Revolution" in 1964[6]. This group consisted of computer experts, economists, business people and other respected observers of technological change. Their document was not left unchallenged. Bell[7] and Silberman[8] were among those who felt that the case had been overdrawn, and pointed out that at that time, ten years and more after computers had started to come into general use, no fully cybernated industry was to be found in the United States. In the several years following the publication of "The Triple Revolution" document, up until 1969, the unemployment rate dropped below 5% (to as low as 3.3% at the end of 1968) suggesting to those who denied any triggering effect of cybernation on unemployment that their position was vindicated. With the unemployment rate increased to well above 5% in 1970 and 1971 the question has surfaced once again. Of course, none of the responsible theorists on either side have relied wholly on data concerning productivity and the use of computers and other machines to reach their conclusions. Many other conditions, including military defense needs and shortage of capital or markets are important if not crucial and must be examined to arrive at a balanced assessment. Moreover, statistical data alone can be misleading, and at the very least ought to be illuminated by descriptive example.

THE WHITE-COLLAR WORKER

It can be instructive to look at how computers and control technology may have the effect of replacing workers; or conversely of generating new kinds of jobs. In industry, the companies which have used computers have done so primarily in the hope of improving profits, and less often to attempt to correct some refractory production or organizational problem (which may effect profits). Profits can hopefully be increased by applying computers for: (1) administrative and accounting purposes, (2) operations control systems, (3) product innovation and customer service, and more indirectly, (4) for improving staff work and management decisions[9]. Up to the present time the first of these applications has been predominant. If computers have begun to change employment patterns, one would suspect that the clerical workers would be the most affected.

Looking at Table 3-2 we see that in the period from 1967 through the first quarter of 1970, the number of employed clerical workers climbed steadily if moderately while the unemployment rate remained near the 3% figure. During the second quarter of 1970 the latter rate began to go substantially higher. The increase in the numbers of employed salesworkers was slight and erratic in this same initial period although their unemployment rate followed a pattern similar to that for clerical workers. As might be expected the other groups in the white-collar class—the professional and technical personnel and the managers, officials and proprietors—showed a consistently lower unemployment rate.

The figures in the table include employment in all businesses including insurance and finance so we can draw no definite conclusions about white-collar employment in the manufacturing industries alone. However since all businesses sharply increased their computer usage in the period before 1970 we can at least infer that there was no comparable marked unemployment trend directly attributable to computers among clerical personnel. Examination of individual cases, rather than statistics, has shown verifiable instances of computers eliminating clerical jobs of certain types. In most of these same cases, (but by no means all), the displaced workers can be absorbed after retraining in another phase of an expanding operation, or fitted into vacancies occurring through retirement or resignation. The same opportunities for retention of employment would obviously not be available under conditions of business downturn, such as that which occurred in 1970 and 1971.

An expanding economy is the *sine qua non* of full employment, and as the United States enters the 1970s no industries promise to expand more

Table 3-2 Employment totals, by occupation, with unemployment rates, seasonally adjusted, quarterly averages [10].

Characteristic	1970 2d	1970 1st	1969 4th	1969 3d	1969 2d	1969 1st	1968 4th	1968 3d	1968 2d	1968 1st	1967 4th	1967 3d	1967 2d	Annual average 1969	Annual average 1968
EMPLOYMENT (in thousands)	78,533	78,992	78,570	78,090	77,550	77,418	76,409	76,017	75,898	75,392	75,121	74,630	73,911	77,902	75,921
White-collar workers	37,981	37,938	37,509	36,923	36,677	36,264	35,906	35,732	35,419	35,140	34,888	34,456	33,943	36,845	35,551
Professional and technical	11,129	11,026	10,936	10,764	10,740	10,638	10,473	10,392	10,295	10,142	10,067	9,952	9,761	10,769	10,325
Managers, officials, and proprietors	8,290	8,215	8,141	7,970	7,993	7,841	7,897	7,827	7,661	7,716	7,633	7,630	7,453	7,987	7,776
Clerical workers	13,748	13,906	13,655	13,478	13,281	13,171	12,876	12,823	12,816	12,694	12,624	12,343	12,250	13,397	12,803
Sales workers	4,815	4,791	4,777	4,711	4,663	4,614	4,660	4,690	4,647	4,588	4,564	4,531	4,479	4,692	4,647
Blue-collar workers	27,663	28,236	28,389	28,425	27,931	28,202	27,774	27,491	27,513	27,297	27,279	27,343	27,175	28,237	27,525
Craftsmen and foremen	10,109	10,264	10,265	10,174	10,044	10,298	10,147	9,972	10,003	9,936	9,827	9,790	9,853	10,193	10,015
Operatives	13,891	14,168	14,412	14,589	14,208	14,264	14,051	13,911	13,956	13,896	13,918	13,999	13,787	14,372	13,955
Non-farm laborers	3,663	3,804	3,712	3,662	3,679	3,640	3,576	3,608	3,554	3,465	3,534	3,554	3,535	3,672	3,555
Service workers	9,589	9,673	9,589	9,493	9,467	9,558	9,411	9,385	9,395	9,337	9,330	9,277	9,276	9,528	9,381
Farm workers	3,234	3,153	3,089	3,231	3,417	3,438	3,346	3,400	3,507	3,649	3,654	3,556	3,448	3,292	3,464
UNEMPLOYMENT RATE	4.8	4.1	3.6	3.6	3.5	3.4	3.4	3.6	3.6	3.7	3.9	3.9	3.9	3.5	3.6
White-collar workers	2.8	2.4	2.2	2.2	2.0	2.0	1.9	2.0	2.0	2.0	2.2	2.2	2.0	2.1	2.0
Professional and technical	1.9	1.9	1.5	1.4	1.3	1.1	1.2	1.3	1.2	1.2	1.3	1.3	1.4	1.3	1.2
Managers, officials, and proprietors	1.3	1.0	0.9	1.0	0.9	0.9	1.0	1.1	0.9	0.9	1.0	0.9	0.9	0.9	1.0
Clerical workers	4.0	3.3	3.2	3.2	2.8	2.9	2.8	2.9	3.0	3.1	3.4	3.3	2.8	3.0	3.0
Sales workers	4.0	3.2	2.8	3.0	2.9	2.9	2.8	2.6	2.7	3.0	3.2	3.6	2.9	2.9	2.8
Blue-collar workers	6.0	4.9	4.3	4.0	3.8	3.7	3.8	4.2	4.0	4.4	4.5	4.5	4.6	3.9	4.1
Craftsmen and foremen	3.9	2.6	2.2	2.2	2.1	2.1	2.2	2.4	2.4	2.5	2.5	2.3	2.8	2.2	2.4
Operatives	6.6	5.7	5.0	4.4	4.3	4.1	4.3	4.5	4.3	4.8	5.1	5.1	5.0	4.4	4.5
Non-farm laborers	9.4	7.9	6.9	7.2	6.5	6.4	6.7	7.4	7.0	7.7	7.8	7.6	8.0	6.7	7.2
Service workers	5.0	4.7	3.9	4.5	4.4	4.0	4.3	4.5	4.6	4.3	4.9	4.5	4.2	4.2	4.5
Farm workers	2.5	2.1	1.8	2.2	1.9	1.6	1.6	2.4	2.3	1.9	2.3	2.4	2.4	1.9	2.1

rapidly under favorable economic conditions than those based on communications technologies. Figures suggesting the growth of the computer industry have already been cited. What do they mean in terms of job type and distribution?

In the manufacturing of computers half of the workers have been estimated to be in the white-collar category and of these only 15% were clerical, the rest being scientists, engineers and related professionals and technicians. These figures may be compared with an estimated less than 30% white-collar workers on the average in other types of manufacturing[11]. The high proportion of white-collar professionals in computer manufacturing is largely a result of research and development requirements. The actual assembling of the equipment involves a surprisingly small work force considering the dollar value of the product.

Looking to the new enterprises which are the offspring of computers, Alt[12] has estimated that if a typical computation laboratory spends in the order of $20,000 per month for machine rental, an additional $80,000 must be spent on personnel for operating, key-punching, programming, supervision, etc. This can add up to an impressive amount considering the annual growth in expenditures either for renting machines or amortizing and maintaining purchased machines. The evidence clearly points to an accelerating need for professional personnel trained for businesses which manufacture and use computers – as long as the economy keeps expanding.

THE BLUE-COLLAR WORKER

The rapid growth of the white-collar businesses compared with those in which the blue-collar worker predominates is again reflected in the statistics in Table 3-2. Among the blue-collar workers, the unemployment figures for the years 1967 through mid-1970 show a consistently high level for operatives and non-farm laborers, appreciably higher than in any of the white-collar categories. The data are not easy to interpret but generally those reflecting unemployment in the period before the recession beginning in 1970 would seem to be more significant. In the context of this examination the next question might be, "To what extent is automation or cybernation responsible for this unemployment level?" Part of the answer can be found by looking more closely at certain industrial applications of those technologies which affect the blue-collar worker. Examples are drawn from steel-making, automobile manufacturing and chemical processing.

Steel-making

Until comparatively recently, the steel industry in the United States was notably reluctant to cast aside habits and practices having roots in the 19th Century. The very terms "works" and "mill," often applied to their individual plants, have about them the flavor of the Industrial Revolution. This kind of conservatism is understandable whenever there are huge amounts of capital tied up in existing equipment. The rapid changes occurring since World War II were forced on American steelmakers by competition from abroad. The basic oxygen process, which in 1969 surpassed the older basic open-hearth as the primary method of steel refining, was developed not in the United States but by a small company in Austria in 1946.

The fundamental innovation was in terms of chemical principles, but the control of the stages of the process is very much dependent on the use of computers. To see why this is so we must observe a few details of the basic oxygen process. First, it refines large quantities of steel—up to 300 tons—in a relatively short time, taking less than one hour as compared with the more than eight hours required for a basic open-hearth to produce about 500 tons. This rapid refining is made possible by blowing oxygen from a lance into a vessel containing impure liquid iron from a blast furnace together with scrap steel and flux. The oxidation of elements —particularly silicon, manganese, phosphorous and carbon—in the molten iron produces enough heat to complete the melting of the charge. The basic oxygen process, unlike the basic open-hearth, requires no additional fuel.

In order to shape up a heat of open-hearth steel and bring it to the desired quality before pouring, the operators have needed a constant check of many state variables but there has always been time to make the required measurements and calculations. This luxury of time is not available to the basic oxygen furnace operator. The requirements of determining the proper weights of materials in the charge and of elements to be added to the ladle or ingot: of assessing the relation of transient temperatures to rapidly changing conditions; of quickly integrating the dozens, even hundreds, of inputs representing gas flows, voltages, spectrographic analyses, etc., would be quite beyond his capabilities. Where control was once effected by having individual workers obtaining samples of metal or slag, reading instruments and reporting observations to the chief melter, now the trend is toward automatic monitors whenever possible. There are parameters for which there is still no practicable way of obtaining a direct measurement, and these must be calculated with data from other instruments measuring related phenomena. Inputs from sensing devices can be supplied to a control system in either analog or binary form. A data converter will feed the data as digits to the computer for programmed calculations or further processing. The calculated information then returns from the computer to the process. In steel-making the information assists the operator in his decision making (*see* Fig. 3-1) although ideally the process would be completely automatic.

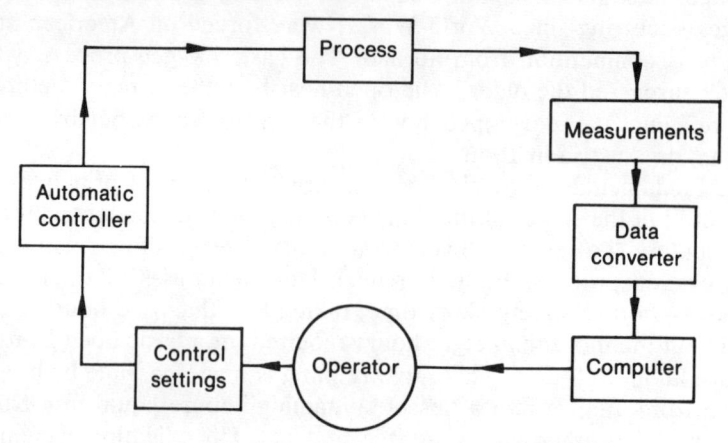

Fig. 3-1. Operator-regulated control.

The change-over from basic open-hearth to basic oxygen practice has meant large capital expenditures. In the fabrication stages of steel-making, many automatic processing machines have been installed. The added cost of computers in the steel industry requires that they be utilized as fully as possible to serve various phases of management and production including scheduling and control of processing from the ingot stage to bar, rod, sheet and wire forms. Productivity increases in the steel industry, mentioned previously, are a natural outgrowth of these kinds of changes. They mean that greater tonnages can be produced faster with fewer semi-skilled laborers; but that there will be a higher knowledge requirement for the operators, and more need for workers competent in instrumentation and computer technology.

Automobile Manufacturing

The two primary production phases in the automobile industry are: (a) the processing of the individual parts by various methods of metal fabrication including automatic machining and (b) the final assembling of those parts.

Machine tools since the times of Whitney and Maudslay have served as a basis for quantity manufacturing, but the machining of complex three-dimensional shapes, such as engine blocks or dies for pressing or forging has always been one of the most costly manufacturing operations, requiring highly skilled personnel. The standard sequence of making a finished part from a conceptual design formerly involved the careful production of an engineering drawing by a draftsman. The machinist, reading a blueprint taken from the drawing, painstakingly performed the specified machining operations. An intermediate stage toward automatic machining of dies was introduced about 1925 with the development of a tracer-controlled milling machine[13] which employed simple electromechanical control systems and electrical sensing devices to trace the surface of models used as templates. The action of the cutting tool on the die was automatically guided in this manner.

These days, multi-stage machining operations are characteristically automated by the method of *numerical control*, which means controlling the action of a machine tool through programmed commands expressed in numbers. The procedure requires a programmer to translate the data from an engineering drawing into a number series, (an automatic programming language may be used), which specifies the coordinates of significant points in a component. After being coded on punched tape the numbers are converted by a special digital computer, capable of serving many machines, into forms acceptable by the control system of the machine tool; or an advanced alternative incorporates computer control into the machine. The instructions, which are on tape, can direct a job requiring complicated machining operations in sequence, or repetitively, once the work has been set up.

The direct control of the cutting tool can be effected by sensing, through the use of diffraction gratings, the actual position of certain moving elements of the machine tool and comparing with command information providing the correct position. The increase in productivity realized by employing numerical control is substantial, and the inevitable result is a lessening in demand for machinists for production work. The position of the draftsman is also vulnerable since there are rapidly evolving methods of specifying designs electronically rather than by drawings.

Automobile assembly lines unlike the transfer lines for machining parts are far less automated than is generally realized. A visit to an assembly plant of one of the large American manufacturers will show an intricate and exquisitely scheduled conveyor system bringing the many parts to the final assembly line at the precise moment needed, but one will also see the assembly operations performed by a large number of human workers. While automatic assembly and inspection in manufacturing processes are rapidly increasing, full automation or cybernation of this stage of automobile-making is proscribed by the variety of body styles and accessories which the American consumers demand. One may watch a modern assembly line for days and never see two identical automobiles produced. It is the essence of cybernated control systems that they be able to respond to constantly changing inputs, but the demands imposed by a process of this complexity would be too great for the most sophisticated systems now available. Instead, the primary use of the computer in this state of manufacturing is for scheduling and inventory.

For the automobile industry as a whole, then, there are jobs in the primary production state which are vulnerable to machine replacement, but fewer such possibilities presently exist in the assembly stage. Since the overall productivity per worker has increased over the years, growth (or even stability) in the work force would seem to depend upon a continued growth in the annual sale of automobiles. In the long run the future of the worker in the automobile industry and in the many industries auxiliary to it may depend on how the average American will react to the fact that the automobile has unacceptably degraded the quality of his environment. It would be foolhardy in the first years of the 1970s to predict the demise of an industry upon which the livelihood of so many Americans depends; nevertheless any responsible forecast of its pattern of evolution would have to recognize the need for wholesale changes in transportation systems. The effects of these changes on the work force could be very great.

Chemical Processing

When looking for applications of unified cybernated systems in industrial production, one begins to find the most conspicuous examples in chemical processing. This is natural because with no need for transfer machines between intermediate stages handling of material in fluid form is relatively uncomplicated. On the other hand, important aspects of the requisite control of the overall processes are effected only through indirect means. Theoretical thermodynamic and kinetic studies can tell a good deal about chemical reactions and changes of state, but since these reactions occur between aggregates of atoms and molecules, only the measurable primary and secondary effects which influence, or are influenced by them, can be utilized for control system design[14]. Ingenious instrumentation was typical of the chemical processing industries long before the advent of electronic computers. Very precise temperature regulation for example has been obtained with controllers which not only detect the difference between the ideal and the actual process temperatures, but also determine the rate of change of the temperature difference as well as how long it has lasted. When a whole chemical process on a very large scale is designed it becomes necessary to integrate the actions of all the monitors and regulators, and this goes beyond the capabilities of any simple electromechanical device. A digital computer must be used containing a mathematical model of the process to which incoming information can be related.

Foster[15] describes the elements of such a model as containing:

(a) descriptions of the basic chemical reactions;
(b) known variations of these reactions with physical changes such as temperature and pressure;
(c) time constants, indicating duration of a process;
(d) mathematical expressions of all the known features of the chemical plant, such as the variations of the characteristics of a heat exchanges with fluid flow, viscosity, etc.;
(e) empirical information obtained through practical experience.

He has prepared a simple schematic of a computer-controlled chemical process as shown in Fig. 3-2.

Synthesizing practical processing and control requirements to make a true cybernated plant is an extraordinarily complex task. When approaches to developing such a system have been implemented in the chemical industries, one has a preview of what manufacturing plants without human operators may become. Yet there are workers still on the job even though perhaps not so obviously involved in the production process as they were at one time. Operating workers are more likely to be stationed in office-like control rooms in buildings separate from the process units, and their concern may have more to do with the computer than the separate controlling instruments. Pipefitters, welders, painters and instrument repairmen are still needed for plant maintenance (although there is evidence that automated plants require less maintenance); engineers and scientists for supervisory jobs; and laboratory workers for tests to maintain product quality. Details of many operations are being

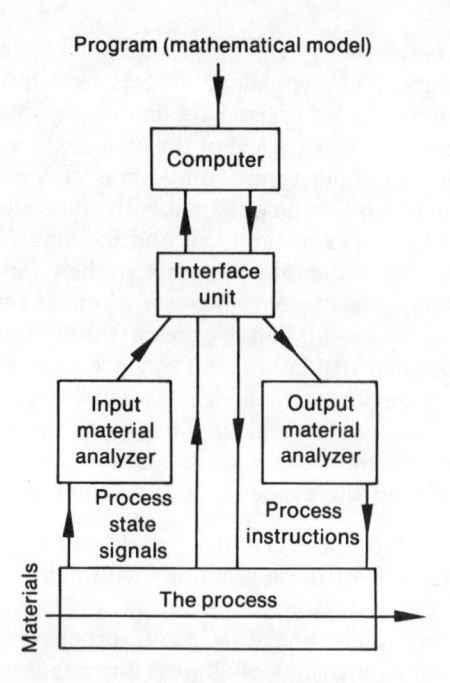

Fig. 3-2. Computer-controlled chemical process[16]. (Reproduced from *Automation in Practice* by David Foster, published by McGraw-Hill Publishing Company, Ltd. © 1968. Used by permission.)

modified constantly, resulting in the elimination of some jobs and the upgrading of others. Even in the laboratory, by devising systems with commercially available instruments, a professional chemist can now develop highly automated analytical schemes based on a continuous process. The designing of these systems provides work of greater complexity and challenge for the industrial chemists at higher levels, but routine subordinate jobs for chemists of lesser training may no longer be needed.

These illustrations of the progress of cybernation and automation in heavy industry ought to add meaning to the numbers shown in Table 3-2. The period represented in the table was one of generally low to moderate unemployment until the economic slump in 1970; nevertheless the figures for blue-collar workers and non-farm laborers are consistently higher than for other categories. One cannot avoid concluding that cybernation, automation, and even relatively simple mechanization have the potential for eliminating many blue-collar jobs. The potential varies from industry to industry, and there are other new jobs resulting from advances in technology, but the workers themselves and their unions see the trends as ominous.

THE TECHNOLOGICAL SOCIETY, UNEMPLOYMENT AND THE POOR

An examination of historical patterns of industrialization strengthens the view that over the long run new technologies have increased the material benefits available to the average citizen in Western nations and have added to the activity options open to him. This should not be allowed to obscure the cruel upset of even a short-run loss of work to those unemployed who are economically and psychologically unprepared for a workless existence, nor the fact that the history of industrialization is yet too brief to offer firm assurance of continuous employment growth.

The problems of unemployment and poverty are inextricably related. It is a continuing irony in an affluent society that poverty remains to reproach wealth quite as challengingly as it did in the flood-tide of the Industrial Revolution. In 1968, the Gross National Product increased 5%—to a level sufficient, if uniformly distributed, to give each man, woman and child in the country an income of roughly $4500. The poor population, measured by the currently accepted Government formula, decreased by four million. Yet, in that same year, according to Heinemann[17], twenty-two million people at a minimum still lived in poverty, although one-third of the poor families had a head who was employed throughout the year. The number of welfare recipients increased by 14%. Dr. Herman Miller of the U.S. Bureau of the Census has questioned the use of a constant standard to measure poverty. He has stated, "By using a constant standard many people are moved above the poverty line each year, but their position relative to others is unaltered... their feelings of deprivation remain the same."[18] That is to say, the definition of poverty is socially determined.

Who are these poor? About one-half of the poor families live in the South, but that does not make it a rural problem primarily. In New York City some 10% of the total population, (one million people), and 20% of the children were on welfare in 1968[19], and the number has grown since then. Of course many of these people came originally from rural areas— from the cotton-growing regions of the South where most of the cotton is now picked by machines, and from the sugar plantations of Puerto Rico—

to crowd into the urban slums just as displaced farm workers did in 19th Century England.

To a large extent the problem is one of race, but by no means exclusively. Although approximately two-thirds of the poor are white, this represents only about 10% of the white population, while over 30% of the non-white population are below the poverty line. The poor are men and women who, by virtue of advancing age, have been eliminated from the industrial working force; or those who are too young or insufficiently educated to have been accepted in it. They are the physically and mentally unfit, and the mothers of fatherless families. Some are working-age males, often heads of families, whose jobs have been essentially wiped out by mechanization. The coal-mining regions of Appalachia have many such men. Many of the poor are a part of what has been called "generational transfer," enmeshed in a repeating cycle of ignorance and despair. Few of the people in any of these categories have ever been a functioning part of the new technological society based on information processing, although they surely represent one of the major problems with which it has failed to contend.

Must there be poverty even in a system of great productivity? It says in the Bible, "The poor shall never cease out of the land" (Deuteronomy 15:11). This has been reason enough for many to assume that there will always be an impoverished class; that perhaps one is necessary to maintain a kind of social equilibrium, even though in the technological society the problem seems less one of scarcity than distribution of wealth. Looking back to the early industrial society it is sobering to realize that in some ways we are not much better equipped to deal with maldistribution of work opportunities and resulting poverty than were our forebears. If it is true that societies in the Western nations now have a firm base in modern science and technology, it is equally true that we have not yet escaped from many habits and consequences of the Industrial Revolution which still strongly persist today.

It was in the 19th Century, as the working masses crowded into festering slums, that a few influential spokesmen began to direct vehement protests against what they considered to be a selfish and oppressive exploitation of labor by the factory owners who were using the machine as their lash. Their arguments were opposed by equally forceful supporters of industrial progress who insisted that the picture of suffering among the laboring poor was vastly overdrawn; and that, in any event, the net result of industrialization must be economic gain for all, over the long term. To understand our own times we can profitably explore the original arguments, clear echoes of which remain with us.

ENGLAND—SEEDS OF SOCIAL DISCORD

Poverty has been an ages-old affliction of the human race. For most of man's time on earth a large proportion of the population has suffered partial or acute want. Since civilizations were predominantly agrarian, a man's success was most often measured by how well he could make the land produce. Not that farming was ever anything but a harsh calling. At best it required unremitting toil, and at worst periodic poor harvests caused unspeakable distress.

England, in medieval times, was an underdeveloped country and as such it found famine no stranger. But by the time of Queen Elizabeth I, the growing wealth from an expanding commerce made widespread poverty less understandable or acceptable. Elizabeth, herself, after a trip through her realm had cried: "The poor, they are everywhere!"

The emerging market system with its emphasis on money rather than payment in kind unleashed certain relentless forces which would seem perpetually out of control. The social stability of the feudal system in which the serf could be assured by his master of at least a subsistence, if one was to be had, had given way. Now labor would become a commodity to be sold at the marketplace for the highest price it would fetch. The buyer would have no responsibility other than the paying of the wage [20]. If times were hard he might not buy at all. Unemployment would become a recurring phenomenon affecting not only the aged or the mentally and physically unfit, but the able-bodied as well. In the 16th Century, however, displacement of workers by labor-saving devices was uncommon. (There were reports of the suppression of the use of ribbon looms about this time.) A bigger problem was a lack of understanding of the relationships between production and consumption in the new market system, by employers bent on making a profit. A new class of landless workers arising from the consolidation of land holdings by the upper classes was starting to add to the difficulties.

There was undoubtedly a combination of complex factors, but whatever the cause of deprivation, the Government felt constrained to take a hand in poor relief. In 1536 a law was passed to provide help for those who could not work. Since the funds were to be from voluntary donations, and the method of administration was not laid out, the effect was minimal. Only the larger towns were able to try to abide by it, and London did so by instituting a compulsory poor rate. In 1601 the parish was named the functioning organization for administering the means of relieving the poor. The enabling law included a poor rate to be applied to householders. Results were uneven. Wealthier parishes found it necessary to maintain barriers against an influx of poor people from other parishes less able to handle the problems, and then in 1662 a law defining the responsibility of each parish was passed.

But there was no attempt to coddle the destitute anywhere. If the poor could not or would not find employment, then something would be found for them to do—in the workhouse. By the early 18th Century, workhouses had become more like penal institutions in which those who were unfortunate enough to be sent there—including the aged or the young, the vagabonds or the hapless—were lumped together under the most pitiable conditions.

Not all of the poor ended up in the workhouse. There were those, (cottagers, squatters and even some of the village artisans) who had managed to eke out a tenous existence by grazing their few beasts or cutting peat or wood on the common, or waste, land. A distinction needs to be made here between the commons and the open fields. Mantoux[21] describes the commons as having been considered of little value except perhaps to the peasants, while the open fields were the property of many owners and were promiscuously dispersed in the form of small plots which could only be farmed in cooperation with all of the other owners. While each owner had unchallenged right of harvest from his plot, during barren times between harvest and sowing, the open fields were treated as a collective pasture.

The acts of enclosure, by means of which the open fields could be fenced, changed all of this. While enclosures had begun before the 16th Century, it was in the 18th Century, with the assistance of a cooperative Parliament, that the rate increased markedly. Less than forty Acts of Enclosure were passed in each decade between 1720 and 1750; from 1770 to 1780 there were 642; and between 1800 and 1810 Parliament passed 906 such acts[22]. Undoubtedly, as the supporters of the enclosures contended, the unenclosed field system had been wasteful and inefficient. The complications it had fostered made the production required to meet the rising demand for agricultural products almost impossible to obtain. But the social torsion which followed in the wake of enclosure was widespread and wrenching. The squatter and the tenant farmer were both hard hit, and the artisans were also affected by the general increase in rural unemployment. There were also stories of a fearsome depopulation of the countryside, but although population figures were unreliable this does not appear to have been the general case. More likely there was a beginning of migration from certain districts which had been particularly affected. Oliver Goldsmith's famous poem, *The*

England — Seeds of Social Discord   81

Deserted Village, written in 1770 tells of the "sweet, smiling village" of Auburn:

> ... Amidst thy bowers the tyrant's hand is seen
> And desolation saddens all they green.
> One only master grasps the whole domain,
> And half a tillage stints thy smiling plain ...
> And trembling, shrinking from the spoiler's hand,
> Far, far away they children leave the land ...
> Where then, ah! where shall poverty reside
> To 'scape the pressures of continuous pride?
> If to some commons fenceless limits strayed,
> He drives his flock to pick the scanty blade,
> These fenceless fields the sons of wealth divide,
> And e'en the bare-worn common is denied ...

The stream of population from most rural areas into the industrial centers would not become a flood until decades after this was written.

In 1793 Great Britain went to war with France and remained in conflict with her almost continuously, except for a two-year break, until 1815. The war was a spur to industry and, as usually is the case, the price of farm products rose sharply. For the great landowners this was an even bigger incentive to move to progressive methods of farming; but for the poor, it made their condition intolerable. The price of food was beyond them, the free access to the land was restricted, and the possibility of supplementary earnings from spinning and weaving in the cottages was being eliminated by the new machines. In the face of a rising discontent which must have been frightening to the upper classes so shortly after the French Revolution, the justices of Berkshire, meeting in Speenhamland in 1795, decided that subsidies in aid of wages should be granted to the needy to provide a minimum total income — a bare minimum — whatever their earnings[23]. The subsidies would be pegged to a sliding scale depending upon the price of bread. It was, in fact, the first clear example of a guaranteed income for the uninstitutionalized poor. The plan was widely adopted, particularly in the south of England, and at first it enjoyed general popularity. In bad times laborers could still eat, and a limit to the deprivation of whole families was provided by supplementary allowances for wife and children. Over the long term, however, the results were disastrous. Those philosophically opposed to such a measure contended that it encouraged large families among the poor, improvidence and idleness. Perhaps there was some truth to this, although a few industrious fellows did very well for themselves by receiving the poor rate and working at other jobs undercover. The primary effect, however, was that it tempted employers to drive wages down to the absolute minimum, secure in the knowledge that the subsidization by the general ratepayer would forestall actual starvation. The temptation was irresistible and the result was that the rural workers were being increasingly pauperized. Let it be recognized that the concentration of the new mechanical industries in the cities had not yet proceeded at the same pace that it would in the early decades of the 19th Century. The war sustained a demand for farm labor, and the difficulties of migration (largely from south to north) prevented a mass movement from the rural areas until it ended.

Nevertheless there was a growth in the class of industrial workers, and in the industrial cities during the French Wars there was also disaffection. Sudden fluctuations in demand for goods resulted from obstacles to expansion to trade caused by the continental blockade. This, added to the mounting use of machines, helped bring about alternating periods of boom and collapse. Now industrial laborers began serious, though scattered, attempts to organize themselves and there were occasional turbulent outbreaks. The most famous riots were those of the Luddites — followers of a mythical King Ludd — who in 1811 and 1812 broke into factories under cover of night and smashed machines; as much, it appears, as a demonstration of a growing bitterness against the unfairness of the employers as it was a fear of the machines. Parliament responded by making the machine-breaking transgressions of the Luddites a capital offense.

In 1815 England was about to enter a long period of national growth and prosperity. Nevertheless it was often erratic and attended by continuing strife between classes trying to adjust to the complexities of an emerging machine-dominated society. The landed class wanted the price of food, particularly corn, kept at a high level; the industrial employers and the workers wanted a low level. On this at least they could agree, but on little else. The goal of the employers was to produce cheaply and in quantity. It was obvious to them that one obstacle to the industrial system's functioning at its destined level was the public subsidization of wages. It took nearly two more decades to accomplish, but finally in 1834 the power of the merchant and employing class in Parliament did away with the previous subsidy for the poor and put in its place the New Poor Law Reform. The three main principles, to be administered through Boards of Guardians, specified: (1) no relief to the able-bodied except in a workhouse, (2) whatever relief given must be less desirable than the most unpleasant means of earning a living outside, and (3) men and women should be kept apart to hold down the population of the pauper class [24].

To put the sponsors of this pitiless law in the fairest light, no doubt some, at least, felt the laborers would, in the long run, be more likely to make out if left to their own initiative. Hard work was the only creed these new "captains of industry" understood. Had they not made their fortunes fighting the most intense competition by keeping their shoulders to the wheel through fourteen hour days? Then there could be no reason why the lower classes should not expect to take care of themselves in the same way. No reason except, of course, that the odds against most of them were staggering.

ECONOMIC THEORIES AND INDUSTRIAL CHANGE

From 1834 there was a full competitive labor market, and this, according to Karl Polanyi, established industrial capitalism as a social system [25]. The currents of industrial and economic growth were now surging more strongly than ever, but they were broken by periodic depressions. The conditions cried out for some kind of explanation which would bring order to complexity. Attempts to comprehend the forces of change, bound as they were by pre-industrial traditions, proved dogmatic and often misguided.

Perhaps the most influential interpreters of the forces molding the times, ultimately, were the economic philosophers who, since before the days of Adam Smith, had been seeking to place the turmoil of economic society in the framework of an all-embracing system. Theorists of mercantilism during the 17th Century had considered that the newer production methods should be restricted to goods for export to limit the displacement of labor. Those who were put out of work by technological change should be re-employed in state-aided enterprises[26]. But technological change in the mercantile era was slow and scattered, and even at the time that Adam Smith wrote his *Wealth of Nations*, (1776), the problems were still mainly agrarian. He could say, without being seriously contradicted by events in his time, that if the market were left alone, with each individual pursuing his own self-interest, society would regulate itself to the best advantage of all of its members. The key to the self-regulatory mechanism was competition which prevented individual greed from causing inordinately high prices. Whenever any element of the system got out of balance—e.g., profits too high or wages too low—then, in the first case, other entrepreneurs would rush into that business and establish competition, or, in the second, workers would simply move to jobs where the pay was better. In his system it was not conceivable that the upper classes could accumulate wealth while workers became poorer than ever; but the flaws and lags in the self-regulatory mechanism he described would not take long to surface. For one thing, collective greed (which he worried about) would find ways to subdue competition. For another, the division of labor, which impressed him so, would trap workers into increasingly narrow industrial specialties from which they could not readily turn.

The stresses of late 18th and early 19th Century industrialism engaged the attention of Smith's successors in the field of economic theory. Among them were David Ricardo and the Reverend Thomas Malthus. They both took a view of impending events much less sanguine than that of Smith. Parson Malthus was sure that the poverty he saw among England's lower classes would hang as a permanent cloud over society for as long as it was able to survive. The bleak facts of fertility and population multiplication would impose an inevitable plague of permanent famine throughout England and the world. In Ricardo's world there was only one class that could look for lasting well-being and prosperity—the landlords. Populations would multiply as Malthus predicted and, increasingly, lands formerly considered marginal would need to be cultivated to feed new mouths. Food costs would rise, land costs would rise sharply; and both the worker (whose wage raises when they came would go to supporting more children at higher prices and rents), and the capitalist (whose profits would be limited by competition and higher wages) could only look to the landlord with envy. The predictions, in fact, did not come to pass, or at least not to the degree nor so soon as they foresaw. Importation of grain into England, later, and the application of scientific methods to agriculture kept supplies sufficiently high to maintain prices within bounds. Also, in some way we do not completely understand, but which seems to have something to do with psychological and sociological factors as people crowd together in industrial environments, the increase of population in England, while rapid, did not maintain the pace that Malthus feared. Emigration was a factor, too.

With regard to the question of the effect of machinery on the condition of the working class, there was a disagreement between Malthus and Ricardo. To the latter the continued expansion of the wealth of the nation obviously depended on the willingness of the businessman to increase production by investing in new equipment. Malthus, on the other hand, feared overinvestment would cause goods to be produced at such a rate that the demand for them, even from an expanding population, could soon be outstripped. Not all the value of what the workers produced, after all, appeared as wages with which to buy the manufactured products. The outcome of such a "general glut" would surely be unemployment. Not possible, rejoined Ricardo. He referred to the proposition advanced by the French economist, Jean Baptiste Say, in 1803 that supply creates its own demand. "Say's Law" contended that there was no limit to the desire for commodities, and therefore, since the cost of goods had to appear as somebody's income—whether worker or capitalist—there could not be such a thing as a general deficiency of effective demand. If income went into savings it made no difference, according to Ricardo, because the only logical way to save was to invest in more production potential, and this was just another form of spending. Later he came around to admitting, in a chapter added to his *Principles of Political Economy*, that the too rapid introduction of machinery could cause distress among the laboring classes.

In London, a generation later, a glowering German scholar living in virtual exile began to work on a theory of the capitalist system which would conclude that capitalism contained the seeds of self-destruction. This was Karl Marx. He concurred with much that David Ricardo had written but not with Ricardo's conclusions about the generally beneficial effects of technology under capitalism on the worker. According to Marx, the recurring periods of unemployment in the Western industrial nations would inevitably become more severe. His theories leading to this conclusion were often murky and self-contradictory but basically he contended that capitalists, hounded by competition, would have to invest in labor-saving equipment. The oppressed workers would then be forced to accept even lower wages as jobs became scarce. Moreover, since the real value of a product depends on the hours of human work required to make it, machines do not actually add to total profits. Profits must fall for the owner as men are thrown out of work. As businesses decline they are bought up cheaply by larger businesses which can flourish for a time, but then the whole gloomy process begins again.

THE LABORING CLASS IN VICTORIAN ENGLAND

Most of the theorists in the 19th Century felt that machinery would improve the lot of the workers (and the owners), if not immediately, then ultimately. Among the more active spokesmen for this position, Lord Brougham and Charles Knight conducted an energetic campaign through the Society for the Diffusion of Useful Knowledge, (nicknamed the "Steam Intellect Society") to demonstrate to the educated, as well as to the industrial classes, the benefits of machinery [27]. But the testimony of the senses, for anyone who wished to look beyond the arid themes of economic theory, made it impossible to ignore human suffering. The reports of government commissions investigating problems of the laboring population vividly described the miserable conditions in the factories and the mines. E. Royston Pike has collected together original documents from the Industrial Revolution which show what it was to be a poor worker in those times.

(From the Parliamentary Papers 1831-32, vol. XV, p. 159) [28]

Joseph Hebergam, aged 17, worked in worsted spinning factories at Huddersfield since he was seven:

When I had worked about half a year, a weakness fell into my knees and ankles; it continued and it has got worse and worse. In the morning I could scarcely walk, and my brother and sister used out of kindness to take me under each arm and run with me, a good mile, to the mill, and my legs dragged on the ground in consequence of the pain; I could not walk. If we were five minutes too late, the overlooker would take a strap, and beat us till we were black and blue...

Did the pain and weakness in your legs increase? Just show the Committee the situation in which your limbs are now. (The witness accordingly stood up, and showed his limbs.)

Were you originally a straight and healthy boy? — Yes, I was as straight and healthy as any when I was seven years and a quarter old...

Your mother being a widow and having but little, could not afford to take you away? — No.

Was she made very unhappy by seeing that you were getting crooked and deformed? — I have seen her weep sometimes, and I have asked her why she was weeping but she would not tell me then, but she has told me since...

(From the Parliamentary Papers, 1831-32, vol. XV, p. 168) [29]

Mark Best, flax-mills overlooker, was asked to describe the sort of straps that are made use of to keep the children at work:

They are about a foot and a half long, and there is a stick at the end; and that end they beat them with is cut in the direction of my fingers, thus, having five or six thongs, some of them. Some of them are set in a handle, some are not.

You say you had one of these delivered to you by a master, who urged you to make use of it, and to lay it on freely? — Yes.

Do you think you could have got the quantity of work out of the children for so great a number of hours (from 6 a.m. to 7 p.m., or 5 a.m. to 9 p.m. when they were "thronged") without that cruel treatment? — For the number of hours, I could not, I think; it is a long time. The speed of the machinery is calculated, and they know how much work it will do; and unless they are driven and flogged up, they cannot get the quantity of work they want from them...

(Mr. Horne's Report, Parliamentary Papers, 1843, vol. XIII, pp. 90-1) [30]

... Little boys and girls are here seen at work at the tip-punching machines (all acting by steam power) with their fingers in constant danger of being punched off once in every second, while at the same time they have their heads between the whirling wheels a few inches distant from each ear. "They seldom lose the hand," said one of the proprietors to me, in explanation; "it only takes off a finger at the first or second joint. Sheer carelessness—looking about them—sheer carelessness!"

The champions of the industrial order had a quite different picture of factory conditions to present. Dr. Andrew Ure, a Professor of Chemistry at Glasgow, had no doubts about the benefits of the factory system. In his book, *The Philosophy of Manufactures* [31], he wrote:

> I have visited many factories, both in Manchester and in the surrounding districts, during a period of several months, entering the spinning rooms unexpectedly and often alone at different times of the day, and I never saw a single instance of corporal chastisement inflicted on a child, nor indeed did I ever see children in ill-humor. They seemed to be always cheerful and alert, taking pleasure in the light play of their muscles, — enjoying the mobility natural to their age. The scene of industry, so far from exciting sad emotions in my mind was always exhilarating... The work of these lively elves seemed to resemble a sport, in which habit gave them a pleasing dexterity...

He was also to write:

> When capital enlists science in her service, the refractory hand of labor will always be taught docility.

Indeed, evidence could be found to support either a positive or a negative view of industrial society because there were now two classes of workers—one, as impoverished as ever, and the other becoming increasingly well off. The picture of 19th Century England was never one of unrelieved tyranny for all in the factory. Men of good conscience in the Parliament sought (at length successfully) to reform the most flagrant evils, and some owners did what they could to humanize industrial relations. Robert Owen was the best known of these latter. His famous cotton-spinning mill at New Lanark in Scotland was a model establishment—orderly, clean and well-ventilated—but beyond this he was equally concerned about the kind of people who worked for him, and what they could become with proper encouragement. New Lanark became, in effect, a planned community. Workers had comfortable living quarters; for the sick and the aged, cooperative funds were available, and the children up to the age of ten received an education which even included dancing lessons. Owen was able to run his mill profitably, but other factory owners did not rush to follow his example.

The workers in England had attempted to organize themselves to guarantee better conditions, but their first efforts were scattered and feeble. In the Victorian era, they succeeded in establishing Trades Unions which provided benefits for needy members and had some effect in shortening the length of the working day. These were, by and large, non-radical organizations. Rather than trying to upset the relation between employer and employee by following syndicalist doctrines (Karl Marx could make little headway with the British working man) they were more interested in keeping the price of labor in the crafts at a high level by restricting overtime and limiting the number of new workers except as apprentices. Charles Babbage had earlier turned his prophetic attention to the problem of industrial unions and had concluded that they would undoubtedly encourage the factory owner, harried by competition, to relocate his factories and to speed up the installation of labor-saving machinery.

SEEDS OF ENVIRONMENTAL DESTRUCTION

Through the rest of the 19th and into the 20th Century wages *did* rise and material standards of living *did* improve for many—slowly and haltingly—as the economy lurched its way to ever higher levels. So the optimists of the early years of the Industrial Revolution were right, to that extent. But the gains were denied some, and they had a heavy cost for all. In the choking cities of England, changes in the physical environment had begun to assail the senses. They had, in truth, even as Elizabethan towns, never been models of cleanliness and good planning but there was now such a difference in scale—with wastes being produced faster than they could be dissipated—that human sensitivity was being fundamentally altered.

One might ask, if pounding engines and smoking chimneys meant a prosperous England, who could object to that? There were, to be sure, not many penetrating voices expressing public concern about the physical changes, but there were some. In 1819 Shelley wrote[32]:

> Hell is a city much like London,
> A populous and smoky city....

and Charles Dicken's Coketown of 1854 is also the picture of 20th Century industrial pollution[33]:

> ...It was a town of machinery and tall chimneys, out of which interminable serpents of smoke trailed themselves forever and ever, and never got uncoiled. It had a black canal in it, and a river that ran purple with ill-smelling dye...

Community health standards were abysmally poor, and it was not until 1848 that repeated cholera epidemics brought about the establishment of a General Board of Health. Even its powers were seriously restricted after the passing of an immediate crisis[34]. Not only were the air and water impure—so was the food. Embalming fluid to keep milk from souring, and red and white lead in sugar confections[35, 36] were just a few of the adulterants used.

Some Victorian writers felt it was more than the body that sickened in this environment. One could now extend the 17th and 18th Century concept of the Universe as a great clocklike mechanism to the notion of the industrial machines having imposed their powerful rhythms on society, such that men had no choice but to make their own natural pace conform. This meant not only a loss of independence, but also, for the worker especially, a deadening effect on the spirit.

If we wished to assess the changes in the conditions of existence for the working man in industrial Great Britain from, say 1850 to 1950, we should have to conclude that many factors could not be quantified, and others only crudely, given the kinds of data available. It would probably be fair to state, however, that public health care, child labor laws, and working hours improved markedly, while the quality of the environment did not get much better, and may in some respects have grown worse; that the worker's pay increased on an absolute scale, but his sense of relative economic position did not evoke a better psychological attitude toward his employer or the machine. The fear of bad times and lack of work has persisted.

EARLY AMERICAN PATTERNS OF INDUSTRIALIZATION

For our purposes, the events during this later period in England need not be detailed because almost from the beginning, although few could have realized it, the mantle of technological superiority had begun to shift toward the United States. Since some of the stages in America's early industrial development have already been discussed in Chapter 2, there is no need to dwell further on this era here, except to point out a few respects in which the American response to industrialization differed from that in England.

The movement in the United States from the farms to the factories came somewhat slower because there was seldom any serious slackening in demand for agricultural products, nor was there such a serious labor shortage in industry. The hordes of immigrants from Europe and other parts of the world—over thirty-seven-and-a-half million between 1820 and 1930—provided an ample manpower source. They crowded into the cities at such a rate that by 1900 over forty percent of the people in the twelve largest cities in the country were first generation immigrants. "Every man is as good as his neighbor" had been a heartening sentiment for the 18th and early 19th Centuries but many—too many—of the native-born had come to feel it might well be amended by excepting those whose language was strange or whose skin was another color. The Wisconsin and Illinois legislatures did not affront prevailing opinion when, in 1889, they outlawed the teaching of foreign languages in the schools[37].

Strife between owners and workers in the United States was slower to develop than in England, but by the last two decades of the 19th Century a growing discontent with the obvious gap between rich and poor helped to produce a number of turbulent strikes, most of which were severely suppressed. At this time the capitalists were strongly in command and on the few occasions that they were publicly challenged, it was more often for suspicion of corrupting public officials or destroying small businesses than for exploiting the laboring class.

Great industrial concentrations were appearing in the East and Midwest but in a land so wide and well-favored the resources were inexhaustible, and the wastes would be quickly purged by the winds and the abundant streams. Or so it seemed.

CONTEMPORARY ECONOMIC CONCEPTS AND UNEMPLOYMENT

Between the early years of this century and World War II, important among the changes affecting the worker were the increased use of mechanical and automatic aids to production, and the expansion of the union movement. The latter was one of the central developments in the bleak, hopeless time of the Great Depression which left its mark on great numbers of Americans now in their middle years.

To many of these people the amazing growth of American industry in the years after World War II could only be a prelude to a coming depression, but others have rejected this fear with the contention that the rapid accumulation of national riches has naturally arisen from two major influences. The first has been the overwhelming commitment to research and development which has spawned and nurtured many new, wealth-producing industries. The second, even more fundamental, influence from which the first actually springs would be the gradual political acceptance, long delayed, of the idea that the full weight of government can and ought to be used to expand and maintain the total demand for goods and services, by manipulation of fiscal and monetary policies.

The concept was first proposed as a full-bodied economic theory in the book, *The General Theory of Employment, Interest and Money*, written in 1936 by John Maynard Keynes. In a letter to George Bernard Shaw in 1935, Keynes stated the following: "I believe myself to be writing a book on economic theory which will largely revolutionize – not, I suppose, at once but in the course of the next ten years – the way the world thinks about economic problems."[38] It did just that. The "Keynesian revolution" overturned the classical free enterprise doctrine that government intervention in the processes of the market was willfully unwise and even dangerous.

Keynes's theory was proposed as a remedy for desperate times, but its originator was not a desperate man nor was he a radical foe of capitalism although he has sometimes been pictured that way. In fact, he found himself repelled by the drab face of communism as he had seen it on a visit to Russia; not surprisingly since he was a man of elegant tastes who moved equally gracefully through the various worlds of government and high finance, the fine and sometimes Bohemian arts, and Cantabrigian scholarship. He once stated that his main regret in life was that he had not drunk more champagne[39]. He cherished no warm affinity for the "boorish proletariat."

The crisis of the 1930s was that the economic cycle seemed to have got stuck inextricably on the down side. There was unemployment on a scale never seen before, (fully twenty-five percent of the work force in the United States in 1933 were out of jobs), and the wisdom of the classical economists could give no satisfactory explanation for it. Say's Law ought to have proven conclusively that full employment was the normal condition. The occasional depressed periods of the trade cycle due to overproduction of capital goods were supposed, automatically to induce falling prices and rates of interest, which would in turn stimulate investment in new enterprises and a higher level of innovation. When this happened jobs should once again become plentiful until the rapidly expanding industries started eventually to overproduce, whereupon the cycle would begin anew. When employment failed to rise, year after dismal year following the 1929 market crash, most economists still insisted the standard response would work in time, though indeed the downtrend was much more severe than anything they had known of before. That meant above all being patient, perhaps cutting wages and prices, and keeping the nation solvent by balancing the budget. Marxists, of course, were sure they were hearing the death rattle of capitalism.

Keynes had felt for many years that Say's Law was somehow inadequate. During the middle 1920s when Great Britain was plagued by unemployment and a generally declining economic position, (well before hard times hit the United States), he advocated that the government begin to spend money on public works without waiting for the recovery which was supposed to follow necessarily after the private sector decided it was once again the right time to invest. The weight of official opinion was quick to reject his suggestion on the grounds that public expenditure would create a greater budget deficit, besides making less money available for investment in wealth producing industries. He had no intellectually convincing answer to this argument at the time. He did agree with the government policy of lowering the interest rates but this seemed to have little effect on unemployment[40].

By 1935 he had found his answer to the puzzle. The key was in his rejection of the classical contention that full employment is the natural state of things—that unemployment could only be a temporary aberration. Rather, there was an intimate causal linkage between consumption, savings, investment and the national income which profoundly affected the availability of jobs. He decided it was possible to have an equilibrium between these elements which resulted in a high level of employment, but it would not occur automatically.

The lifeblood of the economy is the national income—everybody's income—which ordinarily flows constantly from one hand to another. Even when someone saves a part of it, the logical way to do so is not by hiding money in a shoebox but in a manner which provides some measure of appreciation such as through an interest-bearing savings account. The bank can afford to pay interest because it in turn can lend the money at a higher rate to a new business or to an existing enterprise which plans to expand its operations. It had been accepted that in a free enterprise economy there would always be enough demand by businesses for new capital to put savings to use, or conversely, enough savings to sustain any desired investment.

Keynes concluded there was no such certainty in a free market. Even though declining rates of interest in times of depressed business activity ought to stimulate the desire to invest what in fact was happening during the Great Depression was that there was precious little saving, by business or by individuals. People were either not working at all, or if they were their wages were at a level too close to subsistence to encourage saving. The factors triggering the slump were many and complex, going back to World War I. Overinvestment, tax, wage and tariff policies, and overproduction in agriculture all had a part in it. In whatever ways they interacted, the devastating result was that the savings deficiency and the low potential yield on investment (when much of the capital equipment already available was not being used) had created a condition of equilibrium at the bottom of the cycle. To Keynes the only way to revitalize all of the idle resources was to have the government step in and take an active hand in the process. This would mean manipulating the interest rate if this were needed to influence businessmen to invest more; reducing taxes; creating public works; and, most heretical of all, allowing the budget to go unbalanced when necessary, on the theory that deficits are not of themselves inflationary. If the economy rose to the top of the cycle and inflation threatened, then the measures could be reversed.

Under President Roosevelt and the New Deal, a policy at least partially related to Keynes's ideas had already been getting under way. The prominent feature was a program of public works financed by the federal government. Most of these projects were visible attempts to improve physical amenities and included the construction of public buildings, parks, roads and dams. Others were less tangible. Artists and theater groups were supported, and so were many other social schemes to improve the condition of those hit hardest by the Depression. The "pump was being primed" and income began to flow, but the unemployment problem, though relieved, did not disappear. Never once during the New Deal years, and actually not until massive preparations for war began in 1941 did unemployment drop below ten percent[41]. Not only that, private investment did not rise very dramatically, and, most unforgivably in the eyes of those who adhered to classical ideas, the National Debt continued to increase. These programs, however, were never a valid test of Keynesian economics. Considering the state of the economy in the early and middle 1930s the policies of the New Deal did not constitute government intervention to the degree or in the several ways required by the General Theory. That they were acceptable at all, in the most capitalist of countries, was a reflection of the gravity of the times, and a tacit recognition that more radical solutions were lying in wait.

Only in the last decade, beginning with the administration of President John F. Kennedy, has the executive branch of the government of the United States tried consistently to apply Keynesian doctrines to the national economy. This is in contrast to major industrial nations in Western Europe which began to do so shortly after the end of World War II. We have already referred to the high level of unemployment in the early 1960s. Keynesian-oriented advisors to Kennedy recommended that it be combatted by a multi-level attack which embraced tax cuts and tax credits, additional allowances for plant depreciation and a higher rate of government spending.

To propose new measures to the Congress of the United States is not always to convince them. There were then and there are still influential members of that body to whom any resorting to deficit spending is anathema. It was Kennedy's successor, Lyndon Johnson, who, wise in the ways of Capitol Hill, succeeded in obtaining tax cuts from a subdued Congress in the period following the tragic assassination of the young president. The economy responded as the theory predicted it ought to. Unemployment dropped and the national income began to rise sharply.

One must add that another most powerful stimulus was acting simultaneously. The war in Vietnam created, as have other wars before it, a strong demand for goods and services. From the middle of 1965 to March 1967, military expenditures increased from 3.3 billion to 6.7 billion dollars, while the troop strength in Vietnam went from 25,000 to approximately 500,000. The total employment produced by the military expenditures, including military personnel and Department of Defense civilian workers, was estimated at about 5.7 million persons in fiscal year 1965 and 7.4 million in fiscal year 1967[42]. Had the Keynesian theory suggested the need for augmenting Government expenditures in the absence of a real or anticipated military emergency there is no certainty that the Administration would have received authorization to make the additional sums available. Because of domestic and foreign political considerations, responses to the need for government monetary and fiscal action are most often imperfect, or at least tardy.

By 1971 it appeared that the Keynesian revolution was indeed complete when President Richard Nixon, whose party has generally resisted most of the implications of the Keynesian theory, indicated in his January message to Congress that he felt it was necessary to increase Government expenditures and loosen the money supply in the face of relatively high unemployment even though it might cause a budget deficit. Yet if there is one thing to be learned from studying the nature of revolutions it is that they are never more than preludes to further revolutions. The General Theory, a carefully wrought product of a superb intelligence, was designed to preserve capitalism rather than to obliterate it. Nevertheless, free enterprise capitalism had to take on a new form, now subsidiary to the role of the national government in the economy. Keynes was aware of dangers inherent in transferring unprecedented control over investment to a centralized bureaucracy, but being a supremely rational individual himself he hoped (not without some misgivings) that a similar rationality, tempered with sensitivity, would prevail among the managers of the economy. He died in April 1946, just at the outset of an era which would be transformed by new technologies. While the consistent application of his ideas appears to have stabilized employment problems, there are indications that the dynamic complexity of contemporary industrial society can resist the carefully-timed fine adjustments required by Keynesian theory. Businessmen and labor leaders may be expected to cooperate voluntarily with a policy of economic management when they feel their own interests will be well served; with less enthusiasm, if at all, when they do not. As the Keynesian approaches have become familiar, the measures taken by private groups to soften or evade their negative effects become less predictable. Higher interest rates and taxes, designed to slow down the economy, can act as a stimulant if investors and purchasers rush to the market hoping to beat even further increases. There are those who say that the Keynesian theory is undependable, and attention to less complex factors, such as controlling the supply of money, gives a more effective regulation of economic trends.

In a later essay entitled, "My Early Beliefs," Keynes wrote:

> ... We were not aware that civilization was a thin and precarious crust erected by the personality and the will of the very few and only maintained by rules and conventions skillfully put across and guilefully preserved.... The attribution of rationality to human nature ignored certain powerful and valuable springs of feeling [43].

In a sense, a basic test of adherence to Keynesian principles in the United States would be the response to a prescribed need for heavy deficit spending at a time when international political tensions were so relaxed as to provide no justification for an augmented defense budget. The system assumes the need for a continuous expansion of the economy with a constantly increasing consumption of goods. With that in mind, perhaps the more important questions to ask about its long term adequacy should deal with the effects of uninhibited growth on resources and the environment. These are questions the theory did not try to answer.

STRUCTURAL VERSUS AGGREGATE-DEMAND UNEMPLOYMENT

Let us now try to bring into perspective the variables relating to the effects of accelerating technological change on the employment status of the American worker.

Prevailing opinions suggest that the kinds of unemployment can roughly be grouped in two general categories called *structural* unemployment and *aggregate-demand* unemployment. Many students of labor problems tend to concentrate on one or the other of these categories to explain the causes of joblessness or to propose solutions for it, although there are many variations of both positions and they are not mutually exclusive.

It seems that identifiable structural factors can be used to define the nature of unemployment. As examples of these factors we could mention the degree of automation or cybernation, the kind of industry and its location, the possible depletion of resources, or perhaps the age or educational level of the workers. Structures change, sometimes slowly, sometimes rapidly, and for different reasons; and when they do, men and women are often thrown out of work. From the point of view of many structuralists the nation has not yet begun to deal with the situation of those for whom unemployment or partial unemployment threatens to become a pattern of life; and now the introduction of computers and associated technologies can change the nature of work not only so rapidly that society will be unable to adjust in the short term but also in fundamental ways in the long term such that large numbers of people will never again be needed for producing goods and services. Orders of magnitude changes can have profound qualitative effects.

Those who favor the aggregate-demand position contend that to concentrate on structural change causes the central point to be missed. They feel that if consumers, business and government spend enough to maintain economic growth this will provide the employment opportunities necessary to keep people working. Although there is a basic contradiction between a full employment policy and the pursuit of stable wages and prices and there will be times when deflationary policies cause jobs to be lost, the government has it within its power to see that the condition does not go too far. Adherents of this position will usually admit that some unemployment through structural change is inevitable but will also argue that it tends to be temporary and isolated and that one of the most effective ways to fight it is to support retraining programs which will improve the worker's versatility. Sponsored relocation is also sometimes recommended.

Some of these conditions and arguments, we have observed, had earlier counterparts although the acceptance of broad-gauged government intervention in the economy was not an idea seriously followed in modern Western nations before Keynes. One specific aspect of the problem that *is* new is that in the knowledge society an increasingly larger proportion of the population is engaged in pursuits whose success cannot be measured directly by tangible units of production. Many of the new jobs are in non-profit areas like government service or education; or if in profit-making enterprises they are often in research and development or other departments peripheral to the production line. The stability of these jobs is to a considerable extent based on faith that something necessary is being done. Not necessary in the sense that, without the performance of the particular functions, physical survival would be made less tenable; but rather that they contribute to the future growth of the enterprise or somehow to its current quality. This is not work as it was known at one time. In an expanding economy, faith abides. In a slowed-down economy, faith falters, and it is possible to establish by executive direction that something useful is *not* being done. Then the tendency is to reduce the cost of doing business. Many highly trained people, including engineers and scientists working on research and development, may suffer loss of jobs in these periods of retrenchment. In 1968 through 1971, as federal spending for science or defense projects was cut back, physicists, chemists, mathematicians, engineers and natural scientists found it much more difficult to get jobs than they had before. Thirty percent of the 1200 to 1300 physics Ph.D.s who obtained the doctorate in June of 1968 were unable to find jobs immediately, and a year later ten percent of them were still unemployed. The job placement service at the annual meeting of the American Chemical Society in 1966 had 1180 employers prepared to interview 544 job applicants. In 1969, 481 employers were present to interview 880 applicants[44]. By 1971 many educators were suggesting that the numbers of students going into graduate training in science and engineering should be reduced.

So if it is true in modern society that highly educated people who work for others are in greater demand and have more job security than manual workers, it still cannot be said that their positions are invulnerable to economic or technological trends. If it is a fact that the organized pursuit and application of knowledge is inherently productive, this fact has yet to be universally accepted as a guide to the retention of professional workers. In the best of times there are many workers at every level who can be considered superfluous in the sense that the system of which they are a part can be assumed to be able to operate without loss of efficiency if they were removed and not immediately replaced. This is just as true for the corporation executive who is "loaned" to the Community Fund for a year as it is for the employee who is permitted to "featherbed" by the terms of his union's contract. In the worst of times the pressures to identify and eliminate this kind of superfluity can be intense.

WORK AND LEISURE

All of these considerations can lead to the conclusion that neither unemployment nor poverty (which are related conditions but not necessarily the same) is close to being solved in our generally affluent society. In sum, we have a sizable body of people who are unemployed—or only partially employed—by virtue of age, physical disability or sociological handicaps; and some are unlikely ever to enter the permanent work force. Many of these are classified as "poor," and thus qualify for some kind of regular government aid. We also have workers who have lost their jobs through certain changes in technology or in the conditions of their work, but who might be quickly reabsorbed if they could be retrained. Opportunities for retraining would of course be more abundant whenever the economy, in a period of upturn, found workers in generally short supply. Sometimes when the government is reducing spending and private business is experiencing a slowdown, some highly educated professional people who are not self-employed are quite unable to find work for long periods; or else they must accept employment which is less professionally challenging than that for which their training prepared them. While new jobs and professions are continually being created in an atmosphere of unparalleled attention to research and development, many of these people are in very narrow specialities which could suddenly lose marketable value for technical or economic reasons. It is sometimes said that highly trained persons are flexible enough to switch to different fields under such circumstances without breaking stride, but actually statements of this kind overlook the considerable psychological problems which can arise when the security that goes with occupying a seemingly assured and needed place in the professional world is swept away.

It is fair enough to say that industrialized technology increases affluence, so the lingering-on of the ages-old plague of poverty cannot be laid at its door. However, it is also true that the abundance of material goods may magnify the discontent of those who have less of them. The sense of relative disparity of economic position is just as active a source of social malaise as is the absolute level of need. Neither can the entire blame for problems of unemployment be ascribed to the machines that replace men. Things are much more complicated, as we have tried to show. For many reasons there is an almost continuous state of imbalance between production and consumption, however neatly they are supposed to complement each other over the long term. Even if the wants of consumers could somehow always be kept abreast of production capabilities it would still be difficult to avoid unacceptable consequences, of which environmental erosion would be one of the worst.

Many writers in recent years have foretold the imminent arrival of a time when the new machines would eliminate the need for men to perform most physical and mental labor. An era of leisure rivalling that of Ancient Greece would supposedly ensue. Actually enough evidence is now on hand to demonstrate that the cybernated world, considered as a place where the engines of production substantially regulate themselves, remains largely a theoretical feasibility. Rapid strides are being made in overcoming technical impediments to the design and operation of large-scale cybernated processes, but there are also strong economic and sociological factors countering the thrust to implement them. To be sure, more persons in Western nations are now able to spend more time at activities apart from their regular jobs than they have since early in the Industrial Revolution. The average length of the official work week for production workers has dropped sharply in the past 100 years or so and decreases seem to be continuing, especially in those occupations which are strongly unionized. There is also an observable trend toward later entry into the work market, and earlier retirement, but there are countervailing tendencies, too. The demand for material luxuries has increased along with widespread affluence, especially among those people who have not been accustomed to them. Many workers will take second jobs if their ends cannot be met by the pay from their primary job. As for executives and many professionals, they often work fifty hours a week and more; yet if offered the choice of a comfortable retirement, few would care to take it. These are people who have worked very hard to achieve their favored positions in the industrial hierarchy and not many of them would willingly relinquish their special prerogatives or sense of authority.

A workless Eden becomes more difficult to conceive when we examine closely the kinds of people who would have to populate it, and the kinds of conditions which would have to obtain to bring it nearer to reality. The Puritan ethic of salvation through work exists today in different guises. The discipline of work still has the powerful effect of permitting man to put aside the larger anxieties and questions about the ultimate nature of his existence. As Sebastian de Grazia has phrased it[45]: "The men who go to work in the morning and come home at night are still the pillars of society and society is still their pillar of support." We can begin to see cracks at the base of those pillars, but the form of the new structure of a leisure-oriented world remains indistinct.

Leisure is a seriously misunderstood word. It ought not to be described as the converse of work. In it there is no compulsion to do or not to do. Neither is it slothfulness for it can mean vigorous engagement, but only in something desirable for its own sake. Above all, most writers on the subject feel an enjoyment of leisure in the technological society can be stimulated by a broad, thorough education. If it merely meant time apart from work, then of course we could look among the ranks of the indigenous poor or the unemployed to find it. Needless to say, the condition of neither of these groups represents the ideal state of leisure. There would be relatively few among their number who would not want to re-enter the world of work, given the opportunity. On the other hand, the average worker on the assembly line—bound to its unsparing monotony and lacking even the compensation of being able to demonstrate superior strength or versatility—feels that he would gladly throw over his job tomorrow and forever if he could be assured of at least his present standard of living. One wonders how workers would really fare if large numbers were suddenly granted an exemption from the necessity of earning a living. Quite apart from the intolerable burdens which would be placed on public and private recreation facilities, torsions in the entire structure of society would be extreme without extraordinary changes in typical human attitudes.

While spokesmen for business, government and labor may acknowledge the notion of the leisure society in public statements, there seems to be little inclination among the majority of them to accept it as a looming possibility. Their professed, primary objective is still to provide jobs for all who are capable of working. The fear of joblessness and the shame attached to it die slowly in our society. If the time should come when cybernation (or other related conditions) would cause a radical and wide-ranging displacement in employment patterns, however, there would be a need for more comprehensive measures than we now have operating to relieve the distress of the workless.

THE GUARANTEED INCOME

The attempts to deal with poverty in this country have antecedents in the Elizabethan Poor Laws and the modifications to those laws in 18th and 19th Century England. For many years in the United States, public relief measures for the sick and needy have been part of a welfare system which during the late 1960s came under such heavy and sustained attack from all sides that it was apparent other methods would have to be devised. The main arguments against the assistance programs are familiar ones: they foster an immense and costly bureaucracy, cause indigents to migrate in large numbers to localities where the welfare support is highest, and discourage thrift and independence. With regard to the last point for instance, in the Aid to Families with Dependent Children Program, any earnings of the family have ordinarily been deducted from their benefits. Defenders of the welfare concept (not necessarily in its familiar form) have contended that the great majority of welfare recipients are literally unable for physical or other reasons to earn a living. (About ninety percent of AFDC families have been headed by women or incapacitated males [46]). The argument then goes that these people have no fewer rights than other segments of the population who obtain public subsidies in one form or another.

That there have been administrative inadequacies in traditional welfare programs is rarely denied. Some other reasons for the general dissatisfaction with the system have been discussed by Wilensky and Lebeaux [47]. Racial and religious fragmentation in the large urban areas particularly inhibits voluntary cooperation on common concerns, and overlapping local, state and federal responsibilities compound the difficulties. A tradition of individualism in this country favors reliance on social insurance programs, although ample evidence exists to show they are relatively ineffective in improving the situation of those at the bottom of the economic ladder.

At the start of the decade of the 1970s, the principle that the federal government must guarantee a minimum annual income to its citizens has received a wide, although far from unanimous acceptance. Opinions differ on what represents a desirable minimum, and on what means should be used. Whether the proposal has been to grant minimum income payments to the needy unequivocally, as suggested by Robert Theobald in 1963 [48], or on the basis of their willingness, if able, to accept "suitable" employment or vocational training, (as advanced by the Nixon Administration first in 1969); whether the support ought to be granted by paying a fraction of unused income-tax exemptions and deductions (the Negative Income Tax proposal of Milton Friedman [49]), or by distributing ownership of new capital to the poor through government-guaranteed loans, (as submitted in similar suggestions by Kelso [50] and Fein [51]), the plans have commonly professed to be able to reduce the cumbersome administrative machinery of traditional welfare programs, to remove demeaning eligibility tests, and to eliminate variations in the minimum payments from state to state. All of these plans recognize that poverty, even in our affluent society, can be so widespread and enduring that a uniform policy of relief is mandatory.

There is yet very little agreement on what the consequences of a federally-funded guaranteed income plan will be in the long run. On the positive side, the hopes of some are that a uniform national rate will encourage impoverished families to leave the crowded industrial areas which have granted the highest level of welfare payments, to return to the poorer states and rural regions from which many originally came. If this reverse migration would require their re-entering an environment where racial discrimination is more actively practiced, the objective would not be reached quickly. William Vogt[52] has suggested that a more probable migration would be to the states with the most attractive climates, like Florida and California, thereby adding to the population growth which has already strained their resources. The important question of the effects of the guaranteed income on the birthrate remains unanswered.

With the requirement that recipients of aid must accept employment if offered, will some employers then try to use this as a device to keep wages low, as happened under the Speenhamland policy? Will a guaranteed income plan affect the average individual's incentive to "succeed" in the traditional American sense? Will young people use the assurance of a steady income as a means of establishing new kinds of communities on a subsidized basis? Perhaps the effect might be to create incentive. As Ayres[53] has pointed out, the poorest are often the most reluctant to change old habits, feeling the need to hold on to what little they have. With a small measure of independence might come security and a will to branch out, even to acquiring a desire to compete in the economic arena (with its rewards and attendant, but different, anxieties). One can conceive of other potential effects if work and wealth are distributed in different ways in a highly cybernated world. The changes will depend very much on evolving concepts and practices in education for work or leisure.

An influential body of opinion holds that in our technology-oriented society there is no fundamental reason why rational, scientific approaches using the methods or devices at our command cannot alleviate most social problems, including these very basic ones we have been discussing. In the next three chapters we shall examine some of the political and structural impediments to such solutions, which still need to be understood and resolved if modern society is to function in a manner closer to what the majority would consider a theoretical ideal.

REFERENCES

1. Jackman, Patrick C., "Unit Labor Costs of Iron and Steel Industries in Five Countries," *Monthly Labor Review*, U.S. Dept. of Labor, Vol. 92, No. 8, August 1969, p. 18.

2. Jackman, Patrick C., *Monthly Labor Review*, U.S. Dept. of Labor, Vol. 93, No. 9, September 1970, Table 11, p. 75.
3. Machlup, Fritz, *Production and Distribution of Knowledge*, Princeton University Press, Princeton, 1962.
4. Drucker, Peter F., *The Age of Discontinuity*, Harper & Row, N.Y., 1968, p. 27.
5. Kleiman, Herbert S.; "The Economic Promise of Computer Time Sharing," *Computers and Automation*, Vol. 18, No. 11, October 1969, p. 47.
6. Ad Hoc Committee, "The Triple Revolution," *Liberation*, April 1964, pp. 1-7.
7. Bell, Daniel, "The Bogey of Automation," *The New York Review of Books*, August 26, 1965.
8. Silberman, C. E., "The Real News About Automation," *Fortune*, January 1965.
9. McKinsey & Co., Inc., "Unlocking the Computers Profit Potential," *Computers and Automation*, Vol. 18, No. 4, April 1969, p. 25.
10. *Monthly Labor Review*, U.S. Dept. of Labor, September 1970, p. 71.
11. Greenspan, Harry, *Monthly Labor Review*, U.S. Dept. of Labor, September 1967, p. 52.
12. Alt, Franz L., "Computers—Past and Future," *Computers and Automation*, Vol. 18, No. 1, January 1969, p. 14.
13. Handel, S., *The Electronic Revolution*, Penguin Books, Harmondsworth, Middlesex, England, 1967, p. 162.
14. Foster, David, *Automation in Practice*, McGraw-Hill, London, 1968, pp. 48-49.
15. Foster, David, ibid., pp. 50-51.
16. Foster, David, ibid., p. 51.
17. Heineman, Ben W., "Guaranteed Annual Income—A Placebo or Cure?" *The Conference Board Record*, Vol. VI, No. 5, May 1969, p. 26.
18. McNamee, Holly, "Adding Up The Problems of the Poor," *The Conference Board Record*, Vol. VI, No. 5, May 1969, p. 23.
19. Heineman, Ben W., op. cit., p. 26.
20. Heilbroner, Robert L., *The Making of Economic Society*, Prentice-Hall, Englewood Cliffs, N.J., 1962, p. 59.
21. Mantoux, Paul, *The Industrial Revolution in the 18th Century*, Harper & Row Torchbooks, N.Y., rev. ed., 1961, p. 142f.
22. Mantoux, Paul, ibid., pp. 141-42.
23. Polanyi, Karl, *The Great Transformation*, Beacon Press, Boston, 1957, p. 78.
24. Cole, G. D. H. and Postgate, Raymond, *The British Common People*, Barnes and Noble, N.Y., 1961, p. 276.
25. Polanyi, Karl, op. cit., p. 83.
26. Jaffe, A. J. and Froomkin, J., *Technology and Jobs*, Frederick A. Praeger, N.Y., 1968, p. 37.
27. Cole, G. D. H. and Postgate, Raymond, op. cit., p. 193.
28. Pike, E. Royston, *Hard Times*, Frederick A. Praeger, N.Y., 1966, p. 124. *Human Documents of the Industrial Revolution*, George Allen & Unwin Ltd.
29. Pike, E. Royston, ibid., pp. 125-26.
30. Pike, E. Royston, ibid., p. 190.
31. Ure, Andrew, *The Philosophy of Manufactures*, (1835), p. 301.
32. Shelley, Percy B., *Peter Bell the Third*, (1819), Part III, Stanza 1.
33. Dickens, Charles, *Hard Times*, (1854), New American Library, Ed., N.Y., 1961, pp. 30-31.
34. Cole, G. D. H. and Postgate, Raymond, op. cit., p. 358.

35. Mumford, Lewis, *Technics and Civilization*, Harcourt, Brace & World, N.Y., rev. ed., 1963, p. 179.
36. Cole, G. D. H. and Postgate, Raymond, op. cit., p. 360.
37. Hays, Samuel P., *The Response to Industrialism 1885-1914*, University of Chicago Press, Chicago, 1957, p. 104.
38. Stewart, Michael, *Keynes and After*, Penguin Books, Baltimore, 1967, p. 70.
39. Heilbroner, Robert, *Worldly Philosophers*, rev. ed., Simon & Schuster, N.Y., 1961, p. 219.
40. Stewart, Michael, op. cit., pp. 66–67.
41. Stewart, Michael, op. cit., p. 144.
42. Oliver, Richard P., "The Employment Effect of Defense Expenditures," *Monthly Labor Review*, Vol. 90, No. 9, September 1967, p. 9.
43. Malkin, Lawrence, "The Economic Consequences of Maynard Keynes," *Horizon*, Vol. XI, No. 4, Autumn 1969, pp. 110–11.
44. Nelson, Bryce, "A Surplus of Scientists? The Job Market is Tightening," *Science*, Vol. 166, No. 3905, October 31, 1969, p. 583.
45. De Grazia, Sebastian, *Of Time, Work and Leisure*, Doubleday-Anchor, Garden City, N.Y., 1963, p. 385.
46. Levitan, Sar A., "The Pitfalls of Guaranteed Income," *The Reporter*, Vol. 36, No. 10, May 18, 1967.
47. Wilensky, Harold L. and Le Beaux, Charles N., *Industrial Society and Social Welfare*, The Free Press, N.Y., 1965, pp. xviii ff.
48. Theobald, Robert, *Free Men and Free Markets*, Doubleday-Anchor, Garden City, N.Y., 1963.
49. Friedman, Milton, "Negative Income Tax," *Newsweek*, September 16, October 7, 1968.
50. Kelso, Lewis and Hetter, Patricia, *Two-Factor Theory: The Economics of Reality*, Random House, N.Y., 1967.
51. Fein, Louis, "The P.I. Bill of Rights," *Change*, Vol. II, No. 5, 1966.
52. Vogt, William, "Conservation and the Guaranteed Income," in *The Guaranteed Income*, Robert Theobald (ed.), Doubleday-Anchor, Garden City, N.Y., 1967, p. 154.
53. Ayres, C. E., "Guaranteed Income: An Institutionalist View," loc. cit., p. 169.

4

Technology and Government

"Government, after all, is a very simple thing."
WARREN G. HARDING

As we enter the 1970s it seems difficult to believe that a President of the United States could, even fifty years ago, have uttered such a fatuous statement. The art (or science) of government has never been at all simple. It was not in Harding's time, and is even less so today as dynamic technologies create new structures to be adjusted to and new roles to be played. Admittedly there is a line of thought that the best government is the least government. This view has its attractions, especially if one assumes that governments of industrial nations are little better than vast, bureaucratic machines in which the individual is forced to become an insignificant cog. Nonetheless, bureaucracies of some sort in the modern world are probably here to stay as far ahead as we can see. Technology has transformed the social order, usually in unanticipated ways, and brought about nagging and explosive conflicts. In the process it has compounded the burdens and responsibilities of governments. Some degree of bureaucracy is undoubtedly necessary to meet these responsibilities in any reasonably effective way.

There are alarming signs that, in some urban areas at least, local governments have reached a point of administrative saturation where lack of funds, inadequate communication, and sheer weight of numbers have produced crises of control. We hear predictions of inevitable social breakdown. Is it possible that many cities are now, or will soon be, ungovernable? The position taken here is that existing forms and tools of government would be inadequate to cope with some situations should they go beyond a certain stage of deterioration. If chaos short of total physical destruction by atomic war should come to a city or region, other governmental structures would surely rise in place of the old because modern man cannot long exist in anarchy, but whether any alternative forms could do much better in the long term without a wholesale reformation in the attitudes of the individual citizens remains to be seen.

As a vigorous agent for social change, technology has always affected the ways in which societies have organized themselves. The previous chapters have offered examples of this. Since World War II, however, technology, science and government have commingled in unprecedented ways. The magnitude of federal expenditures for research, development and production largely in indirect or direct connection with the war effort defies comparison with any earlier times. Referring to this relationship, the term, "military-industrial complex," which President Eisenhower included in his farewell address, has become a widely, and often pejoratively, used expression. (Professor Melvin Kranzberg has coined a more descriptive acronym, MAGIC, standing for the military-academic-government-industrial complex.) To be sure, other affairs than the strictly military also link technology and government together. Our concern in this chapter will be to explore several aspects of these affinities.

The previously raised question, whether governments ought to do more or less for their citizens has been a central point of interest in political discussions at least since the 18th Century. However, as Shirley Letwin [1] observes, the arguments which stem from this issue may often lack an important dimension—that of a clear measure of the function in question. In a dispute about how much support the government should provide in the field of social welfare, the protagonists are unlikely to begin with any close agreement about the real meaning of welfare. That kind of debate can quickly dissolve into ambiguity. We must then ask whether there are yardsticks which can be applied to political and social issues; whether the art of political action can be reduced in any degree to a science; and, in a collateral vein, if technology is a prime initiator of social change to what extent must a political leader be able to understand it and have it utilized to best advantage?

The inquiry will run along several lines in an attempt to outline certain key problems which press in on a political leader in a technological society. It will begin with the election process itself and examine the ways in which technology may be affecting the choosing of our representatives in government. Then, an area of persistent concern will be studied in appropriate detail. Urban transportation is selected as an example of an issue which was too often approached in the limited framework of technological and economic policy, but which is now recognized as having broad socio-political ramifications, spilling across many jurisdictions. While some areas are more seriously affected by transportation difficulties than others, the problems are basically of national dimensions and represent a major challenge for man and his technologies. A complete book would be required to treat the many levels of the transportation dilemma. Within the limits of space here, we can only try to show how technological factors must undergird key decisions, and how necessary it is for the decision maker to have a clear and sensitive understanding of these factors. This is not to imply in any way that technology can be considered out of context — its context being the whole national and international social milieu. It only means that no appropriate scientific or technological tool can be overlooked in man's struggle for survival, nor should any be used indiscriminately. We are only now beginning to understand that the use of technology always has a cost, and the cost cannot always be counted in monetary terms.

One observation, of concern to many students of government, is that important political decisions have been and still are being made in an atmosphere of incomprehension of relevant technological variables and their possible outfall. Long gone is the image of government as a group of men trained in the law, meeting apart from the mass of citizens to make new laws when necessary, and seeing that the laws are obeyed. The theory of a governing elite goes back to Plato, and the arguments about its advantages and disadvantages troubled our own founding fathers no end. Thomas Jefferson, the libertarian, said, "State a moral case to a plowman and a professor. The former will decide it as well, and often better than the latter, because he has not been led astray by artificial rules." Alexander Hamilton, his constant adversary, contended on the other hand that the common people could "seldom judge or determine right." It is not clear whether Jefferson with his keen knowledge of the science and technology of the times would have made the same statement about technical cases — probably not — but then, neither is it unlikely he felt that the moral problems were the critical ones. In any event, the question of governance by a technical and scientific elite, rather than an economic elite, now recurs insistently. We shall want to discuss features of some of the most common suggestions, namely that: (a) the primary decisions are already being made by technicians in a huge mechanical bureaucracy, (as writers such as Jacques Ellul[2] would contend); (b) that we do not now have, but should have, a ruling hierarchy of people highly trained in the sciences and technology; or (c) that the first contention is only marginally true, the second would be unfortunate if not disastrous, and so a preferable alternative would be to have broadly trained government leaders who are accessible to the advisory services of distinguished scientists and engineers of various political persuasions.

No matter how a government in any industrial society prepares itself to deal with problems of social change and technology, it will have to recognize a growing disenchantment with the concept of limitless technological progress. Opposition to uninhibited and unconsidered growth is better organized and more well informed now than it has ever been. The Congress of the United States has sponsored several studies to address systematic attention to the probing questions that have been raised. The primary objective has been to analyze possibilities for continuous assessment of the potential that technology possesses for harming or benefitting man's environment. The last section of this chapter will seek to evaluate the prospects for an operational scheme of technology assessment which might serve as an early warning system to help avoid the most pernicious consequences of technology out of control.

THE WINNING OF ELECTIONS

The first concern of a politician is to get elected. To an idealist this often-cited maxim smacks of unpardonable cynicism, but to a pragmatic political worker it has the ring of vital truth about it. Neither idealists nor pragmatists can easily avoid concluding that the winning of elections is a more complicated and expensive procedure than it used to be, and that the communications technologies are playing seminal roles, nowhere more noticeably than at the national level where skillful presidents can manipulate communications media for their own political advantage.

One trend which some political scientists feel could subvert our political system is the use of computers to forecast the outcome of elections. In the immediate past, candidates and their supporters have charged that when television or radio commentators have made public computer-based predictions of returns on election night before polls in the Western states have closed, the result has been to influence the choice of the voters or in some instances to discourage them entirely from casting their ballots. Empirical evidence to support or to refute these assertions is not easy to secure, and in any case conclusions drawn from the available evidence need not be expected to have permanent validity. Predictions of final election outcomes based on early returns reported the same evening still tend to be highly reliable, though not infallible, but polls taken a few days before the election have occasionally been seriously misleading. The most recent example of note occurred in the British elections of 1970 when the Conservative Party, quite contrary to the consensus of expert forecasters, swept the Labor Party out of power. This only means, of course, that the human assumptions made in collecting the data and utilizing it were at fault; not that the computer erred. There is a possibility that as ordinary people through exposure to the media become more conscious of the potential significance of their replies to the outcome of an election, they will turn more evasive, if not deceptive. Polling theory has rather assumed that the average voter, pleased to have his ideas attended to, will respond truthfully, if at all, to unintimidating questions but this need not always be so.

Some political strategists are intrigued by the thought of being able to develop a large data bank containing masses of categorized information about voter attitudes on various issues by means of which a successful campaign could be developed and carried out. The best known early example of the use of computer techniques to anticipate the political behavior of voters occurred during the national presidential campaign of 1960. This project, which was reported by Pool and Abelson[3], was an attempt to simulate the probable voting response of large groups of individuals to specific issues on the basis of a limited amount of selected data. The idea of the simulation was to have a computer program test certain assumptions of political theory with regard to group voting behavior and compare the results with the actual election outcome.

Data available to the research team came from public opinion polls conducted in a period from 1952 to 1958. These included responses from about 130,000 individuals from various localities. The problem was to settle on certain issues, some fifty in all, which could be expected to dominate the campaign, and to organize the information so that it would have strategic value for one side, in this case that of Senator John Kennedy. The voters were classified into a series of 480 voter types each defined according to specific combinations of socio-economic characteristics, e.g., "upper-income, Protestant, rural, white, Republican, female." Included in the data were the number of people of that type living in each state, and the way they had divided in their reaction to poll questions on the key topics. It was assumed that the people classed in an individual type would react similarly no matter where they came from.

Few individuals are willing to accept without reservation every part of the political platform of their party, or all of the characteristics of its candidates. For this reason social scientists must try to assess the significances of relative pressures resulting from multiple group loyalties. In the 1960 election one of the foremost of these cross-pressures appeared to be related to the religious issue. How could anyone determine whether a life-long anti-Catholic Democrat would be more inclined to vote for the Catholic candidate of the Democratic Party or the Protestant candidate of the Republican? The research team devised a set of equations for computer solution which took into account several factors such as how one type of voter had divided on a given issue previously and whether that type was inclined to vote in all elections. They made certain additional assumptions to help them prepare an estimate, based on a hypothetical campaign, of what proportion of the Catholic group in each party would vote for Kennedy. The results of the simulated campaign were a factor in the strategy of the Democratic Party. Senator Kennedy determined to meet the religious issue head on, and it is interesting to note that the outcome of the simulation correlated 0.82 (out of a possible 1.00) with the final results in 1960.

One can only surmise how much different, if any, the results of the election would have been without the making of certain decisions influenced by the simulation study—the researchers themselves do not claim that Kennedy would have lost without it. Still, it is natural to wonder how influential similar studies have been in subsequent elections. The answer appears to be not so influential as one might expect, considering the more advanced state of development of methods of data taking and analysis compared with 1960, and the greater capacity of the newer machines. Not that political parties do not use computer analyses for various purposes—they do. For example, when congressional district boundaries were redrawn in a large state a few years ago the parties used computers to decide what shape and size of district, given a certain population distribution, would be most apt to favor their own candidates in future elections. This is nothing more than old-fashioned "gerry-mandering" of course, but accomplished in a more complete and efficient way than by pre-computer methods. The results, from the standpoint of aiding the fortunes of the party, are a function of whether the political theories on which the computer programs were based, were correct. This last factor is the key.

If it is economically practical for political groups to have this kind of job done for them, or any other that requires analysis of masses of data, then there is little question that they would gain by doing so. Yet this is not the same as basing an entire campaign on political science propositions incorporated in a simulation program. There are few indications that practicing politicians are ready to accept this as a pattern. Numerous reasons exist for this. Their attitudes are surely influenced by a suspicion that the complexities of programming could limit flexibility in a rapidly moving campaign. Although political simulations would ideally take into account the most recent relevant events as well as long term trends, this is most difficult to do. Long term data tend to confer stability on a study but individual and group attitudes change with time, often very rapidly in this age of mass communications. A sudden international crisis, an unexpected scandal or an economic slump could require tedious and complicated changes in programming. There is a surprising paucity of reliable data regularly compiled over any considerable periods of time tracing the changing attitudes of people on various issues.

The traditional politician attributes a good part of his claim of expertise and influence to experience and a steady intuition. He probably also holds strong feelings about certain political tenets to which his party subscribes. If he is a successful politician he is no doubt capable of compromise. Still we can wonder about his response if a computer simulation should point to a course of action judged most likely to influence voters toward his side, but one which is directly contrary to long-cherished articles of faith. Under these circumstances if he is a man of principle he might be expected at most to agree to muted references to the issue. This is nothing like having a political campaign planned, packaged and sold with the computer as the principal agent.

Most evidence points to the computer as playing a secondary role in the direct process of winning, as opposed to recording, elections. Not many would question that the primary role has been taken over at this time by another branch of communications technology-television (and, radio, to a lesser extent). The two major national political parties have spent many millions of dollars in recent campaigns to try to get their candidates elected. A large proportion of the expenditures has gone for television and radio time. The trend has disturbed students of political processes for a number of reasons, not the least being that the need to mount a ruinously expensive campaign has prevented many deserving candidates of modest wealth from seeking office. It certainly favors the incumbents in most cases because they have more leverage when seeking campaign contributions, but that is nothing really new.

Legislation intending to limit total campaign expenditures might create conditions to help rectify the imbalance in favor of the incumbent—unless the challenger should happen to be an exceptionally wealthy person. In that case his chances would be favored by a media-saturating campaign to give him name-identity.

There is little cause to doubt that many political figures now in office owe their elections to their television personalities more than anything else, but trading on personality to gain a political victory is nothing new either. If having moving picture actors elected to serve in the high offices of governor or senator seems like a relatively new phenomenon, we must remember that long before television, at least one governor was known more for his ability to sing to the accompaniment of his guitar than for any improvement he wrought in the administration of his state. The importance of television is undeniable but its long term effectiveness in influencing elections or in helping elected officials retain their popularity is not fully predictable. We can be sure that enormous effort will be expended to make any major candidate appear more attractive before the television cameras. However, the image produced may be ephemeral. Television is an extraordinarily sensitive medium which quickly picks up falsity or lack of candor. It has to be used imaginatively and, over the long run, sparingly. Even the most popular programs and entertainment figures with some notable exceptions begin to pall on the average viewer after a few years. When the script cannot always be controlled, as it cannot for political personalities under a democratic government, fall from grace could come very quickly.

THE MAINTAINING OF GOVERNMENT

The winning of elections is not the only important political process to be affected by technology. Victor Ferkiss, in his book, *Technological Man: The Myth and the Reality* [4] has provided a perceptive summary of a few of the other effects. He points out, for instance, that the general affluence which has accompanied the growth of the technological society has shifted the focus of attention away from strictly economic issues, toward concerns about values or ideologies. What is more, television sharpens the competition between opposing ideologies by bringing conflict as a reality into the living room instead of on the printed page or in the debating hall where it most often existed before for the average citizen. The result is a threat to the two-party system of government as it has been known for generations. The constantly stimulated hostilities or loyalties accompanying commitment to a host of conflicting ideologies cannot easily be confined within the framework of even a minimally consistent party program. Style then becomes more important than substance, and the difficulties of marshalling a coherent majority opinion to support a rationally planned program become immense. In other words, although modern communications technologies have often been considered to be unifying forces working to produce a smooth social mix when certain common denominators of language and behavior prevail, the reverse has been equally true – perhaps even more so. An awareness of what is going on makes everyone want to get into the act, as the saying goes, but very often he wants to do so by asserting his individuality as a person or as part of a cultural subgroup. Political fragmentation is the consequence. Fragmentation carried too far becomes chaos for which the available antidote nearly always appears as a repressive political monolith. In theory then that should bring about the overwhelming mechanical bureaucracy of an Ellul or an Orwell, dominated by an all-seeing central electronic command post. There are good reasons to believe that this kind of system is basically impermanent. For one thing, conditions of communication-overload quickly obtain in any centralized bureaucracy, and the reverse trend toward decentralization becomes inevitable. Equally important, however, is the powerful interplay between special interests – business, the military, labor, etc. – which collaborate sometimes closely and sometimes to the extent of grudging compromise when they must. In so doing they often seem to reduce the central government to the position of broker rather than guide or master of the public destiny. This situation bothers some people who have experienced, first hand, a sense of impotence in trying to bring the governmental structure to bear on major problems. McGeorge Bundy, who served as an influential advisor to Presidents Kennedy and Johnson, has written that the explosions of technology, purpose and human need demand a stronger executive branch of government, not to do what can be done better by private agencies, but to engage itself more decisively in the regulation of processes in which the present and future interests of the public are being ill-served [6].

These are questions and concepts which have long exercised political scientists and philosophers. Although they must be very much on the mind of a government leader, he is also heavily beset with the day-to-day problems of trying to run a government as effectively as it can be made to operate. In fact it is a common complaint among heads of all large organizations (and this includes governments) that the unexpected crises of the moment and the continual reappearance of unresolved problems crowd out any possibility of taking time to plan and make rational decisions on the basis of studious assessment of long term consequences.

In one way or another every problem faced by governments in industrialized nations bears a relation to the technological revolution. This implies no necessary order of precedence in the sense that technology always causes social change – the reverse sequence might be as true. It means rather that any problem we wish to consider – war, poverty, pollution, overpopulation, disease or alienation – would have a different character had certain technologies never appeared or had others been brought to bear in some different manner. Furthermore, there is an incredibly complex web which connects every segment of the social structure with every other. No matter how tenuous the relationship may be, we can therefore expect that when a new technology is developed or an existing one redeployed all social links will be affected, perhaps massively, perhaps minutely. These last two terms suggest measurability, but we are not at all sure that the major perturbations will occur in ways that can be quantitatively measured with our present knowledge.

One of the abiding dilemmas for a political leader, then, is to find out if and how the relationships between the coexisting technological and social structures can be brought into clear focus. If they can be understood, can technology then be controlled to produce effects more beneficial than dangerous according to unchallengeable criteria?

URBAN TRANSPORTATION

The real magnitude of the task of effectively using technological remedies for socio-technical problems can best be appreciated through illustration. The example we shall turn to is that of urban transportation, which represents a situation at crisis stage in many areas.

Wilfred Owen[7] states that, historically, there are five stages a country will pass through to arrive at what could be considered a good standard of transportation: (a) a traditional, relatively immobile society, (b) a period of internal improvement and trade growth, (c) mechanization and industrialization, (d) motorized mobility and (e) conquest of distance by air. But it seems that this turns out to be more of a circular than a linear progression in that the most affluent nations, as well as the highly urbanized areas of the less affluent, show indications of entering a new stage of relative immobility. This is not entirely a phenomenon of modern industrialization, (ancient Rome is said to have barred all but public vehicles from the central city streets during the daytime because of congestion[8]) but the automobile has given it a special intensity. One reason is that the individual automobile is relatively large considering its payload, and thousands of them about at one time consume an incredible amount of space, as do the facilities for servicing them. Another is that the relative freedom of movement allowed by the private car confers a randomness to the congestion which compounds the difficulty of finding a basis for orderly traffic flow at certain times of the day. There are concomitant ills—pollution of the air, visual blight, baneful accidents, noise and related socio-psychological traumas—to which the automobile contributes in unique ways.

But a transportation network is essential to a nation or to a city. It is the bloodstream of any society, and critically so in a modern industrial society for without it the requirements for basic survival would not exist. In underdeveloped nations poor transport is one of the major factors inhibiting growth and technical advancement, directly contributing as it does to the high cost of goods. We must also recognize that transportation can be a vital form of communication.

The undeniable need for transportation in itself does not necessarily demand that the automobile be the primary mode but there are powerful reasons why it became so originally in the United States, and why it thus remains at the present time. We have already touched on historical considerations affecting the acceptance of the automobile. Henry Ford's vision of it as the transportation of the common man was undoubtedly influenced by growing problems of maintaining, servicing and financing public urban transportation in the first decade of the 20th Century. When the vision had become reality, there was a great leveling, as Marshall McLuhan[9] puts it, of physical space, and especially of social distance. Every rider had become Superman and some of the more desirable parts of the nation that had been denied him as a pedestrian were now as available as they had previously been to the wealthy few. Even the segregation between the races in the South began to be affected.

There is a growing tendency now to believe that this factor of the car as a symbol of status and achievement may be on the wane. Even before spokesmen for environmental causes had begun to give the automobile industry a bad name, signs of boredom with expensive, over-powered motor cars had begun to appear, and new status symbols such as boats and large houses seemed to be taking their place. But it is easy to overstate this case. There are certain economic verities that cannot be ignored. Millions of new automobiles and mobile homes, not a few of them quite expensive, are sold each year in good times or bad. In California alone enough jobs are provided through the manufacture, sale, servicing and use of motor vehicles to populate a city as large as the San Francisco–Oakland metropolitan area. Over a quarter-million Californians are stockholders in automotive firms. These are potent reasons why any attempt to redesign the systems by which people and materials are moved about cannot cavalierly dismiss the automobile. When in the spring of 1970, a large group of students and faculty at a large California college collected money among themselves for the purchase of a new Maverick car and then destroyed and buried it in a colorful public ceremony, they displayed an undeniable earnestness in their rejection of the car as the materialist symbol of the technological age, as well as an artful sense of the publicity value of their action. The press gave wide coverage to the event, and while there was some affirmative support there was also outraged denunciation of the action from high government officials in the state and ordinary citizens throughout the country. The labelling of the stunt as stupid, if not almost criminally irresponsible, revealed a depth of emotion which, while it certainly bore a relation to the attitude toward college students and faculty in general, also said much about the feeling of the average person toward the automobile. To him it is not only a skillfully and intricately wrought piece of machinery—a monument to American engineering genius—but also a symbol of his livelihood and of a better life than he could ever have expected to know without it.

Transportation is so integral a feature of modern living that it cannot be considered strictly as a service function. It is a bulwark of commerce and an instrument of government policy. The government interest in the United States has traditionally been expressed in the form of subsidies and through certain regulatory agencies or laws. At one time or another during their development virtually all domestic transport agencies, and all air and shipping lines owned by U.S. firms, but engaged in international commerce, have received public subsidies of some kind[10]. Today we could still say that the overall system in the United States is mixed in that the local, state or federal governments share with private enterprise at least some of the costs and risks of furnishing transport. Highways are publicly built and maintained, (although toll-roads are in a somewhat different category), airways are parceled out and airport facilities are supported, shipping is subsidized, and results of government research and development are made available to private carriers essentially without charge. Railways and their terminals are privately owned although they were aided greatly in the 19th Century by federal land grants; but contemporary problems of major rail carriers, dramatized by the bankruptcy of the Penn-Central railroad in 1970, have stirred renewed requests for study of possible nationalization of railroads. Regulation of commercial rail transport through government commissions has been benign or burdensome (depending upon whether the supplier or user of the service is doing the assessing). Many carriers would no doubt prefer to have restrictions removed entirely and to trust to competitive market forces for regulation to maintain economic and efficient service.

All of this is stated to show the scope of government involvement in the whole transportation scheme. Still there is an overall basic lack of coordination. Under present circumstances there is such a large factor of private enterprise, particularly in the case of urban transportation where the essentially unregulated automobile plays a major part, that restructuring the transportation system to try to make it function at an optimum level is bound to be a monumental task.

Questions about the technical coordination of various modes of transport, though significant as well as troubling, draw attention to only a small segment of the overall picture which in fact needs to encompass a concept of intricately-related, competing social priorities. Transportation problems are different in developing and in industrialized nations; they are different in one section of a given nation, or a given state, from another. It is not at all simple even to decide what the ultimate purposes of a satisfactory transportation system ought to be. This discussion, to be kept within manageable proportions, will have to concentrate on certain basic needs for a transportation system in an urban area. To suggest a sampling of the kinds of questions that must be asked we can list just a few.

What ought to be the relative roles of the private automobile and the public carriers? What potential technological innovations should be anticipated in planning for the future? What are the effects of new transportation technologies on patterns of land use? How does housing segregation affect the nature of urban transportation or vice versa? How should relative costs of competing methods be assessed? If parking charges are increased to reduce auto traffic in a central business district will this hurt downtown businesses? Will air pollution become a dominant factor in transportation decisions? Can one seek out psychological means, using mass communication techniques, to influence the average person's feeling toward using public urban transport? If increasing numbers of people should begin to use public transport, will it compound or lessen the difficulties which existing systems face?

Planning for Transportation

To develop answers to these kinds of questions nothing less than a comprehensive planning scheme will suffice. Some of the essential elements needed for broad-gauged planning are summarized as follows along lines suggested by an instructional planning memorandum prepared by the Bureau of Public Roads[11].

Population studies

The measure of the transportation problem in any large metropolitan area in the United States is also a measure of an astounding growth and redistribution of population. The number of urban residents in the United States increased by about 30 million from 1955 to 1965 and in the same period the number of motor vehicles more than trebled from 24 million to 85 million. Census figures in 1970 suggest that the trend (which at this rate would place more people in urban areas by 1985 than the total 1965 population in the United States) is continuing. Orange County, in California, had a population of 703,925 people in 1960 and this had increased to 1,408,969 by 1970. The major growth has occurred in what might technically be called the suburbs, but the problems of the suburbs are in fact the problems of an urban continuum. In this area, which relies almost exclusively on the automobile for transportation, three and four car families are not uncommon. While it is true that many older cities have shown a decline in population in the central core, the need for moving people in and out of the central business district has not declined. Other surrounding communities have as much as quadrupled in size in this decade. There are also indications, however, that a complex of causes which include the contraceptive pill and liberalized abortion laws in many states may now be reducing the population growth rate in this country. The distribution factor may prove to be more important than the overall growth factor in the next ten years. How should the transportation planner decide? There is great difficulty keeping the large volume of data current and relevant. Historical data, population distributions in both large and small areas, migration, changing birth and death rates all need to be surveyed. Moreover, he can be sure that whatever transportation method or mix is developed it will have its own marked effect on the resulting distribution of people and new building construction. The computer is an indispensable aid to demographic studies both in the gathering and interpreting of masses of data, but control of computer procedures has not reached the point where researchers can translate the information for optimum utilization in transport planning[12].

Economic factors related to community development

Other useful types of information, sometimes available from government census figures or from other sources but also generally difficult to correlate, are certain economic factors required for realistic planning. For example:
(a) Employment data and projections including the nature of industries in the study area, future industrial potential as a function of geographic and climatic conditions, characteristics of the labor force, and relation of the local economic unit to the national economy.
(b) Income-consumption patterns for the area.
(c) Car ownership per household.

Land use

A dominant factor in the festering of the transportation crisis and the urban crisis in general is that land in the urban areas has only incidentally been considered as a precious resource to be utilized to meet human needs and purpose except when that purpose has been to wring a handsome profit from it. C. A. Doxiadis[13] states that human needs are governed by the four principles of: (a) maximum of human contacts; (b) minimum of effort; (c) optimum of space, and (d) quality of environment. There is probably no inherent reason these principles could not be maintained, consistent with a fair return on private investment, far better than they have. Students of the problems would presumably agree that a necessary precondition, aside from the good will of private developers, would be a regional planning agency empowered to override fragmented political jurisdictions and narrow considerations of any special interests.

The transfer of power to a large bureaucracy presents its own dangers. Paternalism can suffocate the variety of living experiences that urban areas ought to foster; but on the other hand one would have to look hard to find evidence that the unrestricted creation of cities in our democratic tradition has resulted in anything but an ugly, uncomfortable, lacklustre homogeneity, its formlessness dictated by the needs of the motor car. Much of the answer lies in the attitude of the people toward their government. If, as in Sweden, they refer to the government as "ours," and act as though they have no doubt it is, then good planning for optimum land utilization is a demonstrated possibility[14].

But again there are relatively limited opportunities for structuring a community and a transportation system in a virgin location. Vast areas of the United States are already committed to urban life-styles which overwhelm the occasional pockets of rurality still existing within them.

From Maine to Virginia, municipalities with a density of more than 100 people per square mile stretch in a continuous 500 mile belt[15]. This means an existing system, or more likely an unplanned chaos, has to be dealt with. Assuming the will to start, and to commit the necessary resources for the development of an overall scheme, data describing the existing conditions of land use should be gathered as completely as possible. The inventory ought to include information on the location, identification and a real measurement of land uses, including the precise nature of individual commercial units. Unless the cost of the survey would make it prohibitive, it would be desirable to count the smallest separate parcels except in homogeneous single family residence areas. Information on characteristics of vacant land which may influence its potential for development should also be included. The tax structure which restricts or determines the land use is another important factor to consider.

Existing transportation facilities

A further important phase of an informational survey of transportation needs for an urban region would have to do with the existing facilities. The relative mix of private core and public transportation facilities varies markedly from one urban region to another. Thus, Chicago relies heavily upon public transportation, but Los Angeles, very little. The kinds of information which need to be obtained about existing transportation methods include the physical aspects of roadways and their links, their classification by function when this can be determined, capacities of streets and intersections, volume or speed of movement at rush and non-rush hours, and location and frequency of accidents. Most drivers are familiar with the experience of having their car counted as they pass over a pressure detector stretched across the roadway. Information about public transportation is ordinarily more readily available, not only because the routes are on a regular schedule, but also because the various operating data concerning such features as running time and seating capacity are needed for accurate accounting purposes and are therefore already on hand. Surveys of this type are necessary but by no means sufficient aids to estimating the nature and volume of future travel.

Travel patterns

The study of patterns of travel overlaps the inventory of transportation facilities. Determination of the hours of maximum demand for facilities is more generally useful than are figures for average daily utilization of various means and routes. The five-day-a-week movement of business employees into the center of the cities from the outlying fringes and the suburbs and back again, plus a sizable amount of travel in the reverse directions is the basis for peak-hour urban traffic congestion. The combination of passenger cars, buses, subways and commuter transit by rail can barely cope with the maximum demand in most large metropolitan areas, but during the hours of light demand an oversupply of facilities and services exacerbates the difficulties of making public transportation systems pay their way. Whether municipally owned or privately owned most of them go through the same depressing cycles of reductions in service, increased costs and fares, and decline in utilization in the off-hours as potential riders take to their cars. Profits are rare and bankruptcy a constant threat. Generally speaking only the systems serving cities that have fairly low population density in comparison with their total size are able to remain relatively profitable[16].

On week-ends and between the morning and afternoon peak-hours the data point strongly to the predominance of the private automobile in urban transportation for recreation-social purposes or irregular personal business trips for which a more flexible mode of transportation is desired. Public carriers, especially those moving by rail, are only lightly used on week-ends even though the need for transportation during those times is very heavy. The bulk of the travel falls in quite different patterns than it does during the work week.

Evolving conditions associated with an apparent increase in crosstown or intra suburban travel (for which comparatively fewer public facilities are available); with urban redevelopment projects; with explosive tensions in ghetto areas which demand positive action; and with potential changes in the nature of the work force may cause new patterns to appear. Whatever changes take place, though, there is little question that environmental requirements will impose limits to the extent of increase of automobile traffic. Then it will be axiomatic that people will be able to continue to move about in their urban environment in vehicles only if they, or society as a whole, are willing to pay the full costs of the privilege of doing so.

Terminal and transfer facilities

One of the serious handicaps that public transit often faces is a lack of adequate terminal and transfer facilities. An urban transportation analysis cannot overlook the fact that the commuter is concerned with the convenience and the cost of his total trip between home and work when he uses public transport. It is certainly true that the person who drives his private automobile between suburb and town is similarly troubled as parking facilities in central business districts become more and more unsuitable, and insurance costs increase. This tends to narrow the competition gap between public and private methods under conditions of extreme congestion; but if the public terminal is located at an inconveniently long distance from work or home, the need for some form of feeder transportation arises. Planning projections must consider the possibilities of integrated systems to handle the post-terminal movement successfully, and make allowances for the incremental costs incurred by providing for local passenger distribution.

Zoning regulations and building codes

Zoning ordinances, building codes and subdivision regulations have been among the primary devices used by local governments for control of community development. The first zoning ordinances were enacted in New York City in 1916, and ever since that time, as the idea has spread to other communities throughout the country its application has been both condemned and upheld. One of the most common charges levelled against zoning contends that the ordinances "Balkanize" communities. Since they are not often conceived with other than parochial interests in mind they inhibit the development of any integrated, area-wide development plans. If zoning establishes pockets of low-density middle and upper income neighborhoods isolated from other types of land use it cannot help, so this argument goes, but perpetuate urban sprawl and ghettoism and multiply traffic problems by requiring people to travel longer distances between home and work.

The contrary arguments insist that without local zoning and subdivision regulations there would be no places left where the amenities of quiet, graceful residential living would be preserved. Having located their homes on the presumed guarantee of a stable and congenial community structure those who take this position are strongly opposed to submitting to changes which would remove local control, imperfect though that may be. They fear that a bureaucracy would probably be influenced by rapacious contractors and businessmen, overthrowing painfully won safeguards and invading the area with traffic and noise-creating enterprises which would destroy their chosen manner of living.

These feelings are strongly held and no study of transport capabilities should be carried on without careful consideration of existing ordinances and the possibilities for change should it appear to be needed.

Financial resources

The kinds of economic resources available or potentially available to a given community, and the ways that they are allocated to finance an urban transportation system are matters of overriding significance. When the federal government under the Eisenhower Administration in 1957 authorized expenditures of twenty-six billion dollars for a highway program, that decision had an incalculable effect on the balance among all transportation modes in the country for decades to come. The mix among alternative modes in future urban systems will also be strongly affected by the financing. A key issue is the extent to which they will be subsidized or self-supporting[17]. Of course any money which is spent must come eventually from the citizens themselves either directly or indirectly but the general acceptance of the need for spending is strongly influenced by whether the greater share is borne by the users themselves or the general public, by business or private citizen, or by rich or poor.

At the present time there seems to be no reasonable chance that local governments will be able to finance urban transportation needs when the systems have gone beyond a certain state of deterioration. The intrusion of the automobile into the central city core is only one of several major causes for a shrinking of the local tax base. Any attempt to create open spaces or to construct new government services to make the lot of the city dweller more equable eliminates even more land from the tax rolls. This suggests that the federal government must take a strong hand, but any conceivable plan for them to take over the whole financial burden, even if the administration in power were so inclined, would probably become a nightmare. Therefore, those who use and profit most directly from the transit facilities, and in particular, those who drive their cars over the streets and freeways, must pay a larger share of the costs. The principle of making transportation systems self-supporting through direct charges on the user can lead to a number of complications. One of the drawbacks is that the poor, who already find most means of public or private transportation too expensive to use, would be restricted to the slums even more than now. The principle of self-support has many ramifications which are beyond the scope of this section but it has been developed at some length by Owen[18]. Along with subsidies from the federal government it promises to provide some answers but it is not offered as an ideal solution for financing urban transit. Obviously none exists.

Technology and urban transportation

Very few people who have studied massive social problems seriously believe that even the most innovative technological devices are capable by themselves of providing complete and permanent cures. The preceding paragraphs in this section on urban transportation point to some reasons why other considerations are so vital. Nevertheless, it is also clear that humans, although normally reluctant to change old patterns of action, *may* do so if the patterns become too uncomfortable or if more attractive or profitable means are demonstrated to them. If there is general agreement that alternative ways of transporting people in urban areas have to be found, then the challenge to the engineer is to employ current science and technology to develop new ways which the public will want to try. He must also accept a parallel commitment. That is, he must use every resource at his disposal to insure, as completely as his professional skill and integrity will allow, that the remedy he proposes will not create other and greater problems either for those he is attempting to serve directly or for others who are sharing the same environment. This is of paramount importance because we are no longer so confident as we once were that what man has done, he can also undo.

The engineer can actually proceed in two ways. He can study existing systems to see what adjustments can be made to render them more suitable to the public need; or he can seek new and different ideas to help people move about in an alternative fashion, or even to supplant the older ways entirely. There are general methods of procedure which the engineer classifies under the heading of such terms as systems engineering or systems analysis[19]. The systems approach has been found helpful for many technical problems and Chapter 5 will examine its elements in some detail, but for the present our discussion will relate to a few general examples showing what the technologists are doing.

Traffic engineering

Traffic engineering is a term referring to the activity which seeks, as a major function, to devise ways of increasing the effective capacity of existing transport facilities. Many of the schemes of the highway traffic engineer are in partial measure intuitive and commonsensical (even though they have been worked out in the context of engineering planning) and are in evidence on many streets and freeways. Changing the pattern of traffic signals, eliminating parking on heavily traveled streets, and introducing through and one-way streets are all examples of commonly observed techniques. Having made the changes, he must then measure their success by determining the new volumes of traffic flow at various times of the day.

The traffic research engineer is not, in fact, ordinarily satisfied with simple trial-and-error attacks on a problem. He has developed empirical, mathematical and simulation approaches to help him understand transport networks, and to allow predictions of change in certain variables such as the time of delay at an intersection. Empirical models relate quantitative observations on the independent variable with effects on the dependent—thus, a dependent variable, such as the lengths of lines of cars queued up at an intersection, changes when a traffic light sequence is changed. Mathematical models make performance predictions on the basis of certain theoretical assumptions. Queuing theory and linear programming are examples of mathematical modeling techniques which have been used in studying traffic problems. They are both mentioned further in Chapter 5. Among the drawbacks of empirical studies is that they may be impractical in some situations, or expensive and time-consuming in that large amounts of data are ordinarily required. On the other hand the complexities of mathematical models make it difficult to study the effects of all the important variables. Simulation models tend to be more flexible and have the advantage of displaying recognizable similarities to real traffic systems.

Computers in transportation design and research

We have already gained some appreciation of the complexity of information gathering in the design of a transportation system. It is obvious that the computer will play an important part in compiling and analyzing these data, but it serves other functions in the transportation field, as well. For instance, scheduling and operating rail transit have become important applications. There is another interesting use of which there are now a number of examples in the United States and other countries — that of regulating automobile flow by digital computer control of a signal light network. The first study project of this kind was begun in 1964 under the joint sponsorship of IBM Corporation and the Public Works Department of the City of San Jose, California. A brief discussion follows.

San Jose Traffic Control Project. The original study aimed initially to determine the functional and economic feasibility of controlling a system of traffic signals by a central digital computer, and has been described in an unpublished report, entitled *San Jose Traffic Control Project* [20].

Basically, the performance of the control system can be evaluated by certain criteria which reflect driver benefit and satisfaction. The required number of stops in the study area, queuing delay, trip time, density throughput of traffic, and speed variation are all useful indicators of this kind. Of course the variables influence each other, e.g., restriction of traffic in the control area could certainly reduce delays there, but it would also reduce the throughput of cars and could adversely affect the situation somewhere else. This study concentrated on measuring stops and delays caused by traffic signals at fifty-nine intersections.

The traffic control system is similar to all other closed-loop control systems[†] in that it involves information gathering, decision making, execution and verification, and evaluation and adjustment. A system for controlling traffic does have specific additional needs. The information gathering must be continuous to handle momentary and long-range changes induced by accidents, weather, parades, new construction, etc., and so, obviously, must be the evaluation. Individual computer control for each intersection also provides for flexible decision making without requiring a physical changeover. The diagram, Fig. 4-1 indicates the control loop relations.

The system, among other things, provides for the following:

1. Positive computer control.
2. Accurate timing for each controller step.
3. Synchronous phasing of all changes when responding to changing conditions or manual instructions.
4. Real time detection and counting of vehicles. Storage of counts. (A real time process is one in which the data are handled rapidly enough so that the results of the processing are available in time to influence the process being monitored.)
5. Detection of special functions such as fire station activities.
6. Automatic generation of traffic statistics.

A block diagram of the master control program is shown in Fig. 4-2.

The useful results claimed for this program at the end of its initial trial were that reliable techniques were developed to reduce average

---

[†]For further discussion of closed-loop control systems *see* Chapters 5 and 6.

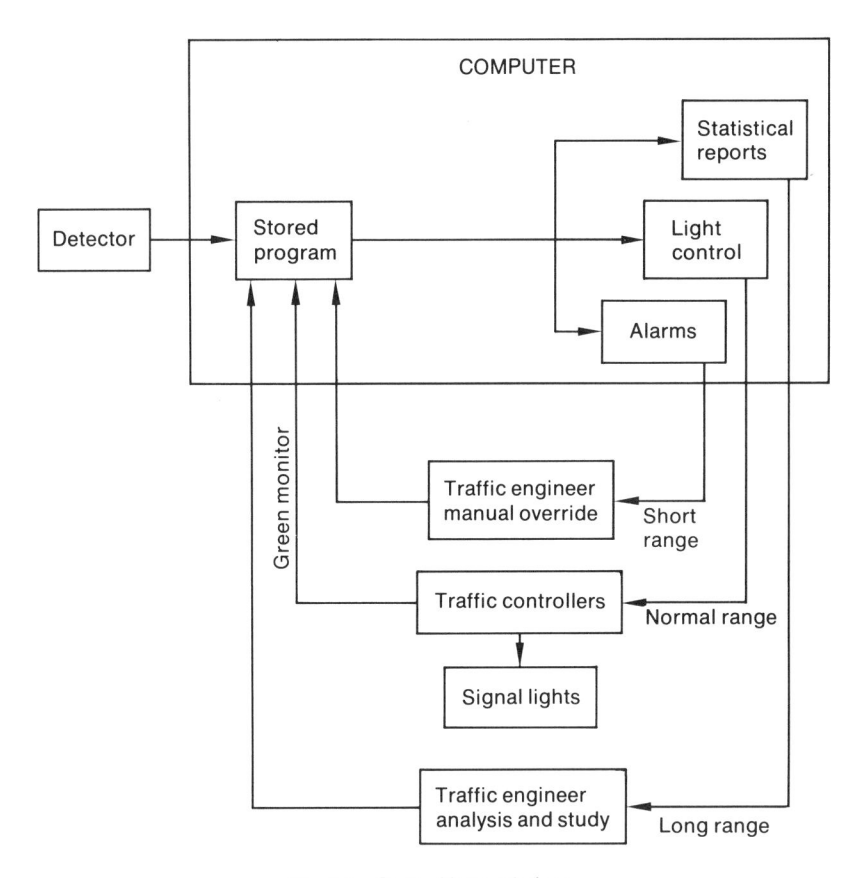

Fig. 4-1. Control loop relations.

vehicle delay, the probability of a stop and the trip time of cars passing through a test area. For a control area containing fifty-nine stop signals the average delay per vehicle was reduced by 12% and the probability of a stop by 7.3%. Cost calculations indicated that these reductions resulted in an approximate total annual cost reduction of over $300,000 for motorists. In addition, the computerization adds control flexibility, evaluation capability and high reliability to the system it serves.

The project has successfully demonstrated over a period of several years that computer control of a traffic signal system is practicable and helpful but it is still a modest contribution to a solution of the overall transportation problem. The expansion of this kind of system to control all the traffic lights in a large metropolitan area would introduce unpredictable

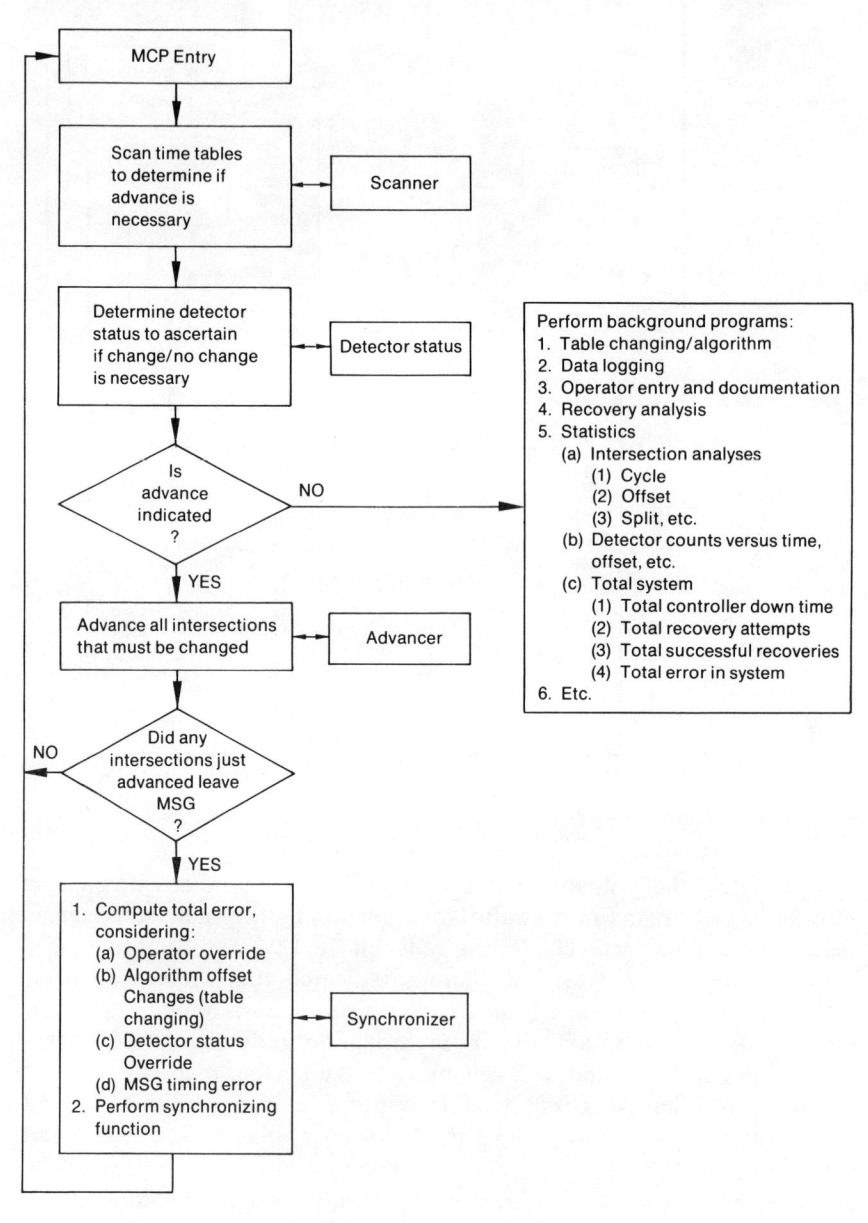

Fig. 4-2. Master control program.

technical complications and might well require larger expenditures than most hard-pressed municipal governments would want to undertake given the fact that the savings to the motorist, and perhaps to the city, are largely indirect and, to an extent, intangible.

Highway location analysis. One other illustration of the use of the computer in transportation design problems will be discussed briefly. The location of a new highway or freeway demands decisions which must take into account a variety of ill-defined options and, as the population spreads, it becomes increasingly more difficult to avoid unfavorable effects on groups or individuals. The design study must seek alternative action plans and then predict and evaluate certain of the consequences of these alternative courses. Manheim[21] and Roberts and Suhrbier[22] have described the development of a theory of problem-solving processes for such a purpose, which utilize several search and selection techniques.

One specific problem might concern the location of a twenty-mile stretch of highway in a given location. For such a problem, twenty-six important requirements were identified by Manheim, each of which could be associated with a preferred route location. For example, the need to consider bridge costs would certainly require the seeking out of a route with a minimum of intersecting ravines and bodies of water. Other requirements would demand quite different locations. Land costs, travel time, services and public financial losses may be more readily assessed in a quantitative fashion than others like air pollution, eye sores, comfort and safety and regional development. For each of the requirements a graphical representation of the area being studied was prepared showing light and dark areas of various shades, such that a black area or point indicated a desired location for building the highway while a white area indicated a location to be avoided with respect to the given requirement. Combining all twenty-six of the diagrams was a hierarchical process in which the designer began with certain small subsets of requirements photographically superimposed on each other to produce one solution. The process is repeated as a series of superimpositions of subsets until a final map is prepared on the basis of the given assumptions showing the preferred location. The advantage of using the computer is that it can handle large amounts of data as input, and can also generate and display the images.

New transportation technologies

Both the popular and the technical literature contain numerous references to imaginative ideas for facilitating the movement of people and materials based on new devices and technologies. Some of those accounts describe changes that are more dramatic than imminent, one suspects, when he observes that it has been many years since there has been a major functional change in our urban transport. Nevertheless one should not underestimate our capacity for engineering ingenuity if social pressures for change become sufficiently demanding.

Just a few among the methods that have been proposed are the monorail and various other forms of high-speed rail transit, hydrofoils and other ground effect machines, short take-off and landing aircraft and electric cars. Different ways of increasing the efficiency of travel by private and public vehicles include computerized guidance systems, expanded leasing arrangements and transit vehicles capable of operating either on rails or on uncontrolled highways. The developments which have progressed furthest toward practical implementation with few exceptions have been those designed to improve or increase high-speed intercity travel.

These, unfortunately, serve to accentuate even more the obsolescence of local urban transit. The turmoil at large metropolitan airports dramatically demonstrates the unbalance.

Using a morphological approach to invention, Howard R. Ross[23] has offered a helpful guide to a classification of potential new technologies. It is possible to identify a transportation method in terms of the three elements of: (a) *support*, (b) *propulsion* and (c) *guidance*. Methods of vehicular support which have been employed or proposed consist of air foils, hydrofoils, air cushions, wheels, magnetics and lubricants. Types of motive power include the use of a propeller, jet, turbine, linear induction motor, linear turbine, gravity or pneumatic force. Guidance may be provided by a flanged wheel on rails, through friction, or by magnetic, aerodynamic or hydrodynamic forces. Other methods are presumably conceivable but combining those listed in various ways essentially covers the present range of possibilities. One might wish to add the concept of the number of degrees of freedom of movement allowed. Thus a railroad or a tracked air cushion vehicle has one degree of freedom; automobiles, ships and air cushion vehicles without tracks have two, and aircraft and submarines have three.

When a choice of system is made on the basis of a given combination of elements, this introduces a certain set of advantages and constraints. For example, an air cushion vehicle with two degrees of freedom must have its own power supply, but a one-degree air cushion vehicle can get its power from the track. Or, to give another illustration, automatic control systems are much more readily adapted to tracked vehicles than to automobiles with their two degrees of freedom. This relationship, however, is more a matter of the peculiar, unregulated conditions of traffic in which the automobile must function, than the number of degrees of freedom. An aircraft, which presumably operates in an environment where the flight paths of the other aircraft are effectively controlled, can be safely guided by an automatic pilot despite its three degrees of freedom. This is true only because the numbers of aircraft in the sky at any one time are relatively few in number compared with automobiles in urban traffic. One-degree systems have the inherent advantage of being able to accommodate large numbers of passengers in a relatively limited space. From 40,000 to 60,000 passengers per hour can be handled by a rail rapid transit system at peak while a single lane of freeway (which in fact permits two degrees of freedom) can carry only 10,000 passengers per hour when the vehicles are fully loaded[24] and under optimum conditions. That, of course, is never the case, so the actual capacity is considerably less.

A study of the practical potential of new technologies for urban transport makes it hard to avoid concluding that designs embodying some of the more radical changes are presently available only as engineering concepts, at best in the model or test car stage. Thorough engineering analysis has long ago discounted many of the attention-getting systems so often hopefully promoted in the popular media. The monorail is a good example. The monorail featured at Disneyland has prompted many to wonder publicly why similar systems could not be adapted to solve transit problems in a larger urban area. The fact is that the monorail is an old idea, the first successful one having been constructed in Ireland in 1888. Whatever its merits in special situations, there are also serious engineering drawbacks (such as the problem of switching because of the need for suspension from massive beams, lack of stability at high speeds, or the difficulty of car-to-car articulation around curves) which have not been overcome.

The one major newly-designed mass transit line in the United States nearest completion is the 75 mile, 1.2 billion dollar Bay Area Rapid Transit system (BART). BART (which appears to have been primarily conceived as a means of rejuvenating downtown San Francisco and increasing property values there) has been designed to operate between San Francisco and cities on the east side of San Francisco Bay, all a part of the San Francisco–Oakland metropolitan area. While the original transport plan envisaged a system circling the bay and serving the metropolitan area south of San Francisco, those counties involved, partly because of strong resistance from taxpayers groups, chose not to be included, so that when the system begins to operate in late 1972 it will not even connect the City of San Francisco with its airport in San Mateo County. BART is a mix of old and newer technologies. Steel wheels still run on steel rails, but the cars are computer controlled for optimum efficiency and safety. Although some of the original designs had to be substantially modified because of financial problems, it appears that the commuters will still ride in unaccustomed comfort. The basic question is whether this will offer the needed incentive to attract enough potential customers out of their private cars to make BART pay. If so, then perhaps the system will be extended to complete the original routing some two decades or more after the plan was originally conceived. If not, then either government will have to introduce measures to make it so expensive for automobile commuters that many will have no alternative to BART, or else any chance for similar systems to be funded in other areas will probably be delayed for years to come.

The BART system is relatively small and uncomplicated compared with the kind of network needed to serve the New York metropolitan region. The present New York system has 237 miles of track, and although the equipment is antiquated and hard to maintain, one cannot imagine the city having survived into the 1970s without it. A proposed expansion and improvement program will cost at least 2.9 billion dollars. Given the cost of new rapid transit systems and the many years needed to plan and build them, it is difficult to foresee any scheme for alleviating urban transport problems which does not provide a balance between rail or some other single degree of freedom service along population-dense corridors; and freeways and roads for public and private vehicles, nor will any scheme to serve an entire area spring into being without including parts of systems already functioning.

This brings us back to one pressing aspect of concern in urban transport studies—how to retain the advantages of the comfortable, versatile automobile while reducing or eliminating its disadvantages, some of which have gone beyond tolerable limits.

As we have indicated before, if the automobile is to be replaced, its substitute under our system of government would have to be acceptable to the mass public. Most of the engineering concepts proposed[25] seek to combine relative freedom of movement with automaticity in what are called "personal transit" vehicles. Some of the features which many designers feel should be incorporated in this kind of system are automatic parking and retrieval, smogless propulsion systems, door-to-door service and, probably, a combination of public and private ownership. A typical system would have small two or four passenger battery-propelled vehicles which could carry a commuter from his home to public stations. There his vehicle would shunt into a network of electric guideways, and be carried to a destination which was read in to a central control computer. When this station is reached, the car uncouples automatically, and the commuter drives to a parking location near his place of work. The basic problem of parking facilities still exists with this method, unless some form of public leasing could be arranged so that these vehicles are available to others for intracity use during work hours. An alternative method to help avoid the parking problem would be to have individuals transported between their homes and guideway stations by means of computer-dispatched minibuses.

Little imagination is required however to see that plans such as these have some inherent drawbacks. Indeed, all transport schemes do. It can be stated as an immutable principle, for instance, that the cost of energy and materials per passenger will be greater for smaller vehicles than for large ones. It surely helps to have a central track or roadbed take over the task of guidance and propulsion from the individual vehicle. Here technology has interesting possibilities to offer, but the fact remains that no individual can attain full freedom of movement and satisfy his desire for privacy and comfort without an additional, sometimes heavy, cost being paid somewhere in the system.

Urban transportation and political realities

How should any individual or public agency taxed with the responsibility for seeking solutions to the pressing problems of urban transportation proceed? How would you?

The preceding sections summarize only a few of the relevant considerations which come readily to mind, in the main those which are technological or economic in nature. It would require no exhaustive study, however, to establish affective links between the urban transportation crisis and any of the other major social problems which beset our cities. Thus we restate the obvious—that any significant new development in an existing transportation system (or any technology with extensive applications in a community) will have effects which can be expected sooner or later to change living patterns of virtually every individual or group in that community. There will be some effects which favor certain groups and abuse others; there may very well be some effects which are universally detrimental, and perhaps others which seem to generate no recognizable undesirable consequences. What happens when a commitment is made to construct, say, a new urban railway? Who has made the basic decisions, and what are the means for effecting them? If we were to try to retrace how existing urban transit systems came into being we would probably have limited success in uncovering the many specific influences and decision points. Even if we succeeded fairly well, the study might shed but little light on a current problem.

A major decision appears in its ultimate form as the inevitable result of many small public actions and private actions taken by autonomous groups. But, as a distinguished committee from the National Academy of Sciences has concluded after studying existing processes of assessment and choice, under our system—"with few exceptions, the central question asked of a technology is what it would do (or is doing) to the economic or institutional interests of those who are deciding whether or how to exploit it."[26] This statement was directed particularly toward the management of private corporations. It also holds true for many government agencies whose ostensible mission has been to promote the public good, but whose view of the public good has been too narrowly conceived, either by virtue of their failing to take into account possible secondary effects of a new technology on initially remote groups, or because of their desire to enhance and perpetuate their traditional function.

Affluence is a warm cradle where apathy flourishes, but hunger, injustice and physical discomfort can quickly turn apathy into angry, active discontent. For some people the urban environment with its traffic, pollution and crime has been an uncomfortable place all of the time. For the more affluent the periods of discomfort have been less sharply felt, but, as the duration and frequency of those periods have increased, pronounced social reactions, both rational and irrational, have been manifested. The last few years have witnessed a marked rise in the numbers of individuals and groups from different social and political backgrounds who have made common cause on environmental issues and have sometimes succeeded in mustering such potent opposition that major projects of the nature of urban freeways have been discontinued or relocated. Actions of this kind would have been rare prior to the latter part of the 1960s. These new pressure groups are often apolitical in the sense of traditional party adherence although they may cultivate a shrewd appreciation for political realities. To the politician or appointed government official they represent one other important variable in the difficult equation he must try to solve before he casts or recommends a vote on an issue affecting the public good.

Given the turbulent climate of American urban life today, any full-scale planning of transport systems has to recognize, as we have already suggested, that new transport may itself be an agent for aesthetic and social transformations. Is it politically possible to work for plans to improve the general social welfare although it means overriding the accepted rules of the market place? The answers are not clear for every situation, but even in the unpredictable social climate of the 1970s, one suspects that any political figure trying to do so will face formidable obstacles. Still it seems certain that pressures from many directions will force him to make an effort at least to consider the second- and third- as well as the first-order consequences of technological change. The first-order consequences would include the observed [27] increase in overall movement of vehicles and people in a given time period, and also changes in property values although the latter result would be less immediate and less prominent in cited statistics. Secondary and tertiary consequences could consist of subtle and sometimes brutal effects on community and family structures — pollution, very likely, but also the mangling of neighborhood patterns through the elimination of older housing or the fostering of "across-the-tracks" ghettoes by inappropriate location of new transport arteries. Historical landmarks or points of natural beauty are often the casualties of transport development. These lower order effects do not *have* to be entirely negative, although on balance this has been the tendency. New transport can be located with regard to needs for open space, and shelter too, and may actually represent an aesthetic improvement in some areas. These kinds of possibilities must begin to get the consideration which they have too often not received in the past.

ECONOMIC AND SOCIAL INDICATORS

What is the public good? For most of two centuries in the industrialized nations this question has been officially answered with primary reference to economic indicators. There has been an abundance of data on the growth of incomes, profits and the value of real property and businesses leaving no doubt about the material progress in the mainstream of society. Economic indicators were institutionalized in a significant way through the Employment Act of 1946 which provided for the President to make an annual Economic Report to Congress. His Council of Economic Advisors now also publishes monthly economic indicators. On the basis of this carefully organized information major public and private decisions are made either to initiate or defer action. The growing concern about the unconsidered consequences of some of these actions has caused different questions to be raised. As early as 1962 presidential advisors, as well as scholars meeting professionally, were beginning to call for the initiation of a system of "social indicators" by which various consequences might be assessed in other than strictly monetary terms[28].

The National Commission on Technology, Automation and Economic Progress in 1966 issued a report[29] proposing the measurement of social costs and net returns of innovation, the measurement of social ills, the development of "performance budgets" where obvious social needs exist, and the development and preparation of indicators of social mobility and economic opportunity. In the spring of that year, President Lyndon Johnson asked the Department of Health, Education and Welfare to develop sets of social indicators to supplement those data already being issued by the Council of Economic Advisors and the Bureau of Labor Statistics.

There have been other studies since that time concerned with the same problem but progress toward an acceptable solution has been slow and tedious. Some fundamental questions remain unanswered. Perhaps we should not be surprised that answers have not come easily. Herbert Spencer, the dogmatic philosopher of the Victorian era, maintained that an orderly accumulation of facts about every aspect of social existence would provide the key to understanding his times. He wrote "we want all the facts which help us to understand how a nation has grown and organized itself.... These facts, given with as much brevity as consistent with clearness and accuracy should be so grouped and arranged that they may be comprehended in their ensemble, and contemplated as mutually-dependent parts of one great whole. The aim should be so to present them that men may readily trace the consensus subsisting among them, with the view of learning what social phenomena coexist with what others."[30] His call has never been met.

It would be too easy an explanation to suggest that is because there was no machine like the computer to handle data in those times although the computer has removed some serious limitations to scholarly activity requiring the accumulation and analysis of data. There are more basic questions to be asked. Biderman[31] has noted that since all social statistics are necessarily the product of certain institutions, they are bound to reflect social and cultural biases peculiar to those institutions. Moreover, the very nature of any institution is impressed on the information it furnishes. What organizations, then should be considered as legitimate collectors and purveyors of social data? The Census Bureau is surely an important source and many other government divisions consider the providing of data one of their prime functions. Legislators must frequently rely on information provided by lobbyists for special interests. The best of the lobbyists try to see that the information they supply is factually correct as far as they can be sure, or they would soon lose their credibility. Still, just as in ancient times when the bearer of bad tidings might be expected to have a hand or a tongue removed by the sovereign, there is even now a natural tendency to temper unpleasant news, if not to misrepresent it. Only the naive would be unaware that even divisions of the government itself are inclined to take a proprietary interest in data and sometimes to repress facts which they feel might be detrimental to their mission.

However, one should avoid trying to throw the worst light on practices of gathering and utilizing social indicators. There are checks and balances in our system which in time, tend to detect and restrain the most flagrant distortions of the free exchange of information. It seems fair to say that most government services do their best to maintain a relatively unbiased stance. But even presupposing a lack of conscious bias on the part of those agencies gathering and furnishing information, there are, as Biderman[32] goes on to point out, other limitations to the use of social indicators. In many cases an indicator, standing alone, is at best only a partially valid measure of the social phenomenon it is intended to represent. In the preceding chapter we spoke of a national income norm by which a family is judged to be poor or not poor. This figure, which is subject to adjustment every few years to reflect such factors as the changing value of the dollar, is supposed to provide a useful picture of the relative success or failure of our system in raising the citizens of the United States above the poverty level. The indicator is not without value for the purpose but it may conceal almost as much as it reveals. Thus, even though one might justly assume that an average child in a family below the poverty line has less chance of receiving an adequate diet than one from a more affluent family, much additional information would be needed to reach any firmer conclusions about serious nutritional deficiencies. Or, we can guess that many families officially adjudged to be impoverished according to the criterion would be denied certain opportunities for social broadening; but some "poor" families manage to acquire a richness and variety of experiences that others above the poverty line may never know. The basic social indicator would not show this. As for statistics relating to the transportation problems in the United States, one might choose to conclude from the numbers of miles of highways completed in the past ten years that the ability to move people from place to place has markedly improved. This would be demonstrably true for certain places at certain times, only occasionally true for other situations, and would tell us little that we really wished to know about some effects of increased traffic on the quality of the environment.

We can pass quickly over the problem of inaccurate data, not because it has no importance, but because it ought to be possible to keep technical errors of sampling and measurement within tolerable limits given adequate resources and a good understanding of statistical techniques and the significance of time-dependent changes. There will always be technical errors but they are at least theoretically manageable, whereas the problem of deriving quite distinct and conflicting interpretations from a given set of indicators is most difficult to deal with. Interpretations may vary for many reasons. Sometimes the judgment of the person examining the data will be affected by philosophical or political preconceptions, or by his professional background. The economist may choose to view a set of data in quite different terms than a sociologist or an engineer.

When a political leader makes a public decision based on the recommendations of various individuals or groups reporting information to him he must bear in mind all of these considerations and more. He must know, for example, that the rhetoric which may have been very useful to his cause during the election campaign will not be conveniently translatable into any concrete plan for social action. Phrases such as "the abundant society," "the general welfare" or even "the full dinner-pail" are very much a part of the oratory of characteristic political figures. They also tend to form the substance of many of the primary recommendations offered by committees convened to establish institutional priorities or goals. It turns out that attempts to match these broad statements of purpose with unambiguous social indicators have met with limited success as Biderman[33] demonstrates in citing the report of the President's Commission on National Goals prepared in 1960. A careful study of available statistical information showed that, out of eighty-two statements of domestic goals to which, according to the Committee, the Nation ought to aspire, indicators that were at least partially relevant were found for only forty-eight. For the remainder of the statements no indicators at all were uncovered. This means that the representative of the people in government is not only confronted with the complexities of political decision making, but having taken a commitment to a specific goal or a form of action he cannot always find a suitable standard against which the wisdom of his judgment can be measured. He would like to be able to show that a concept which he supported say, five years ago, has had certain tangible and enumerable results which have benefitted the majority of his constituents. There may, and almost certainly will, be some qualitative signs enough to convince him intuitively that on balance his position was the right one, but his subjective feelings will not necessarily be shared by many others. It is this relation between measurable factors and qualitative impressions which is so elusive, yet which needs somehow to be approached if the well-being of the society is to be successfully monitored. Some years ago, John Dewey touched upon the scope of the problem which concerns us. He wrote, "It has long been a moot question how civilization is to be measured. What is the gauge of its status and degree of advance? Shall it be judged by its elite, by its artistic and scientific products, by the depth and fervor of its religious devotion? Or by the level of the masses, by the amount of ease and security attained by the common man? . . ." [34].

The question is still moot. The development of a system of social indicators requires certain surrogates to represent attributes or concepts. No matter how carefully the surrogates are selected, at each remove some distortion of the original idea is bound to occur.

TECHNOLOGY ASSESSMENT

Technology assessment, a phrase which appears first to have been introduced by former Congressman Emilio Q. Daddario of Connecticut, refers to the examination of existing and emerging technologies for the purpose of evaluating the potential gains or dangers which might attend their further development and application. This is a limited definition of an activity which carries somewhat broader implications for many of those who have been trying to relate technology to human needs. In one sense the assessment of technology as a harbinger of social change is the same kind of process as the development and use of social indicators. However, social indicators do not have to be considered exclusively in the context of technological determinants. We have chosen to do so for our own purposes. Relating probable economic and social benefit to cost is very much a part of the aim of technology assessment and so is the hope of improving the efficiency and impact of technologically based programs by defining policies and objectives more precisely [35]. There is also a close relationship to short term "technological forecasting" which will be taken up in greater detail in Chapter 10.

It should be obvious that the kind of activity or concept we are outlining is not entirely new. Prophets, both euphoric and dismaying, have greeted the birth of most new technologies with predictions of probable consequences. Sometimes these assessments have been ludicrously wide of the mark, and at other times they have been keenly perceptive. As one might expect, the more accurate assessments have usually been those made on maturing technologies which had already begun to produce effects on the physical or social environment. In these cases the assessors have been not so much prophetic as they have been sensitive observers keenly attuned to the changes taking place long before most of their contemporaries had reached the same awareness. Rachel Carson was one of these. Her book, *Silent Spring* [36] was a clear call of alarm drawing attention to the insidious effects of pesticides on living creatures other than the lower insects they were intended to eradicate. At some other time her voice might well have been drowned out by the chorus of rebuttals from various representatives of the scientific and industrial communities who insisted that she had improperly ignored the positive benefits of these products of chemical technology, especially the improving of agricultural yields. But her impassioned, lyrical appeal touched a public nerve and although results came slowly in the face of entrenched practices, disorganized concern and apathy, the unrestricted use of dangerous chemicals which will enter the food chain may now hopefully soon become a thing of the past in this country. Many other individuals have sent out early warning signals of technological misapplication — Lewis Mumford, Aldous Huxley, S. P. R. Charter (or much earlier, George Perkins Marsh) among them, but these have not been organized assessments on a large scale. The earliest attempts to assess effects of technology by institutional means were usually initiated in response to some single disaster or a recurring series of technical crises. One of the first examples was a request to the Franklin Institute in Philadelphia to look for the causes of the numerous catastrophic explosions of high-pressure boilers which were plaguing steamboat travel in the early years of the 19th Century [37]. There have been other instances of similar studies in recent times. The U.S. Navy has conducted research over several decades to determine the reasons for the sudden, brittle failure of steel plate which has resulted in the loss of a large number of ships. But this is technology assessment viewed in a very narrow sense. When existing research organizations or *ad hoc* groups are asked to look into problems of this kind the charge to them does not ordinarily include any license to go beyond an examination of the engineering and scientific variables even though there is a chance that regulatory laws may be based on an interpretation of the results.

The U.S. Bureau of Standards, the Food and Drug Administration and other similar agencies of the government for years have had a responsibility for evaluating new engineering and scientific developments and establishing standards of quality. They have not been inclined to probe very far into non-technical consequences or central political issues but when on occasion a recommendation has upset a powerful Congressional constituency the pressures exerted to repress or invalidate the results of a study have been substantial. In the Eisenhower Administration, the respected head of the Bureau of Standards was threatened with the loss of his job for refusing to withdraw a report which questioned the effectiveness of a simple battery additive. Only the intervention of national scientific organizations prevented his dismissal. More recently, in 1969, after a government scientist made public the results of a study demonstrating that cyclamates, which were being used as sugar substitutes, had caused tumors in heavily-dosed rats there was a strong reaction from commercial interests against the decision of the Secretary of Health, Education and Welfare to have cyclamate-bearing products taken off the market. That the decision was sustained might be considered as a reason for believing that the climate of opinion has changed with respect to conflicts between advancing technologies and the well-being of the people. Of course things are not quite that uncomplicated. No one cares to find out that a common food product which he has been consuming might cause cancer. Since there was an existing law proscribing the marketing of such products, the government's necessary course of action was rather clearly defined. How much different and more involved are the problems of assessing the need for and the probable consequences of developing the supersonic transport (SST) for example?

It was apparent from the start that there would be many thorny engineering and scientific obstacles to be cleared prior to a successful completion of the project, but by 1970 the builders left no doubt that they felt they would be able to have a large fleet of these aircraft in commercial service by the early 1980s capable of flying 1800 miles per hour at 70,000 feet above the earth. Also before this time, however, serious questions had been advanced about certain environmental implications of the SST — the unavoidable producing of thunderous sonic booms in flights over continental land masses, and the possible modification of the environments through the expelling of radiation-absorbing combustion products into the stratosphere. While sonic booms would be inherent, results of research on potential stratospheric and atmospheric effects have remained inconclusive except that earlier concern about shifts in the ozone balance of the stratosphere appears to have been unwarranted. Other effects considered included the human psycho-biological response to time-zone disorientation.

Two distinctly identifiable positions inevitably develop in situations of this kind with various shades of opinion in between. One side points to the possible undesirable side effects of the technology and insists that further developments be deferred until unmistakeable evidence is produced to show that irreversible damage will not occur to the environment. The other side tends to assume that meaningful assessment of the effects can be made in actual operation of the system and then, in the presumably unlikely event that the consequences prove damaging, technology can come up with palliative measures. Given the lack of consensus among nominally competent technical advisors, the decision makers—the leading members of the executive and legislative branches of the government—are historically disinclined to make these technical factors ultimately decisive. Instead, the nature of their position prompts them to give heavy weight to the basic political and economic issues as they see them. In the case of the SST, millions of dollars were involved in government subsidies for development work which the industries could not afford to pay for by themselves. Many more millions seemed to be at stake in the future in the competition for international markets and in the ever-present balance-of-payment problem, not to mention the more imponderable consideration of maintaining national prestige in the technological competition with other countries. In addition to these concerns, the decision makers tried to balance a desire to limit government expenditures in a long period of inflation against the assumed need to bolster the fortunes of aerospace industries hard hit by declining profits and heavy unemployment. In so doing they had to balance one sectional interest against another.

How important, then, is the contribution that technology assessment can make under circumstances similar to these? The Committee on Public Engineering Policy of the National Academy of Engineering concluded in a report to the Committee on Science and Astronautics of the U.S. House of Representatives that a program of technology assessment could make a significant contribution under certain conditions. These conditions were outlined in a list of fourteen principal points of which the following summary attempts to bring out what seem to be the most salient features: [38]

    (a) Technology assessments should be prepared by independent, *ad hoc* groups chosen from public and private organizations having an interest in and knowledge about the subject considered.

(b) There should be a small, permanent management organization, answerable to Congress, to arrange for the preparation of the assessments, but it should not depend upon in-house expertise for the actual conducting of the study.

(c) As much as possible, the assessment environment should be free from bias or political influence. Preferred courses of action, among alternative possibilities presented, should be chosen by the legislature after the assessment has been made and not by the technology assessment group.

(d) The group should take pains to solicit presentations from all interested parties, including those not normally well-organized in their own interest.

(e) Many professional specialties should be represented in the assessment group—behavioral and political scientists as well as engineers, economists and physical scientists.

(f) The study should be completed in about one year.

(g) Assessment of possible consequences looking more than five years into the future are apt to be unreliable, and this should be recognized.

(h) There appear to be two main classes of approach to technology assessment: problem-initiated analysis and technology-initiated analysis. Problem-initiated analysis, in which the variables are pointed in a cause-effect sequence toward a common concern, is closely related to the method of systems analysis described in greater detail in Chapter 5. The Committee concluded that more progress could be made by having assessment groups use the problem-initiated approach focussing on converging events leading to a specific problem, rather than beginning with a technology and trying to foresee the enormous number of possibilities in the necessarily diverging cause-effect array. The drawback with the technology-initiated approach, of course, is that no individual or group no matter how well-informed or intuitive is likely to be able to anticipate all potential consequences when new technologies are invented.

It is possible to study these recommendations and ask how the suggested program differs from the time-honored practice of appointing government commissions composed of distinguished experts in certain fields to report back to Congress recommendations for appropriate government action on pressing national problems. The Committee did not ignore this question. They, in fact, implied that there was an aspect of *déja vu* in their own study by referring to a report prepared in 1937 by a special federal agency, the Subcommittee on Technology of the National Resources Committee. This report, entitled *Technological Trends and National Policy, Including the Social Implications of New Inventions* [39], concerned itself with the possible impacts of technology on society, and attempted to anticipate and evaluate several distinct areas of technological development. Some of the assessments were prescient but some of the most potent technological achievements of the World War II period (and of course their effects) were not foreseen. One can also infer that little notable federal action followed the recommendations of the report. Technology assessment as it is currently conceived does not differ very much in its aims from earlier efforts similar to the one just mentioned. Instead the differences seem to lie in the concept of carefully institutionalizing the mechanism for managing the assessments so that there will be a competent body continuously functioning to link Congress with the scientific and technical communities rather than relying upon specific crises to alert the government to a need. The *ad hoc* committees which the management staff recommends for short term service would constantly re-evaluate the state of technology and the deliberations of previous study groups, (assuming the optimum resources for data generation); would hopefully operate with more independence than in-house departments; and would be sufficiently broadly constituted to emphasize social and environmental concerns in their cost-benefit perspectives. If there is reason to hope that such an arrangement would have more effect now on promoting corrective legislation than any similar attempts made previously, it exists in the focussing of national attention by the communications media, and by the ubiquitous conditions themselves, on environmental issues.

One of the most significant contemporary developments has been the emergence of loose organizations of concerned citizens unconnected with the government who, in the changing climate of opinion, have been able to exert an exceptional influence on legislation affecting the spread of technologies. Perhaps the most publicized of these groups has been the one led by Ralph Nader. Nader, operating almost single-handedly at first, succeeded in bringing together a group of people who have actually served as a kind of unofficial technology assessment organization. There have been charges that these kinds of organizations function with an anti-technology bias which ignores or threatens to inhibit the beneficial aspects of technology. Among the younger members of the groups, at least, it seems there would be more than a few who would freely admit to such a bias and who would spiritedly defend their attitude. Be that as it may, there appear to be advantages and disadvantages to both types of technology assessments, either publicly or privately supported, and they could be mutually reinforcing. Independent groups can presumably operate free from political or institutional pressures and they may incline to give more weight to fresh, if unseasoned and inconoclastic, views of younger participants. Publicly sponsored groups should be expected to have more resources of information at their command, and with some sort of symbiotic—if distant—relationship to the legislative bodies ought to be able to preserve a sense of political realities.

DATA CENTERS AND THE ISSUE OF PRIVACY

We must turn once again to the problem of information gathering and usage which recurs so insistently. Implicit in the assessment of technology and in the determination and evaluation of social indicators is the premise that very large masses of data need to be collected, tabulated and analyzed. Beyond that, however, there is an additional difficulty in the seeming paradox that the information which can be expressed best in precise quantitative terms most often has the least to say about basic human needs and aspirations, and decisions which affect the well-being of the people are often made on the basis of inadequate or misleading statistical guides. One way to try to rectify the problem is to gather a great deal of qualitative information (i.e., religion, political affiliations, etc.) about individuals and groups to give more substance to the data files. This kind of information can be stored in some standardized format. Nuances cannot, but they can be inferred when the information is retrieved. Social scientists do this and public and private agencies do it also, sometimes for purposes that have no relation to the interests of the subject whose file is being maintained. Businesses of various kinds not only collect data about individuals for their own uses, but they are sometimes able to purchase or otherwise obtain information from public data files to supplement their own. The basic problem is not new—dossiers on individuals are probably almost as old as record-keeping in government—but the computer has given it a special urgency. In recent years the possibility of establishing a National Data Center has been broached a number of times but until now it has been rather decisively beaten back primarily out of concern by conservative and liberal legislators alike for possible abrogation of the individual's right to privacy. However many separate public and private "banks" exist and there is strong incentive to link them together to provide a richer data base.

Central to the issue is the right of individuals or organizations to decide what information about themselves they can withhold or permit to be disseminated publicly. Many of the relevant considerations have been carefully reviewed by Alan Westin[40]. The Constitution of the United States has provided long-standing guarantees against authoritarian inquiries into the beliefs of the individual, although it does not specifically mention the right to privacy. However, time has eroded some of the social restraints, and technology many of the physical restraints on information gathering which existed when the Consitution was written. Acts and conversations of individuals can be recorded by long-range cameras and sensitive eavesdropping devices. Their habits and even their ideas can be probed through questionnaires and psychological testing and through analysis of data which they are required by law to submit to government agencies. Then computerized communications networks can digest and store these data where they can be made quickly available for a variety of purposes.

The right of privacy is not absolute in any society. Indeed there are no definitions of privacy which would fully serve as a basis for developing policies to insure it. To what extent is each of us a private person? Surely an individual has an inherent desire to conceal some knowledge about himself from anyone else and most knowledge from possible unfriendly surveillance; just as he often has an impulse to share other matters with whomever he chooses. At the same time his own curiosity about the actions of others knows few bounds. When the information he seeks has to do with political decisions which may affect his future his assumption of a right to know is well-founded, and yet in the affairs of government there are activities whose public disclosure could, at least sometimes, do more harm than good. Who is to decide what they are?

A government could not function without certain kinds of information about its citizens. We have always kept public records about private persons. Some of the records stored in manual files have been available on anyone's request; others, in principle at least, only to authorized members of certain agencies, but there have been few central guiding policies to safeguard the individual. With centralized data processing units, there would be a temptation to gather more information than is actually needed and any errors made may be more damaging and difficult to rectify. This has led to a growing awareness among legislators, the public and the designers of information systems themselves that there are still important questions to be resolved. First, besides decisions about what kinds of information are really needed, the determination should be made whether there are some types which must be excluded by law, such as hearsay or personal opinion. Next, for each kind of information, one needs to know to whom the data can be disclosed, and what right the individual has to specify the details of disclosure of material in his file. His right to examine the information in his own file at any time ought to be inviolate, as should be his opportunity to challenge details contained therein and to have erroneous or obsolete material eliminated. A neutral auditing agency could serve a very useful function in this regard.

There are many like concerns which apply equally to the smaller data files presently maintained by credit bureaus and legal information agencies. In fact Sprague[41] has indicated that the formation of a national information utility would be preferable to the continued existence of those scattered agencies because its operations could be better controlled by regulatory laws. There are theoretical means by which computerized information can be made available only to qualified persons but it is very difficult to insure absolute terminal user verification. Besides, all such safeguards add to the cost and lower the efficiency of the system. Absolute secrecy is an illusion. In treacherous times even legal safeguards are overridden. The real threat to the privacy of the individual is also a challenge. He must exercise constant vigilance through his own activities and those of his representatives in government to see that his rights are understood and protected.

THE TECHNOLOGICAL ELITE

The concept of the national data center has been one symbol of a trend which in the view of some writers presages a dangerous tendency toward concentration of political power in the hands of those who best understand the workings of complex technical systems—i.e., the technological elite. Jacques Ellul is one of the foremost spokesmen of those who foresee a controlling bureaucracy guided by "technocrats" a term which in this sense refers not to scientists and engineers exclusively but to the entire managerial class in government or in positions influencing the government. However the special role played by the scientists and engineers in an advanced technological society causes the spotlight to shine most strongly on them.

This projection of a governing technical elite has affected the outlook of many thinkers since the start of the scientific age. At first the prospect was regarded generally as a desirable consummation, rather than threatening. Francis Bacon was one of the most influential early writers to speculate on an ethical society guided by science, one in which ordinary political institutions would not even be necessary. Bacon was considered a leading scientist of the Elizabethan period in England even if sometimes misinformed about contemporary theories. In his dual capacity as statesman he was in a special position to consider the relation between science and government. He left an unfinished work called *New Atlantis* which described a scientific utopia. Its activities centered around a great laboratory, Salomon's House, in which scientists and philosophers searched for "the Knowledge of Causes and secret motions of things; and the enlarging of the bounds of Human Empire, to the effecting of all things possible." These searches would be conducted in a thoroughly organized, empirical way unhampered by misleading theoretical preconceptions. Mathematics would be the servant of physical science, with classification more important than measurement.

Across the channel in France, the Enlightenment produced René Descartes, who took quite a different view of science. He was prepared to deduce in his brilliant mind the whole system of the Universe. His influence remained very strong through the 17th Century and the early 18th. This period saw the emergence of the "philosophes," an acute clear-thinking group of men who were literary figures as much as they were scientists or philosophers. They were able to take the abstractions of 17th Century science and bring them to the world as a new view of the Universe and everything that composed it. Deism became a new religion and for the Deists it has been said that Reason was God, Newton's Principles, the Bible, and Voltaire, the Prophet [42]. Succeeding generations in scientific, engineering and mathematical pursuits were imbued with the spirit of rationalism as France forged into leadership in these fields with such names as Laplace, Lagrange, Condorcet, Carnot, Poisson, Ampere, Gay-Lussac and Berthollet heading the lists.

From the standpoint of our interest in the relationships between science, technology, industrialism and government, however, we must mention another individual who, though not a scientist was among the most influential—certainly the most bizarre and interesting—of the successors to the philosophes in the period including the French Revolution and after. This was Claude Henri de Rouvroy, Comte de Saint-Simon. In his writings we find the bases for ideas of a controlled industrial economy, international cooperation, and the employment of scientific experts at the decision making levels of government. He was a nobleman with

Imperial Germany, and the *Industrial Revolution*. In all likelihood, he had misread the signs of attempts of engineers to draw together, during and just after World War I, to try to advance their professional status. It is true that a few prominent engineers, mostly followers of the late Frederick W. Taylor who had introduced "Scientific Management" to American industry, sought to apply Taylor's principles in a wider political sphere but they never made any substantial progress in this direction. During the 1920s the theme was kept barely alive by a few devotees with an occasional serious effort to bring the concepts to public view. One study of industrial waste resulted in the publishing of the book, *The Tragedy of Waste*, by Stuart Chase, in 1925.

It was not until the crisis atmosphere of the early 1930s, with the industrial machine prostrate, that Veblen's ideas were resurrected and adapted to fit the formation of the new movement called Technocracy [45]. Technocracy quickly came into national prominence with many voices interpreting its message, but the dominant one was to become that of Howard Scott, a strong-willed man with uncertain technical credentials who had been one of the faithful during the period of the 1920s. The concepts of Technocracy, borrowed from many sources, held that because of the increased use of machines consuming unprecedented amounts of energy, the traditional link connecting income and labor had been shattered. Thus, no scheme for distributing the abundance of the new industrial system based on man-hours of human labor could possibly survive.

Since true material progress was deemed to depend on the natural energy resources available to society and the efficiency with which they could be converted to useful purposes, one of the first tasks of the technocracy group was to try to conduct a survey of all the energy resources of the entire North American continent. With a knowledge of the annual quantity of energy available, technicians in charge of the government could then supposedly plan for its use through a central accounting system bringing supply into close balance with actual demand. Energy certificates representing a certain portion of the available supply would be issued in equal shares to all citizens and used for goods in terms of the exact cost to society. The cost would be based on a calculation of energy expended to produce the goods. Information from cancelled certificates would provide the planners with an accurate knowledge of production needs.

Many of the criticisms which supporters of the Technocracy movement advanced against the existing price system were reasonable, and their idea of an energy resource survey is being carried on today by other groups with vastly more effective means. However, much of their projected design for a scientifically planned system was politically naive — some observers were even concerned about their potential fascistic tendencies — and enthusiasm for the movement, particularly with the abrasive Scott as leader, quickly dissipated. He remained at the head of the society until his death on January 1, 1970 at the age of eighty. Only a few scattered chapters of the organization remain. It is perhaps ironical that at almost the peak period of the movement which had a considerable following in 1932 and 1933, Herbert Hoover, an engineer who had at one time been called the engineering method personified, was soundly beaten in his bid for re-election to the Presidency.

So the prospect of a ruling technological elite flared up briefly and then nearly flickered out during the remaining years of the 1930s. Government support for research in science and technology was small and those few federal agencies devoted to testing and development were seriously limited in their ability to move in new directions. At the end of the decade, however, with armed combat already raging in Europe, it was apparent that this nation's research establishment was in disarray. Strenuous measures would have to be taken to meet the challenge of technological innovation which the looming war would surely bring. Hitler's Panzer division had given a rude shock to those strategists who had predicted that the tanks would bog down in the mud and be no match for the superb Polish cavalry. The Luftwaffe served abrupt notice that this would be a technological war to be fought with weapons which few of the career military people would have the background to understand.

It is not necessary to repeat here the familiar story of the huge research and development effort that was eventually mounted, particularly in nuclear weaponry and in the electronic, chemical and biological fields. Our particular interest is in the relation between the technological and the political establishments. During the war many well-known representatives of the scientific and technological communities such as Vannevar Bush and Karl T. Compton in this country and P. M. S. Blackett and John Cockcroft in Great Britain served as administrators and advisors to their governments in the matter of development and use of new methods of warfare. It was, of course, largely through the efforts of Albert Einstein, Leo Szilard and other scientifically renowned refugees from Europe that the United States led in the race to develop the atomic bomb. There is evidence enough that the scientists and engineers were able to influence the direction of research and development as it coincided with strategic policy but very little to show that they took a major part in the making of high-level political decisions. C. P. Snow has cited one prominent exception to this last generalization. He writes of the influence which F. A. Lindemann (Lord Cherwell) had on the British war effort as Winston Churchill's friend and personal scientific advisor. Snow believes that Lindemann had more power than any other scientist in the war. This had been dangerous not only because opinions contrary to his had been given little weight, but especially because he had used bad scientific judgment and recommended strategic policies based on faulty technical analysis [46].

In any event, concern about a concentration of technologists was heard very little during World War II in spite of some expressed unease about a managerial elite. In more recent years, especially since the Korean War, this kind of speculation has been renewed. It has accompanied a staggering level of federal expenditures for research, development and production, most of it militarily oriented, in the atmosphere of continuously unstable relations with the Communist nations. This, without doubt, has profoundly affected the practice of science and technology and irreversibly influenced other institutions as well. From the point of view of those who favor a high level of federal support for research and development there is ample justification for continuation of the trend. Some of the reasons given include the belief that research generates new production and prosperity, that grants to schools and colleges insure a continuous supply of trained young people, that there are many projects necessary for the future growth and prestige of the nation which would cost too much for private enterprise to develop, that disease and social ills will only be eliminated through the insights and new techniques which result from research, and that our defense must be insured in the face of undiminished military growth by potentially aggressive countries. There are also some implications that basic research is worthwhile even when there is no reason to expect practical benefits in any foreseeable period, because it tends to lead to a general uplift of the culture.

Contrary views cover a wide range. Undoubtedly a minority position, but one strongly held by some younger people especially, would be since science and technology have gone so far out of control, tainted by concentration on destructive and venal objectives, the only useful course is to restrain their development by avoiding any participation in activities which would contribute to their advancement. Others, still convinced of the need for maintaining technological leadership, are more concerned that the major support for sponsored research in recent years has been through the Department of Defense—while conceding that not all of the research has been directly related to war purposes. The inference drawn is still that the increased productivity and greater affluence in the United States in post-World War II years (even if it can indeed be attributed in significant measure to a higher level of research and development) has been the result of lop-sided expenditures on war-related missions and hardware. Consequently, by diverting funds best spent on education, social welfare and environmental research the nation has not added to its fundamental wealth, and its long term stability has been poorly served.

These arguments represent a basic political diversity of opinion and it is not likely that they will soon be answered to everyone's satisfaction. What is reasonably sure, however, as we enter the 1970s is that the fear of a technological-military-industrial elite taking over the United States is overdrawn or at least premature. This is stated without denying the power which these forces have in the political realm, or that in other countries, France especially, technical experts have long taken dominant positions in the government process. Both Ferkiss[47] and Meynaud[48] have come to similar conclusions. The reasons for thinking so can be summarized briefly. Scientists and engineers have become more influential not just because their numbers have rapidly increased but also because only they can fully understand how their complex technologies work. When the political leader is scientifically illiterate, as is so often the case, he must rely on explanations given him by his technological advisors. Yet as the long-standing controversy about the development of the anti-ballistic missile (ABM) has demonstrated, the political beliefs of the advisor can strongly influence his interpretation of the potential of a new technology. He may draw different conclusions from the same quantitative data, or choose his statistics out of a different segment of the general data base from one who opposes his point of view. The nationally televised Congressional hearings in 1969 featuring the conflicting testimonies of Dr. Edward Teller and Dr. Jerome Wiesner illustrate the point. Dr. Teller spoke from a background of an implacable determination to resist the Communist menace. Dr. Wiesner spoke as one who had earlier contributed a great deal to the effort of the United States to keep its technology in advance of the Russians but who had come to feel that widespread introduction of such weapons as the ABM could lead to what Dr. Herbert York has called "the ultimate absurdity."[49] This means not only that the arms spiral has led to a decrease rather than an increase of national security, but that the use of these weapons puts such a premium on speed of response that, even assuming the decision for action could be properly programmed, the political leader would either have to delegate reaction responsibility to a computer or "pre program" his own responses. So the technical aspects of the decision making process in similar cases do not tend to be paramount but rather secondary to the initial leaning of the ultimate decision maker toward one of the positions presented, and also toward what he considers the prime economic and political realities.

It is also true that technologists and scientists have operated neither from a unique political nor economic base of power. Nothing in their nature inclines them to speak as one in political matters, nor are those of their number who are best at their profession normally enthusiastic about moving into managerial positions. If they do, they find it difficult to keep in touch with the most current developments in their technical field. We must also mention that various fields of science and technology are not free from antagonisms toward each other not only because of intense competition for limited government support, but also because of fundamental disagreements about ways of approaching problems.

Nothing in the immediately preceding paragraphs is intended to minimize the profound influence of modern technology on the practices and institutions of government. The technologist himself is indispensable to the system because he brings the technology into being, whether for reasons of monetary gain or professional accomplishment; the businessman, because, recognizing that technology undergirds the industrial system, he provides the environment within which the technologist can work; the politician, because through the policies he supports, the magnitude and direction of the science and technology effort are largely determined; and the layman, because the services and material possessions he has learned to demand provide the motivation to sustain the forces of development which some call the "technological imperative." Technology is immanently inert, but paradoxically it feeds on itself and grows. Once in existence it has a way of demanding that it be attended to —its dendrites enmesh and become an integral part of human activity. Man has reached a point where he must examine this relationship with more care and wisdom than he has in the past.

REFERENCES

1. Letwin, Shirley R., *The Pursuit of Certainty*, Cambridge University Press, London, 1965, p. 1.
2. Ellul, Jacques, *The Technological Society*, translated by John Wilkinson, New York: A. Knofp, 1964.

3. Pool, Ithiel de Sola and Abelson, Robert, "The Simulmatics Project," *Public Opinion Quarterly*, Vol. 25, No. 2, Summer 1961, pp. 167–83.
4. Ferkiss, Victor, *Technological Man: The Myth and the Reality*, George Braziller, N.Y., 1969, pp. 182–96.
5. Ferkiss, Victor, ibid, p. 176.
6. Bundy, McGeorge, *The Strength of Government*, Harvard University Press, Cambridge, Mass., 1968, 111 pp.
7. Owen, Wilfred, *Strategy for Mobility*, Washington, D.C., The Brookings Institution, 1967, p. 35.
8. Smerk, George (ed.), *Readings in Urban Transportation*, Indiana University Press, Bloomington, 1968, p. 3.
9. McLuhan, Marshall, *Understanding Media*, Signet Books, N.Y., 1964, p. 199.
10. Nelson, James C., "Governments' Role Toward Transportation" in *Modern Transportation: Selected Readings*, edited by Farris, Martin T. and McElhoney, Paul T., Houghton Mifflin Co., Boston, 1967, p. 383. Reproduced from *Transportation Journal*, Summer 1962.
11. *Instruction Memorandum* 50-2-63, Bureau of Public Roads, September, 1963.
12. Beshers, James M. (ed.), *Computer Methods in the Analysis of Large-Scale Social Systems*, Joint Center for Urban Studies of the Massachusetts Institute of Technology and Harvard University, Cambridge, Mass., 1965.
13. Doxiadis, C. A., "Ekistics, An Attempt for a Scientific Approach to the Problems of Human Settlement" in *Science and Technology and the Cities*, Committee on Science and Astronautics, U.S. House of Representatives, U.S. Government Printing Office, Washington, D.C., 1969, p. 9.
14. Hirten, John F., "Urban Planning and the Political Process," *California Monthly*, December 1968, p. 41.
15. Tunnard, Christopher, "America's Super Cities" from *Taming Megalopolis*, Vol. 1, edited by Eldredge, H. Wentworth, Anchor Books, Doubleday & Co., Garden City, N.Y., 1967, p. 6.
16. Owen, Wilfred, *The Metropolitan Transportation Problem*, rev. ed., Anchor Books, Doubleday & Co., Garden City, N.Y., 1966, p. 97.
17. Owen, Wilfred, ibid, p. 144.
18. Owen, Wilfred, ibid, Chapter V.
19. Hamilton II, William F. and Nance, Dana K., "Systems Analysis of Urban Transportation," *Scientific American*, Vol. 221, No. 1, July 1969, pp. 19–27.
20. San Jose Traffic Control Project: Final Report, San Jose, Calif., City of San Jose – IBM Corp., 1967. (Courtesy of Mr. Gene Mahoney, Traffic Engineer.)
21. Manheim, Marvin L., "Problem Solving Processes in Planning and Design," *Design Quarterly*, 66/67, 1966, p. 31.
22. Roberts, Paul O. and Suhrbier, John H., *Highway Location Analysis: An Example Problem*, Rep. No. 5, Cambridge, Mass., M.I.T. Press, 1967, 224 pp.
23. Ross, Howard R., "New Transportation Technology," *International Science and Technology*, November 1966, pp. 26–37.
24. Ross, Howard R., ibid, p. 31.
25. Ross, Howard R., "The Future of the Automobile," *Science and Technology*, August 1968, pp. 14–24.
26. *Technology: Processes of Assessment and Choice*, report of the National Academy of Sciences, Committee on Science and Astronautics, U.S. House of Representatives, July 1969, p. 26.

27. McHale, John, "Science, Technology and Change," *Annals. Amer. Acad. of Political and Social Science*, May 1967, pp. 128-39.
28. Gross, Bertram and Springer, Michael, "A New Orientation in American Government," *The Annals of the Amer. Acad. of Pol. and Soc. Sci.*, Vol. 371, May 1967, pp. 7, 8.
29. *Report of National Commission on Technology, Automation, and Economic Progress*, Washington, D.C., U.S. Government Printing Office, 1966.
30. Spencer, Herbert, "What Knowledge is of Most Worth," *Westminster Review*, 1869. Quoted in Spencer, H, *The Study of Sociology*, N.Y., D. Appleton & Co., 1896, pp. IV, V.
31. Biderman, Albert D., "Social Indicators and Goals" in *Social Indicators* edited by Raymond A. Bauer, M.I.T. Press Paperback Edition, Cambridge, Mass., 1967, Chapter 2.
32. Biderman, Albert, ibid, p. 80.
33. Biderman, Albert, ibid, pp. 87-94.
34. Dewey, John, *World Tomorrow*, October 1928, p. 392.
35. *Technology: Processes of Assessment* and Choice, op. cit., pp. 3-6.
36. Carson, Rachel, *Silent Spring*, Houghton-Mifflin, Boston, 1962.
37. Daddario, Emilio A., *Technology Assessment*, Committee on Science and Astronautics, U.S. House of Representatives, U.S. Government Printing Office, Washington, D.C., 1967, pp. 7-10.
38. Daddario, Emilio A., *A Study of Technology Assessment*, Report of the Committe on Public Engineering Policy of the National Academy of Engineering to the Committee on Science and Astronautics, U.S. House of Representatives, U.S. Government Printing Office, July 1969.
39. U.S. Congress, House Document 360, *Technological Trends and National Policy, Including the Social Implications of New Inventions*, Report of the Subcommittee on Technology to the National Resources Committee, 75th Congress, June 1937.
40. Westin, Alan F., *Privacy and Freedom*, Atheneum, N.Y., 1967.
41. Sprague, Richard E., "The Invasion of Privacy and a National Information Utility for Individuals," *Computers and Automation*, January 1970, pp. 48, 49.
42. Kline, Morris, *Mathematics in Western Culture*, Oxford University Press, N.Y., Galaxy Edition, 1964, p. 264.
43. Saint-Simon, Henri de, *Social Organization, The Science of Man and Other Writings*, (edited by Felix Markham), Harper and Row Torchbooks, N.Y., 1964, p. xi.
44. Veblen, Thorstein, *The Engineers and the Price System*, (with introduction by Daniel Bell). Harcourt, Brace & World, Harbinger (ed.), N.Y., 1921, (1963).
45. Elsner, Henry, Jr., *The Technocrats, Prophets of Automation*, N.Y., Syracuse University Press, 1967.
46. Snow, C. P., *Science & Government*, Mentor Books, Harvard University Press, 1960, 128 pp.
47. Ferkiss, Victor F., op. cit, pp. 174-79.
48. Meynaud, Jean, *Technocracy*, translated by Paul Burnes, The Free Press, N.Y., 1968, pp. 131-38.
49. York, Herbert, *Race to Oblivion*, Simon and Schuster, N.Y., 1970, 256 pp.

5
The Systems Planners

"Let fanciful men do as they will, depend upon it, it is difficult to disturb the system of life."

SAMUEL JOHNSON, Boswell's *Life*

One phase of post-World War II technology reached a dramatic culmination on July 16, 1969 when an American astronaut first stepped on the moon. It is perhaps more accurate to recognize that ancient dreams fulfilled by this space adventure had become theoretical possibilities with the science of Galileo and of Newton: however the possibilities had little credibility (despite the pioneering work of Robert Goddard and other pre-war rocket experimenters) until two powerful industrialized nations decided to concentrate awesome resources on their competing space programs. Money alone would have been insufficient, but the experience of organizing highly trained men and their equipment to deal with the growing complexity of military requirements had provided the base from which an attack on the imposing problem of making a successful moon landing before 1970 (as President Kennedy had promised) could be mounted.

It has been pointed out[1] that in the decade from 1948 to 1958 the U.S. Air Force progressed from the B-50 bomber, with some 10,000 electronic components to the B-52 bomber, which utilized approximately 100,000 electronic components. Successful development and production of such complicated systems would never have been served by haphazard means. Within the aerospace industries, certain orderly patterns of analysis and synthesis have evolved to help make many of the decisions necessary to reach a difficult objective, and the processes of development have been similarly aided in other industries. Together, these patterns are termed the "systems approach."

As an aid to engineering and management the systems approach has had some important applications, with the result that its potential for ameliorating the ills of the technological society is being explored on many fronts. In some ways these efforts had earlier counterparts whenever quantitative methodologies using the resources of science and technology were looked to for social utility. This chapter will describe the general nature of the systems approach and will examine the implications of attempting to apply it to a variety of systems.

DEFINITION OF A SYSTEM

Often a phrase will achieve wide currency although there is limited collective agreement about the idea it is supposed to represent. This is the case with the term, systems approach. To say that there are nearly as many interpretations of "system" as there are classes of systems would ignore its generality as a concept, but it is true that a variety of definitions have been proposed.

Derived from the Greek, *systema*, meaning "to place together," the word, system, contains the notion of an assemblage of interdependent components. This is not, however, a sufficient definition. It could apply to anything more involved than the most elementary particles, even though the interdependency is not always readily traceable. We also want to include the idea of purpose, and to this end the following is proposed:

> A system is an array of related principles or parts given coherence by focussing on the achievement of an objective.

Analyzed, the definition signifies that a collection of individual components expresses the whole, but only by virtue of the mode of interrelationships. That is to say, the system, as it seeks to fulfill an objective, requires that the elements of the whole assume a particular function or place within certain limits. At any instant the state of the system is described by the components and their attributes.

CLASSES OF SYSTEMS

Classifying can be helpful. One can list many categories of systems. For example, Ellis and Ludwig[2] suggest three: (a) *natural* systems; either physical or organic; (b) *devised* systems, which men have helped to design and create and (c) *hybrid* systems which are combinations of the natural and the devised. Useful analogies may be drawn between natural and devised systems. It is also common to speak of *open* and *closed* systems, the difference being that closed systems have no interchange of matter, energy or information with their environments, while open systems do. Some engineering textbooks describe closed systems as those permitting energy but not matter to cross system boundaries, reserving the term, *isolated*, for systems not interacting with their environments. The distinction is not ordinarily maintained in most fields nor will it be here. The concept of closed systems is useful for physical analysis but open systems are the ones encountered in practice and are the more challenging to consider. A difficult but important problem is to determine where the optimal boundary between the environment and an open system lies.

Figure 5-1 is a simple schematic diagram to illustrate an open system operating to produce an output changed in some form from the input.

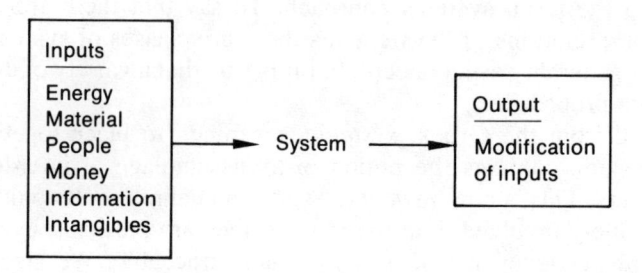

Fig. 5-1. An open system.

This does not, of course, show anything at all about the way that the elements within the system combine to effect the operational process. The interconnections most characteristically occur as networks for which various methods of analysis exist. Often the inner relationships of a system are inaccessible to the observer. It must then be treated as a "black box" which can only be studied by manipulating various inputs to see which combinations lead to coherent outputs.

There is also a basis for identifying systems according to the ways they respond to their environments. Natural systems in particular are often *adaptive*, which means they can adjust to external conditions, even harmful ones, in a manner that supports continued existence although their state is changed[3]. Social groups, such as political organizations, are man-made systems which must demonstrate a substantial degree of adaptive behavior to survive. While man-made machines are not ordinarily considered to be able to adjust to wholly unexpected changes in environment, one of the most intriguing questions about computerized systems is: How adaptive are they or may they become?

Sometimes a small part of a process output — enough to carry information — is conveyed back to the input side to keep the output continuously controlled within prescribed limits. These systems are said to have *feedback*. The home thermostat is a simple example. When a rising temperature in the room causes the indicator to go above a predetermined setting, this information is fed back to a valve to cause the fuel supply to be reduced. As the room cools, the indicator drops to a point below the setting and the fuel supply is again increased. It is the departure from the desired temperature which brings about the correction. In this case the current output helps determine future output. Feedback inheres in automatic systems.

The addition of a computer can provide a tighter degree of control to the performance of a system with feedback. In the system illustrated in Fig. 5-2, the computer can analyze the input data and from the output

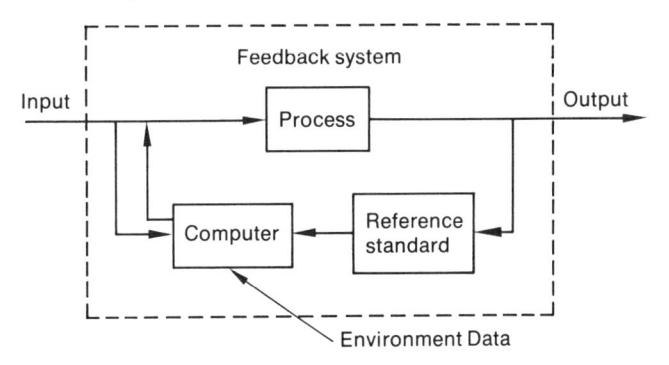

Fig. 5-2. Computerized feedback system.

information and data concerning the environment, as well as from the storage of past events and predictive elements such as statistical tables, it can adjust the input to maintain the desired control. Further examples of feedback processes are described in Chapter 6.

We have listed only some of the ways by which systems may be described. They are also — among other things — *concrete* or *abstract*, *predictable* (static or mechanistic) or *unpredictable* (variable or self-organizing) and any one system may fall under several classifications.

Some time ago, Kenneth Boulding proposed the following hierarchical classification of systems[4].

Table 5-1   Hierarchy of Systems.

Key Word	Characteristic	Example of System at Level
1. Framework	Static structure	Geography of earth, anatomy of cell
2. Clockwork	Predetermined motions	Steam engine, solar system
3. Thermostat	Transmission of information	Furnace, homeostasis†
4. Cell	Open system throughput of material and energy, system maintained in the face of throughput	Flame, river, cell
5. Plant	Division of labor among parts but not at sensory level	Any plant
6. Animal	Self-awareness, Specialized receptors, increased intake of information to form image (different from the information)	Any animal
7. Human	Self-consciousness, knows that it knows, language and symbols	Human being
8. Social Organizations	"Role" and communication, history, messages, music and art	Human societies
9. Transcendental systems	Unknowables	Unknowables

†Homeostasis is the tendency of an organism or mechanism to maintain a relatively stable condition within itself, e.g., blood temperature.

SYSTEMS ENGINEERING AND OPERATIONS RESEARCH

Thomas Tredgold, a British architect of the 19th Century, is credited with being the first to define engineering in one sentence. He said it was "the art of directing the great sources of power in nature for the use and convenience of man."[5] Although the definition did not refer to the *benefit* of man we can reasonably assume that for him a convenience was to be considered beneficial. In any event, many subsequent statements of the purposes of engineering, some by official bodies, included phrases relating to the engineer's responsibility for improving the well-being of mankind.

The point has already been made that in the industrial nations, at least until quite recently, those who contended that the results of engineering and technology were often anything but beneficial remained distinctly in the minority. This is no longer the case. While it may still be a minority view that technological enterprises ought to be sharply cut back because social and biological systems have been unable to adapt to the rapid changes they cause, there is a prevailing sentiment for having social values somehow incorporated into the goals of a technical system. We shall try, eventually to show what kind of task this must be, but first it is necessary to see what systems engineering means.

According to Schlager[6], the Bell Telephone laboratories used the phrase, systems engineering, first in the early part of the 1940s. That this organization should be one of the earliest to think in these terms is hardly surprising since the operation they serve epitomizes a complex network system. Nevertheless, the original growth of the telephone company was not—could not be—along the lines of ideal systems planning. The necessary technical resources for anticipating the future demands for telephone service did not exist in the early years; and so the expansion of the system, while as orderly as engineering knowledge and other circumstances would permit, was essentially a step-by-step process of accretion and conversion. Many systems must develop in this way. By 1940 however, as Hall has described[7], there had been enough progress in the understanding of radio and wire transmission of signals and enough interest developed in the potential of television to support a study on the possibilities of designing a microwave radio system which would handle long-distance telephone communication as well as television programs. The first stage of the study was devoted to the technical feasibility of this new kind of system. The conclusion reached was that, although the project was theoretically sound, the current technology was not sufficiently advanced to chance designing a fully operational system without further research. With the coming of World War II the need for radio and radar communications for the military effort gave an indirect impetus to the development of the technology, and by the end of the war it was possible to renew planning on a much firmer engineering basis.

The studies were extended to include routing, selection of sites for repeater stations and other factors related to the enhanced knowledge, but in addition, economic considerations broadened the scope of activities beyond the merely technological. The procedure then became, as it so often does, one of carrying on research and development planning at the same time as trial systems were being constructed and tested. With the general needs well established, and most of the more specific objectives expressible in quantitative terms, it was now possible to compare the economic and technical aspects of alternative approaches in order to determine the additional development work required for an acceptable working system. When these decisions were made, development could begin on the first stage of the basic project—in this case providing a radio relay system between New York and Chicago. It was completed in initial form in 1950 and almost immediately, on the basis of observed performance, a program effecting corrective changes began.

This is a brief description of what was a very large systems project bringing together engineers, scientists and other professional people in an integrated attack on a many-faceted problem. No job of this magnitude, involving hundreds of man-years of highly skilled effort could be undertaken successfully by any single organization (or consortium of smaller groups) without resources somewhat comparable to those of the American Telephone and Telegraph Company. And yet, once the initial development planning was completed and the engineering decision was made to proceed to the design and construction of the desired system, the chances of successfully attaining the objectives were very good. This is stated not to minimize the difficulties that remained, but rather to make the point that they were largely technical and economic, and the exploratory stage had provided rather firm assurance of their being manageable. Had indications been otherwise it is unlikely that work on a final system would have proceeded. The alternatives of carrying on further research or abandoning the project would in that case have been preferable.

This kind of choice is not always available. When there appears to be a question of security or survival, or other subtle or unwieldy political considerations affect a decision, the final selection of an approach may have to be among relatively undesirable competing alternatives. Many times the potential outcome may be in serious doubt. In periods of war or preparation for war these kinds of pressures often beset decision makers. The tendency has been to move increasingly toward a systems approach to help determine a course of action and carry it through.

No one would wish to claim that a science of problem solving was conceived at a particular identifiable moment. Men have tried to apply knowledge to systematic classifications of thought and action for as long as there has been a notion of science, but in the modern context this kind of development was first identified by name about 1938. Page [8] credits P. M. S. Blackett, the British physicist, with giving the name *Operational Research* to the scientific and mathematical study of man-machine systems which have a purpose. The method was employed on many military problems during World War II, and when the war was over its use spread quickly to problems in business and government. In the United States, it was called Operations Research, and the term, Management Science is sometimes used to mean essentially the same thing.

Many illustrations have been given of the ways in which teams of scientists, studying military tactics and strategy from the viewpoint of mathematical analysis, successfully resolved dilemmas which had been locked in by traditional thinking. The tense contest between the Royal Air Force and the German submarine fleet during the Battle of Britain offers one example[9]. It had long been accepted naval practice to set anti-submarine depth charges to detonate at about 100 feet below the surface of the water, on the theory that at this depth the water pressure produced the greatest explosive power. When aircraft were brought into use to assist the Navy in hunting submarines, most squadrons continued at first to use depth-charge settings which would result in explosions far beneath the surface, sometimes at 150 feet. Others felt that the rapid approach of airplanes would be more likely to catch submarines near the surface so that settings of, say, 50 feet were more logical. Besides, the submarines could not be seen at 150 feet. When operational researchers were given the problem, they quickly determined through the data available to them that the most significant consideration was the location of the submarine when the depth charge was being released. If most attacks were made on submarines which were still surfaced or only partially submerged, then settings of 50 feet, which were effective when a dive had already begun, would be too deep. This was the kind of information they needed, and when enough had been gathered and analyzed statistically the results indicated that a setting of 25 feet ought to be optimum. It so proved. Reported sinkings of submarines more than doubled after the new settings were adopted. In this instance, the method of problem solving owed its success to fresh insights contributed by scientists who had been trained to work with quantitative information and to recognize where it could be applied. The scientist, engineer or mathematician also has a background which enables him to grasp the essential details of a subject new to him when it is expressed in mathematical language or in a technical way.

To the question of whether there is an identifiable methodology involved in Operations Research, the answer would be that, while the procedures described by many practitioners differ in certain details, there is reasonable concurrence on the basic approach. The major steps are:

1. Formulation of The Objective

The objective needs to be one which can be defined in some way so that a quantitative measure of the success in attaining it can be applied. Sometimes the person or group for whom the study is being conducted will propose the problem in a form which will require modification before an attempt at a solution begins. As decision maker he must agree to the modification. This stage requires establishing a whole scheme of reference into which physical, economic and human constraints on any solution must be incorporated. The formulation of objectives is determined by value systems.

2. Building and Testing a Model

The model, whether mathematical, physical or graphical is a symbolic and partial representation of the system being studied. Models are necessarily simplifications, and sometimes misleading illusions of reality attend simplification carried too far. Nevertheless, the advantages of model-building are undeniable. Time and money can be saved, and even the failure of a model can suggest the need for additional information in some areas, which in turn can spark fruitful courses of action.

3. Developing a Controlled Solution, or Alternatives

At the completion of the study, the operations researcher should be able to provide the decision maker with an accurate description of the situation, and an indication of an optimum policy based on the risks and benefits which would presumably derive from alternative courses of action.

Each of these three stages can be broken down into substages and the order of application does not have to follow a definite sequence. In fact some may take place simultaneously. Data can be analyzed and collected at any stage.

The use of science and mathematics as aids to decision making is surely nothing new, and taken by themselves the steps to a decision listed above offer little in the way of innovation. The difference is that a higher order of complexity in our technological society has created systems problems against which unidisciplinary attacks have proven inadequate. The urgency of war-bred crises has carried over into the periods of uneasy peace to force interdisciplinary approaches to take on some shape and coherence. Now, more advanced techniques have been developed to analyze the interrelationships between the elements of complex systems, including those about which an irreducible uncertainty remains. The computer is an essential tool for some of these techniques but not all. A recognition that the operation of systems could be handled as a problem in research was accompanied by an integration of talents and techniques, whether as operations research or systems engineering, which has permitted higher orders of systems to be studied.

In what ways does systems engineering differ from operations research? Once again, since the practitioners of both carry on their activities from so many backgrounds and points of view, no universally accepted statement describing either is available. An examination of the literature shows that there are many areas of convergence. Much of the research methodology is common to both, and some contend that little purpose is served in separating the two. The most generally favored distinction gives to systems engineering the role of designing new systems, and leaves to operations research the problems of finding optimum patterns for existing systems. That is to say, systems engineering might be considered more creative, and operations research, more analytical. Since the radio relay system described earlier was only an idea when the Bell Laboratories study began, conceivably the resulting operation could have ended as something quite different from what the company executives had in mind in 1940, and so this is considered a part of systems engineering. On the other hand, the contest between submarines and airplanes during the Battle of Britain was already an actuality and the clearly specified objective for the team of operational researchers was to maximize the destruction of the submarines, or, more accurately, the tonnage of allied shipping which would get safely through.

Our attention will be directed primarily to the processes of devising systems to help achieve a desired goal. The exclusive use of the term, systems engineering, to describe these processes may however be unnecessarily restrictive even though widely employed. Since it became popular largely in the context of "hardware-oriented" systems, to many it connotes an activity in which traditional engineering practices play the only important roles. For this reason, the more general expression, systems approach, (which may include aspects of analysis, engineering and management), has found favor in man-centered systems.

SYSTEMS PROBLEMS AND TECHNIQUES

Some problems, especially those involving "hardware" are limited enough in scope that applied intelligence from one engineering or scientific point of view is sufficient to give an answer. In large, modern organizations the problems are more often multidimensional and if the study is limited to a single viewpoint a crucial variable is apt to be overlooked. Even with an interdisciplinary team study this could happen, but it is less likely. An orderly systems approach, by focussing on the structure of the problem should help to emphasize the relevancies.

A few key questions such as the following can serve as a framework for the attack on the problem:

1. What is the real objective of the system?
2. Why are you trying to achieve that objective?
3. What are the constraints on any solution in terms of allowed cost, effort, etc.?
4. What are the tools available?
5. How long is the system to last?
6. What are the alternative means of attaining the objective?

These questions, given here in broad terms, do not have to follow each other in this order. There must be a process of recycling or iteration. The asking and answering of one question may require the reconsidering of another.

Designing Systems

Designing new systems demands a high level of sensitivity to conditions of the environment. According to Hall[10], three kinds of environmental conditions with which systems people ought to be familiar are: (1) physical and technical, (2) economic and business and (3) social. They often overlap, of course, and all are usually present with a problem of any magnitude, influencing the systems they enfold to various degrees. Although much contemporary attention is directed toward social conditions, most detailed studies of actual systems design in the literature emphasize the effects of environments in the first two categories. The social problems are very refractory for reasons which have already been discussed in the previous chapter. We shall take up additional aspects of the treatment of socially-related problems later in this chapter. First, however, we shall examine the development of two systems—a large-scale government information system and a common type of industrial system capable of being simulated by computer methods—and then suggest how certain analytical tools may be used to help understand and attack systems problems.

An Information Handling System

In Chapter 3 we spoke of the growth of the knowledge industries, and implied that the designing of systems for the processing of information is of prime concern in large organizations. With the help of the computer, it is possible to develop systems which at least provide historical and current cost and performance data out of which controls of some sort can be fashioned. At best they can do much more. The sheer physical details of paper-handling and the need for constant manual intervention have destroyed the credibility of traditional paper-work oriented methods as suitable for administrative control for large organizations, and this is increasingly proving to be the case for medium-sized and even some small organizations.

To illustrate the design of an information-handling system on a large scale we have chosen to examine the Local Government Information Control (LOGIC) System of Santa Clara County, California.† This example was picked because, although it is very encompassing and complex, the system objective was clearly perceived at the outset by those who would have the responsibility for designing it. We shall want to defer considering until later what can happen when there is no strong consensus either about the nature of the problem or the methods of approaching it. In the given instance no one questioned that the first concern was how to provide adequate governmental services for processing and providing information in one of the fastest growing counties in the United States.

In 1940, the population of Santa Clara County was 175,000. By 1950 it had increased 71% to 300,000. In the beginning of 1970 it was approximately 1,100,000 and has been increasing on the order of about 45,000 persons per year. In the 1950s, processing of information was still dependent on manpower, and the manpower requirements were unconscionable. The County Management began in 1953 to consider new procedures to replace the manual typing of the assessment role, tax roll and tax bills, and decided to go to data processing with punched-card equipment. Soon after the equipment was installed, eight other areas were converted to punched-card data processing: assessment roll, tax roll, tax bills, welfare roll, budget, payrolls, appropriations and roads; but very quickly demand for information exceeded equipment capacity. By 1961,

---

†Grateful acknowledgment is made to Mr. Howard W. Campen, County Executive, for permission to use this material, and to Mr. Thomas Johnson for assistance in preparing it for this section.

the costs of medium-sized computers had dropped sufficiently that the purchase of an IBM 1401 computer system was approved. A second one was installed in 1963. By 1964, during peak periods both computers were on a three-shift, seven days a week basis.

Although a great deal of data was being handled, there was still no overall systems plan—for the most part individual departments still had their own programs. A systems study group was set up on November 1964 to analyze and study all department operations as the base for establishing a county-wide information system. The major objective of the system would be to provide centralized record keeping for the County, (but with decentralized control), and to make those records quickly available to anyone who had a right to examine them. The system would then in effect be a facility for maintaining coordinated sets of records, a means of analyzing the contents of those records to provide statements of pertinent current facts, and it would also be a tool for problem solving.

The realization of the system, it was clear, could only be achieved by means of a reorganization which, from the standpoint of information flow, would picture each departmental subsystem in a new light. Organization and geographic boundaries would have to be redefined and in some cases ignored or crossed. In any large system such as this there are some obvious constraints and others less obvious, but a major constraint has to be the money available for designing the system and making it work. The charges of some political figures and taxpayer groups to the contrary, very few county or city governments are ever able to command overabundant resources to provide needed services for local residents. Another major constraint placed on the system was time. It was intended that the system should be completed and in operation in six years, that is by July 1971. There would be continual updating to reflect changing needs.

Analysis of the system

Apart from primary system objectives, subsystems may also have objectives which need to be considered in the overall analysis. The LOGIC system was originally designed to encompass ten information subsystems:

Property	Supply
People/General	Engineering
People/Law Enforcement	Hospital
Accounting	Library
Personnel	Advanced Analysis

One of the first steps in the project was to make a detailed study of each of the departments to be involved in LOGIC. This meant collecting all facts about current operations including descriptions of methods of data gathering, processing evaluating and disseminating; accounts of the volume of transactions at each stage; and an indication of storage needs. An important consideration was the need to identify all potential recipients of an item of information, and to elicit from the departments the clearest possible indication of what other kinds of information they might find useful. Routinely, individual county departments gather information which is of value to other offices. Both the Assessor's Office and the Planning Department are interested in information concerning real property; departments involved with inventory keep individual records of stock status, and so does purchasing; and many different departments collect information about people, much of it redundant. While each department should be able to get the information it needs from a central source, it is not desirable to process or display more than is required or can be assimilated. With the information about the existing departments collected and evaluated, it was possible to prepare a preliminary outline of information flow for the proposed system. This can be done in the form of descriptive statements, or flow diagrams or perhaps cross-referenced tables. Figure 5-3 gives a general schematic indication of elements of the former data processing system compared with the projected LOGIC system.

A steering committee composed of various County executives had to decide whether it would be best to implement one subsystem at a time, sequentially, or each subsystem concurrently in phases. The alternatives are compared in Table 5-2. The steering committee elected to recommend the segmented method of implementation and as a result a phase schedule for each section of the overall system was developed as shown in Table 5-3.

Costs

Obviously work on a system of this magnitude could not proceed very far without developing cost estimates. A detailed estimated cost schedule was prepared for a six-year implementation plan which included employees, equipment, systems design and programming, and telephone lines. It was intended that at the end of each fiscal year, modifications could be made on the basis of fresh evaluations of work volume and new developments in machine storage or terminals.

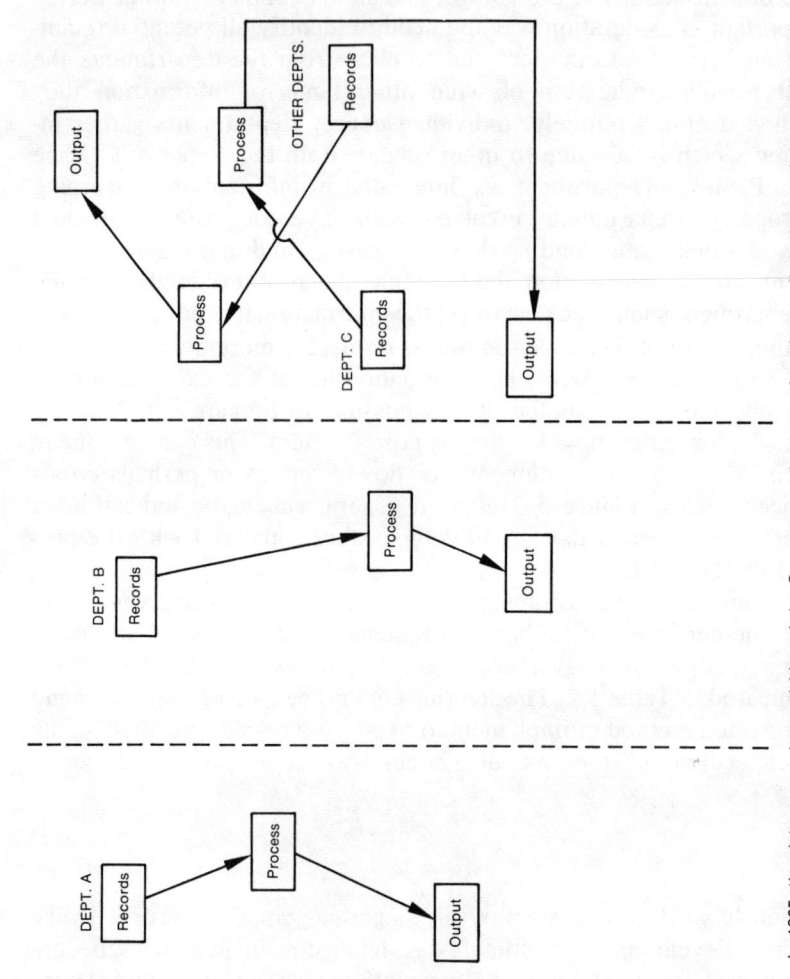

In 1965, the data processing activities of the County were (because of equipment limitations and decentralization) performed for and/or by departments on an independent basis. Separate records were maintained by many departments. Departments were unaware of information of value to them stored routinely by other departments. Little, if any, information of common interest could be readily shared.

172

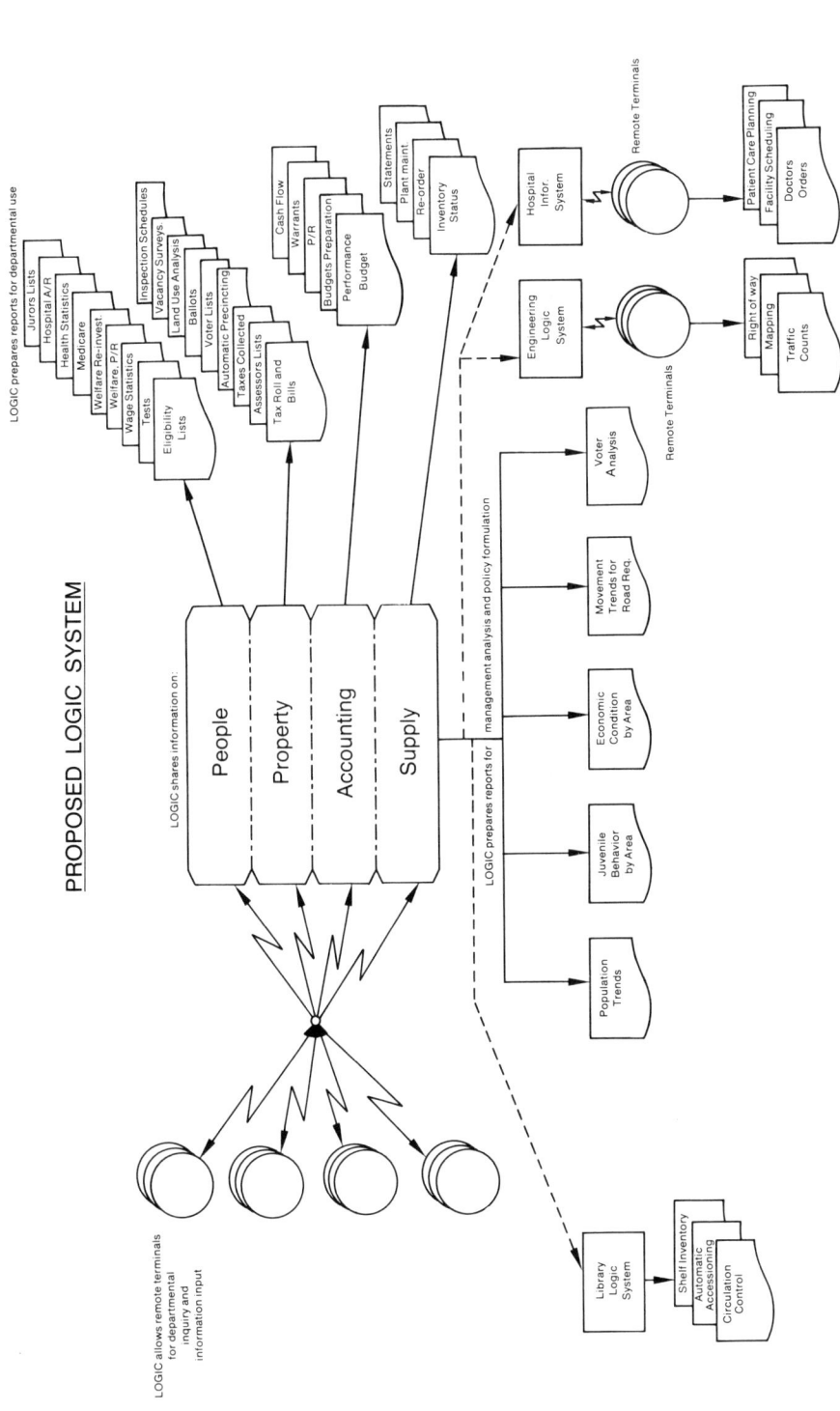

Fig. 5-3. Schematic diagram of 1965 data processing activities compared with proposed logic system (1965).

Table 5-2 Alternative ways of implementing LOGIC system.

Sequential Assignment	Segmented Priority Assignment
Each subsystem would be assigned a priority number. The couple subsystem would be implemented before beginning the next subsystem.	A segment of each subsystem would be implemented concurrently with segments of other subsystems. Total implementation of all subsystems would be completed in several phases.

A. *Advantages*
  1. A complete subsystem would become operative in the shortest possible time.
  2. Experience gained from problems encountered in tying segments of the first subsystem together into a complete subsystem would aid in assembling later subsystems.
	A. *Advantages*
	  1. High "pay-off" areas in *each* subsystem could be made operational first.
	  2. Programming staff could be immediately assigned to areas in which they had experience.
	  3. As a portion of a subsystem became operational information could flow between it and operating portions of other subsystems.
	  4. Segments of smaller subsystems could be scheduled and placed in various implementation phases as desired. This would enable operational progress to coincide with that of related subsystems and could also aid scheduling of programming.

B. *Disadvantages*
  1. Work would begin on lowest priority subsystems several years after the first was started with a possible undesirable effect on morale of employees of some departments.
  2. Changes in operating procedures, regulations, etc. Could make total system as approved obsolete by the time implementation was started on lowest priority subsystems.
  3. Some areas of low "pay off" (reduced clerical costs) such as the last segment of priority # 1 would be implemented before certain higher pay-off elements of other subsystems.
	B. *Disadvantages*
	  1. Complete implementation of any one subsystem would take several years.

Subsystems of LOGIC

Each subsystem of LOGIC is in itself a highly complicated system as we can briefly observe by taking the Law Enforcement Information Subsystem as an example. The rough schematic chart, (Fig. 5-4) shows the kinds of data generated by County law enforcement agencies, on the

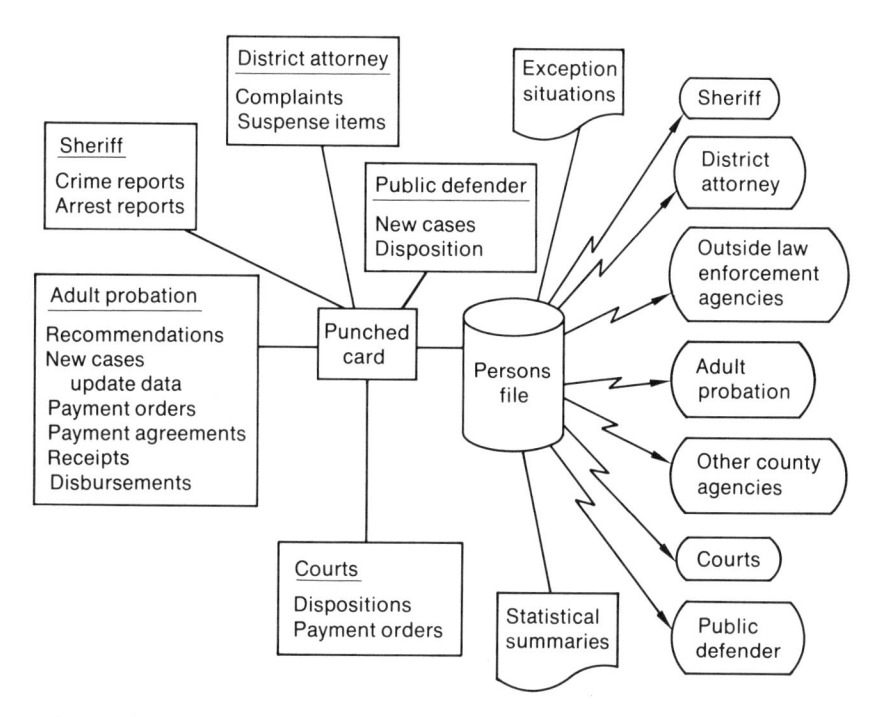

Fig. 5-4. Schematic chart of LOGIC information flow law enforcement persons section.

left, which are envisioned to be filed in the People (Law Enforcement) Subsystem and made available on demand to the agencies on the right. It is important to realize also that law enforcement problems do not remain confined within local boundaries and so any modern criminal justice information system must attempt to interface with regional, state and national systems.

The Law Enforcement System (or subsystem) has been conceived as two functional groups. The first group, the *Person Case Information Subsystem* is concerned with the information required for the processing of persons through the Criminal Justice procedure. The second group, the

Table 5-3 Segmented priority assignment phase schedule of LOGIC implementation.

	Phase 1	Phase 2	Phase 3	Phase 4	Phase 5	Phase 6
Property	Alphabetic Index	Assessor Controller Tax Collector	Planning	Assessor Unsecured	Registrar of Voters	
People "General"	Alpha Index Welfare Hospital	Welfare Budget & File	Hospital Accounts Receivable, Health Batch Processing	Health File	Registrar of Voters	
People "Law Enforcement"	Alpha Index Sheriff Adult Probation, Juvenile Probation	Citation Processing & A/R, Juvenile A/R, Adult A/R, Sheriff A/R	Sheriff file District Attorney	Juvenile Probation, Population Control Statistical Reports	Adult Probation File	Court Calendars, Attorney Scheduling
Personnel	Alpha Index	County Personnel File Position Code File	Eligibility File, Inactive Personnel	Examination Bank	Application Skills Inventory File	

	Supply	Purchasing Sheriff Hospital Reproduction Data Processing	Juvenile Communication Public Works Misc. Fixed Assets	Equipment maintenance schedules & Analysis	Statistical studies Vendor analysis, Item analysis
Accounting		Time recording and costs	Appropriation accounting Budget preparation	Revenue and fund file Outstanding warrants	School Accounting
Engineering		Earth work and traverse Right of way	Flood control Computations and control	Critical path and costs	Graphic data processing
Library		Book select. Book order, cataloguing	Circulation control Registration		
Hospital info System					H.I.S.
Advanced Analysis		Management Analysis of Data Bank Information			

Management Information Subsystem provides the decision making information needed for planning, organization structuring, staffing, allocating of resources, directing of activities, and obtaining feedback from those activities. Under each of these two functional groups there are a number of specific activities, called *applications*. The *District Attorney Application* for example, supporting the Person Case Information Subsystem, utilizes an Alpha card file that contains over 350,000 names and he handles about 40,000 cases a year. The great majority of the names in his file are duplicated in other files.

An information system of the scope of LOGIC is a highly complex undertaking simply by virtue of the large number of sub-functions required of it. Service for inquiry terminals at the Welfare Department was begun on January 15, 1968 and since that time other departments have been brought into the Alphabetic Persons Index according to an amended schedule. The original total systems study took 10,500 man-hours and resulted in reports containing detailed sets of recommendations. The partial descriptions in the foregoing sections can only suggest the complexity of the total system. Neither two-dimensional sketches nor descriptive narratives can convey the rich variety of interactions which must take place.

In any community which has grown as rapidly as this one has, the demand for governmental services is destined to outstrip the ability to provide them on the basis of records kept as individual card files. Quite apart from the duplications of effort, the mounting cost of continually adding buildings and personnel, (some 2.5 million dollars a year were used for data processing before LOGIC), and the awesome task of managing departments which become, in effect, multi-storied "paper factories" ordain the adoption of modern data processing methods. An alternative which has been proposed by a few is to reduce most government services drastically, but there is little evidence to suggest that such an approach would be effective or lasting in our urbanized, technological society.

Not that computer systems are inexpensive. They require initial financial commitments of such magnitude that public officials must examine their potential sources of revenue very carefully before giving approval for installation, even though there may be promise of future savings to go along with improved effectiveness of service. In some cases where planning has been inadequate there can be unexpected operational difficulties. Existing codes and regulations may inhibit innovation and within departments there may often be entrenched bureaucratic resistance to change, not all of it merely obstructive and doctrinaire.

One factor which made the LOGIC system feasible was the general acceptance of the need at County management levels, and the lack of any organized opposition from groups inside or outside the county government. It is worth mentioning that the issue of privacy, which was discussed in the previous chapter, has been carefully reviewed by the designers of the system, and safeguards are being continually evaluated and updated. For example, a "code of ethics" is being developed for the Law Enforcement Information subsystem to insure that steps will be taken to protect both privacy rights of individuals and the confidentiality of data in the system.

In information systems of this magnitude or larger, which have been proposed for other county or state governments or for service on the federal level, it is fairly evident that the data being handled must be of a type which offers minimum ambiguity for programmers and systems designers, or the overall difficulties of the systems problem would be compounded. It seems fair to say that, at the present state of development, very large information systems function at their best when processing relatively routine, if massive, data transactions. Sophisticated systems designed to facilitate higher level decision making and to produce fundamental changes in the governmental processes still have quite some way to go to match their initial promise.

In the planning of large information systems there can be no question at all that the computer should occupy a central place in the designing as well as the implication of the system. In the next example we shall look at an industrial situation where computer simulation techniques are used to solve a materials handling problem. The computer might be used in the resulting process, but not necessarily.

A Materials Handling Problem

A ubiquitous kind of industrial problem is one in which units of production arrive at a service facility to be processed or distributed to other stations. One particular difficulty with these cases comes from the irregular nature of requests for service. If the input units were identical and were to appear at constant intervals; and the service facilities were also predictable in operation and ample as well, the system would be relatively simple. Real situations rarely occur this way and so there is a problem of designing a system which will keep to a minimum the sum of the cost of both bottlenecks and idle facilities.

In Great Britain, a waiting line is called a "queue." To Americans who were there during the 1940s, the patience with which the British people stood in long lines for goods and services in short supply seemed quite remarkable. It is not surprising that operations researchers of that time referred to problems involving bottlenecks in a service line as queuing problems.

The original mathematical theory of waiting lines goes back to the early years of this century when A. K. Erlang, a Danish engineer, conceived the fundamental approach while designing a telephone system. His objective was to reduce waiting time for telephone customers, but the theory and later extensions of it relate to a host of situations where there are people or units of production waiting to be served or processed, or where the components in the processing system are idle part of the time. Traffic tie-ups at a freeway off-ramp, aircraft unable to land at a busy airport, or an imbalance between checkers and customers at the check-out station of a supermarket are examples of this type of problem. The need to wait may result in nagging or explosive frustrations, serious monetary losses or even physical danger. One experience so common as to be almost a stock comedy situation takes place in some banks when a patron wishing to cash a check enters what appears to be the shortest line before a teller's window only to find that someone in front of him has a day's business receipts to be tallied and entered. The patron then switches lines and perhaps the process is repeated. He sees that other patrons in other lines who have come in after he arrived are already being served, and his blood pressure rises. A simple remedy to help alleviate this situation has been adopted by some banks. As each patron enters the bank he gets into a single "vestibule" line. Here he stays until he reaches the head of the line at which time he may then go to the window of the first teller who is free. It is a variation of a method which has long been used in busy restaurants, although some, to be sure, have found that slow service in the dining room can be converted to a profitable business at the bar.

Some problems of this sort are attacked by intuitive means. When they are more complex, however, it may be necessary to turn to mathematical approaches for analysis, or to simulation wherein a functioning model of the actual system can be observed over many different input conditions. There are systems engineers who feel that simulation should be tried only when a problem cannot be handled analytically. An example of designing, with the help of computer simulation, a materials handling system with waiting-line characteristics has been described by J. F. Sundstrom[11], and it is discussed here. The problem was to devise for a new materials distribution center the most efficient means of moving packages from arriving trucks through counting, repackaging and inspection stages, and on to storage. Provision had to be made for special handling of certain packages needing to be treated more expeditiously than others. In considering this kind of system at the outset, a designer would normally have little difficulty in determining the principal objective to his satisfaction. It would be to insure the smoothest correct flow of material for the least overall cost.

On the basis of preliminary data and experience, a pre-analysis stage will often make it possible to sift through alternative systems possibilities and eliminate the least promising, at least tentatively, while settling on one for methodical study. In this case, for instance, the determination was made that forklift trucks would be less satisfactory overall than an automatic conveyor system. These kinds of decisions need to be reached before simulation begins because the technique of simulation by computer modeling is one of trial and error and is not well suited to optimizing between distinctly different possibilities.

There were seven major steps used in this process of problem solving: problem definition, data collection, model definition, computer programming, debugging and validation, actual computer runs and analysis of output.

Problem definition here refers to the asking of questions concerning the proposed system to help determine what information a simulation can provide, e.g., is the suggested design satisfactory with regard to sequential arrangement of the several stations, the length of the accumulators (accumulating conveyors), etc., or will redesigning make the system more effective? The first proposed design is shown in Fig. 5-5.

Gathering and analyzing the data were among the critical and time-consuming elements of the problem. It was necessary to know the characteristic distribution pattern of arrival rates, peak load periods, operating characteristics of proposed components of the conveyor system, operator times and costs, etc.

After these two steps, model definition and computer programming could begin. Figure 5-6 shows a partial model of the system. Part (a) in the figure represents one of the accumulating conveyors from the original design as a logic flow diagram, and in part (b) the logic diagram has been converted into simulation program blocks.

The simulation program used one of the special languages—IBM's General Purpose Systems Simulator (GPSS) which was specifically designed for modeling queuing systems, and the various GPSS sub-models were joined together in a single model containing about 100 logic

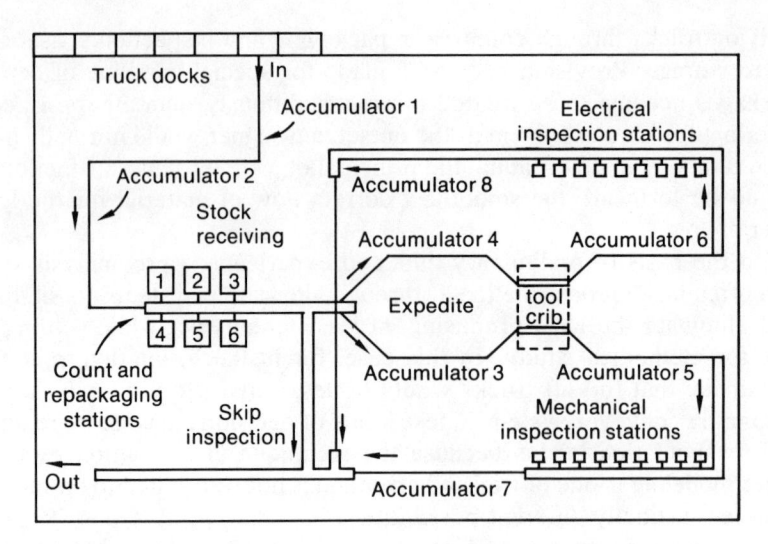

Fig. 5-5. Proposed conveyor system design before application of simulation (after Sundstrom [12]).

blocks. The model was now ready for the next step, debugging, in which coding and simple logical errors were detected and corrected in initial computer runs. As Sundstrom points out, as soon as any output is forthcoming from a computer run on a model one must be concerned with the validity of the results, which at this point cannot be tested physically. There are certain calculations such as those comparing inputs and outputs in the computer simulation which can help point out errors.

The simulation was performed on a medium-sized computer, (IBM 360/40), which was capable of handling a model with 100 logic blocks. About thirty runs were made simulating conveyor operations over a six-month period, to test the effect of various changes in conditions. The outputs consisted of statistics on variables such as queue length and conveyor utilization, and analysis of early runs indicated that the original design would give less than satisfactory results. Subsequent design changes were modeled and then run to see whether they produced improvement. Two accumulators which were shown to be capable of causing bottlenecks were eliminated in the final design selected, and one other was relocated. Another method was utilized to supply tools.

The use of simulation uncovered a great deal of useful information for management before the system was built, and permitted the avoidance of costly trial-and-error experimentation. It also offered an alternative to a

Fig. 5-6. (a) Logic diagram of a functional element of the materials handling system. (b) Simulation program converted from logic diagram in (a). (After Sundstrom[13].)

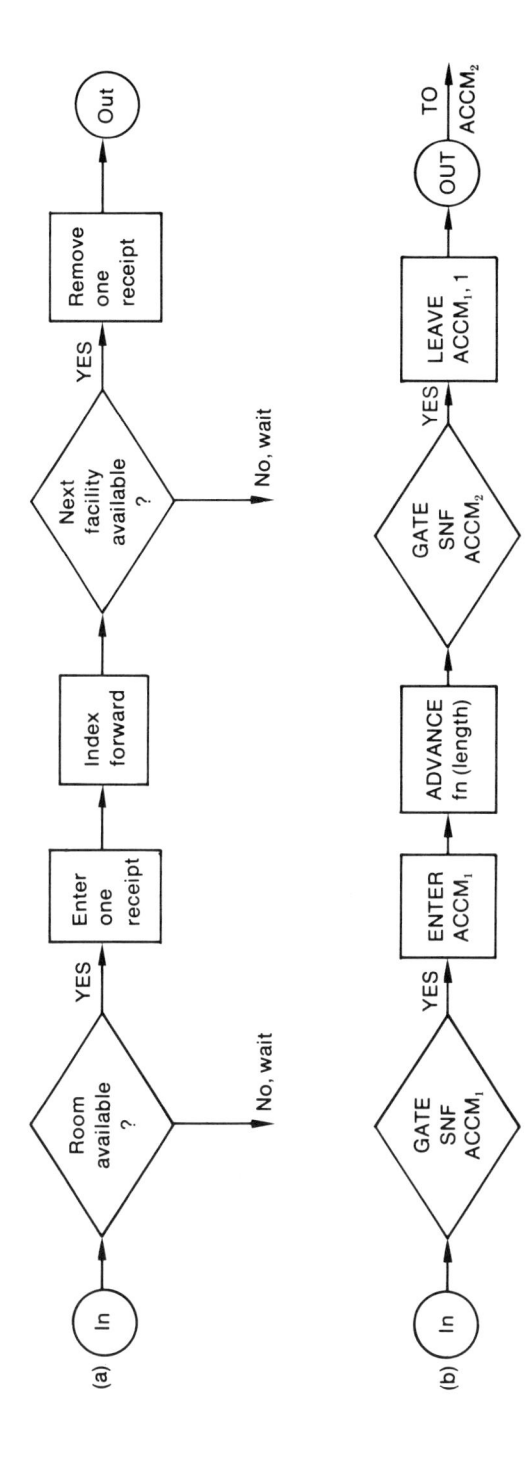

Note: $ACCM_1$ and $ACCM_2$ refer to accumulators 1 and 2; GATE SNF means "hold up packages until conveyor has room," ENTER means place one package on conveyor; ADVANCE fn(length) is the ratio of conveyor length to conveyor speed; and LEAVE means remove package.

formal mathematical approach which can be, at the very least, formidable for many practical problems.

Forms of Systems Problems

The two systems problems discussed in the preceding pages appear, on first inspection, to have little in common either in character and complexity. They are, in fact, different in important respects but there are also similarities. The simulation example involved the flow of material. The goal was to design a facility so that, as packages were serviced through a processor, the waiting time before and on the conveyors would be minimized. In this kind of systems problem, the primary concern presumably is to reduce costs of operation while maintaining service at a satisfactory level. In the information system, providing a needed service to individuals and departments was a basic objective, and keeping the cost of the new system down was also a major consideration. Government officials are ordinarily reluctant to approve new methods of operation which burden the taxpayer additionally. The previous manual information system had had some of the characteristics of an older type materials handling process. (Actually, all complex systems are information processors to some degree.) Information was collected, transmitted and stored in material form, i.e., as paper documents, and at each stage there was a large element of human involvement. The LOGIC system handles information electronically—data can be collected by punched cards, optical character readers or direct-access typewriters; carried by transmission lines and through automatic switching centers; and stored on magnetic tapes, disks, drums or cores. In addition the new system provides for data conversion to more suitable forms for processing, and for automatic calculating in a central processing unit as needed.

Systems engineers and operations researchers find it helpful to group problems in a generic way. Ackoff and Rivett[14] have developed a classification of the form and content of problems. Sometimes the basic task is to keep inventories at an optimum level. Sometimes it is to try to find the best match between a variety of jobs to be performed and the resources on hand to do them. Most individuals continually encounter allocation problems of this sort when they attempt to budget their finances to meet their expenses or arrange their time to satisfy pressing obligations. Government agencies and large corporations and individuals, too, are often required to make routing decisions such as when a location must be found for a new freeway, or a summer trip mapped out. The minimum distance between points is only one of many factors to be taken into account.

There is another type of problem having to do with replacements. When, for instance, parts deteriorate with time, decisions must be made either to purchase new parts at regular intervals or when trouble appears, or to make repairs. Even if the latter option is chosen there comes a time when the cost of constant maintenance, the annoyance, and perhaps the increased danger to people if breakage should occur suggest that some kind of optimum policy ought to be followed. Ordinarily, it is found best to keep the sum of operating and investment costs at a minimum. Businessmen and engineers have known this for a long time, and so some of them can be pardoned for expressing varying degrees of disbelief when they hear some officials proclaim that they are saving money without limiting essential services by ordering substantial "across-the-board" reductions in the purchase of new and necessary equipment.

Again, there are other kinds of problems involving organizations or individuals wishing to achieve objectives to which competing parties are partially or diametrically opposed. The decision of each competitor is affected by the activity or anticipated actions of the other. Corporations place a high priority on planning which takes into account the strategies and tactics of their competitors, and among nations the stakes between adversaries can be terrifyingly high.

Some Techniques for an Approach to Systems Problems

In each of the aforementioned classes of problems, at some point a decision, or decisions, will have to be made. The results of systems analysis are used as an important contribution toward making these decisions. If the interactions are simple and deterministic and expressible in quantifiable terms, the decision, given a limited objective, can usually be reached expeditiously by using good sense aided by common mathematical methods of analysis. In complex systems, too, analytical tools can be useful (and often indispensable) adjuncts to the experience, judgment and synthesizing ability which lie at the root of a successful systems approach. Systems problems have so many parameters it is natural for the scientifically trained person to divide each part into manageable segments which show a recognizable relationship to other sub-problems for which certain mathematical methods have been shown to be effective.

At the beginning of most analyses lies the need to predict what the systems requirements will be. Not only do the inputs tend to fluctuate irregularly over time, but so does the operation of the elements of the system itself. In designing a hospital complex or a post office nobody can be certain what the demand for service will be at any future moment of time. There are countless variables, some of which can be anticipated precisely, some in a general way, and some not at all. Epidemics, natural disasters, strikes, new government health plans or population shifts could each play a part.

Sometimes there is a large body of prior information which seems to bear on the problem, and sometimes there is very little. If the "facts" are to be of any help to the systems planner he must first try to assess their relevance to the objective. He needs to have experience with the best ways of gathering data, and then he must know how to arrange them to give him a useful indication of the likelihood of certain events occurring or not occurring. Statistical analysis and probability theory offer ways to serve him. Probability theory has to do with the mathematical determination of the relative frequencies with which events may be presumed to occur when a given set of conditions is repeated. Some of the important applications of probability theory are those dealing with the gathering, analyzing and interpreting of observed numerical data, i.e., mathematical statistics.

The basic task is how best to take into account the random behavior of physical phenomena. Many natural or contrived events are, of course, readily predictable, but through most of history Nature has been judged as often capricious. In the framework of probability theory, however, it turns out that randomness is not truly capricious; that natural phenomena, considered random, can have their own kinds of regularity which, when recognized, can be incorporated into a design[15]. What, then, do we mean by randomness? According to Parzen[16], a random phenomenon is characterized by the property that its observation under a given set of circumstances does not always lead to the same observed outcome but rather to different outcomes in such a way that there is statistical regularity. (It is only fair to state that no definitions of randomness are universally regarded as satisfactory.) Coin-tossing would be such a phenomenon, as would be the rate of failure of transistors or the flow of traffic on a freeway. If one should toss a perfectly balanced coin a number of times, the sequence of heads or tails appearing would be random—irregular and unpredictable over the short run but regular over the long run in the sense that in any extended sequence of coin tosses heads will be observed to occur about half the time.

The theory of probability is actually the study of randomness and in its many manifestations it provides ways of describing the randomness found in systems and in their inputs and outputs. The distributions of statistical data can be graphed by plotting the number of occurrences of an event in a given interval against the magnitude of the event. For example, the pattern of patient load for individuals needing intensive care in a single hospital ward was found to follow the distribution shown in Fig. 5-7.

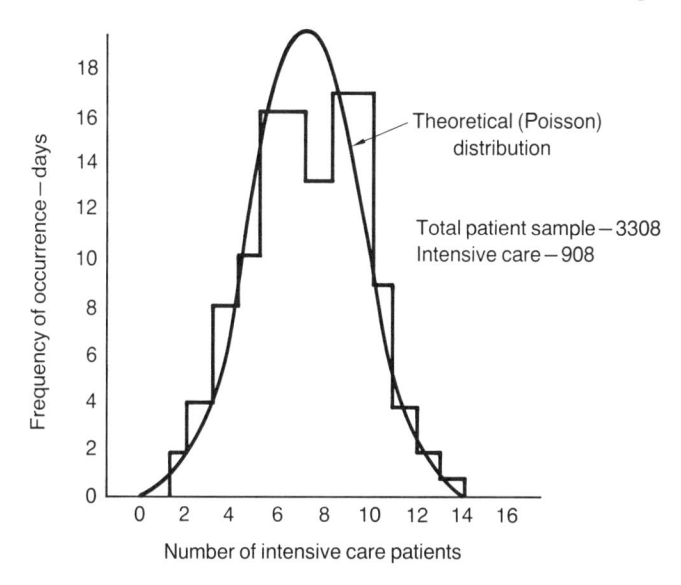

Fig. 5-7. Distribution of population of intensive care patients (after Flagle[17]).

Typical distributions can generally be found from the available data and the special nature of the problem. This one is a close approximation to a theoretical Poisson distribution which characterizes the pattern of irregular customers arriving for service in many real situations. There are other common distributions. For the given example, analysis of probabilistic aspects of patient needs whether for intensive, regular or minimal care could give administrators opportunity to consider what kind of system would be compatible with expected loads. More data over longer periods of time add to the likelihood of random events being taken into account.

Systems problems, by their basic nature rarely offer an easy choice between a "good" and a "bad" course of action. Unless he begins by strictly limiting his objective and essentially ignoring the larger environment, the systems planner is almost sure to encounter incompatible criteria among those he is counting upon to guide him. He then has no alternative if he claims to be rational and objective but to try to seek some kind of compromise solution. The problem, in short, must be *optimized*, i.e., its most desirable solution must be found, given the problem criteria and constraints, and a useful way is to describe it by means of a mathematical formulation and then manipulate the model. Classical mathematics, which is designed for analyzing continuous relations among very few quantifiable variables is not too well suited for treating large systems, so other methods, (some of which may actually help decide what the decisive variables *are*) have been devised or reapplied. Linear programming, dynamic programming and game theory are a few examples of the branches of mathematical research which systems designers have applied or tried to apply to complex problems.

Mathematical programming is used to plan activities for optimum benefits. With the assumption of linear input-output relationships and an objective which can be expressed as a linear function, an optimum solution, given limited resources, can sometimes be obtained through linear programming. (A linear relationship is postulated, for example, in the assumption that it takes four times as long to produce four cars as it does to produce one.) Many common examples of situations in which linear programming can be used for optimizing solutions are given in the literature. One simple illustration is offered here.

Let us assume that a lady's dressmaker has on hand 14 square yards of cotton and 11 square yards of wool. She knows that to make a cocktail dress that sells for $30 she will need 3 square yards of cotton and 1 square yard of wool, while a pants suit selling for $45 will require 1 square yard of cotton and 4 square yards of wool. Her problem is to decide how many of each she should make if she wishes to gain the maximum amount of money from the available material.

We can tabulate the information as follows:

	Pants suits	Cocktail dresses	Available material
Cotton	1	3	14
Wool	4	1	11
Price	45	30	

If we let x represent the number of suits to be made and y, the number of dresses, we then can write an objective function:

$$f = 45x + 30y$$

An optimum solution to the problem is one in which f reaches a maximum value subject to the conditions:

$$x + 3y \leq 14, \quad 4x + y \leq 11,$$

and the additional constraint that neither x nor y are less than zero (i.e., negative). The first inequality signifies that no more than 14 square yards of cotton is available for making suits which use 1 square yard and dresses which use 3 square yards, and the second inequality gives similar information about the wool.

We can now graph a convex polygon whose corner points are:

$$0 = (0,0), \quad P = (2,0), \quad Q = (2,3), \quad R = (1,4), \quad S = (0,4)$$

These are the respective values of x and y which will satisfy the inequalities.

Substituting the x and y coordinates of each of the corner points gives the following values of f for: $f = 45x + 30y$. Then,

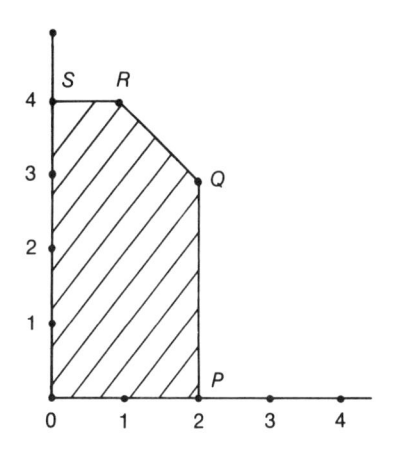

Pt. P, $\quad x = 2 \quad f = 90$
$\qquad\qquad y = 0 \quad$ 2 yd. cotton
$\qquad\qquad\qquad\quad$ 8 yd. wool

Pt. Q, $\quad x = 2 \quad f = 180$
$\qquad\qquad y = 3 \quad$ 11 yd. cotton
$\qquad\qquad\qquad\quad$ 11 yd. wool

Pt. R, $\quad x = 1 \quad f = 165$
$\qquad\qquad y = 4 \quad$ 13 yd. cotton
$\qquad\qquad\qquad\quad$ 8 yd. wool

Pt. S, $\quad x = 0 \quad f = 120$
$\qquad\qquad y = 4 \quad$ 12 yd. cotton
$\qquad\qquad\qquad\quad$ 4 yd. wool

so that it can be seen that the maximum profit will be obtained by making 2 suits and 3 dresses.

Problems with one or two variables can often be solved with paper and pencil, or even instinctively — although instinct is not always reliable — but as the numbers of interactions increase, the computer must be used if solutions are to be obtained in any reasonable time. Providing optimal routing for freight cars, awarding contracts, assigning personnel and allocating electronic equipment to military aircraft are just a few cases where variations of linear programming might be employed. It has had broad use in military planning and control.

However useful, the presumption of linearity is restrictive. Non-linear relationships are far more common in practice, whether we are dealing with animate or inanimate objects. An output, such as the production of a saleable object will almost certainly not double just because the size of the plant is doubled. The non-linear programming methods of analysis which seek to deal with these relationships are ordinarily more difficult than the linear to apply or carry out, although certain equations which are nearly linear may be less manageable than some which are entirely non-linear.

Higher order decision making in real situations is very often sequential. Over a period of time a series of decisions may have to be made, each depending upon the previous one. This being the case it would seem that the individual decisions ought to be made with a view to their future effect, even if it means foregoing what appears to be an immediate gain. Dynamic programming, originated by Richard Bellman[18], is a method of studying multi-stage decision processes under uncertainty on the basis of a policy. The policy is a guide for determining the optimum decision to be made at each stage according to the position of the system at that time, or the conditions currently prevailing. In the case of the plotting of the flight of a space rocket, for instance, even though an initial flight path had been calculated, many difficulties could intervene to send the rocket off its course. Then a program of multi-stage decision making might permit calculating a new trajectory (based on a policy of, say, minimum elapsed time for the total journey) rather than attempting to re-establish the original course. The advantage of limiting the number of variables to be considered by taking one stage at a time is more than just nominal, but some inherent difficulties are not to be overlooked. The assumption that the true state of a system can be determined at any moment, or that the effect of a change in course of action can be accurately assessed is bound to be an idealization for most systems. The numbers of possible control decisions are huge, and although the information storage capacities of large computers are helpful in this regard there are many levels of decision making which are beyond the present reach of the approach.

Systems planners would agree that the randomness of physical events, while often exasperatingly difficult to provide for, at least does not appear to be guided by a cunning adversary. There are systems problems, however, where the contending actions of dedicated and skillful opponents must indeed be anticipated. Thus does the world of social interactions differ from the inanimate physical world.

When business or military strategies are being developed, the decision maker cannot count on his opponent's behaving according to given patterns. Rather, what evolves is a kind of game in which each adversary seeks for the maximum benefit for himself, while keeping constantly in mind that the actions of others, whose goals may be opposed to his, will affect the outcome. John Von Neumann and Oskar Morgenstern offered a mathematical method to try to explain optimum strategic behavior for social situations in their *Theory of Games and Economic Behavior* published in 1944[19]. The theory extended the scope of Von Neumann's earlier work and became the basis for models of rational behavior under competition.

In Part I of *The Man Made World*, the innovative textbook edited by E. E. David, Jr. and J. G. Truxal[20] there appears an interesting example of optimization through game theory in the form of the well-known story, *The Lady or the Tiger* by Frank Stockton.

A young commoner in love with a princess is found out by her father, the king, who decides that an appropriate punishment for the youth's audacity will be to place him in a public arena in front of two closed doors. Behind one of the doors is a ferocious tiger, and behind the other, a beautiful maiden. The young man must open one of the doors and his choice will determine whether he will die a cruel death or win the hand of the maiden. The princess, however, has learned which door leads to the tiger, and from her position in the stands she catches the eye of the youth and motions to the right-hand door. His dilemma then, of course, is to decide whether her stronger impulse is to save him from the tiger, or keep him from marrying someone else. If he should attempt to come to a decision in a logical way he can see that he and the princess each have two alternatives, and, in a sense, it is like a serious game played between them.

To describe the model completely he must decide what will be his payoff from each of four possible combinations of decisions. He needs, then, to place arbitrary values on the several outcomes to make an optimum choice. Almost certainly he would conclude that opening the door to the tiger would have a maximum negative value no matter which door the princess had designated. If, however, he opened the other door his pleasure would be maximum only if the princess had lied, but diminished if he realized her love was strong enough that she would prefer to give him up to another rather than have him come to harm. A matrix can be constructed showing the two possible decisions for each player and the payoffs to the man. His choices are shown in the two horizontal rows, and hers in the vertical columns.

		Princess's Choice	
		A. Maiden's door	B. Tiger's door
Man's choice	C. Door that Princess designates	20	−20
	D. Other door	−20	40

The numbers chosen are, of course, arbitrary, but presumably reasonable in a relative sense.

If he is perfectly convinced of the unselfish purity of the princess's love one might expect him to open the door she pointed to and accept a bittersweet reward. If, as is more common in games of love, war and business, there are doubts about the motives of the other participant in the game, he cannot then be sure what the princess will do. She may be considering the same possibilities as he, and be concerned mostly about her own reactions to his decision.

When he tries to put himself in her place, he soon finds himself entangled in a bewildering succession of guesses as to her intentions on the basis of her anticipations of his intentions as a result of trying to figure her motives, and so on He could then decide just not to think at all and flip a coin to make a choice, but this is not the best way, according to game theory. Without going into the proof, it can be shown that his chances for a favorable decision are improved by doing the following:

(a) Subtract the respective terms of column B from A to form a third column, R, consisting of the entries 40 and −60

	A	B	R
C	20	−20	40
D	−20	40	−60

(b) Eliminate the minus signs in column R

	A	B	R
C	20	−20	40
D	−20	40	60

(c) Reverse the entries in column R

	A	B	R
C	20	−20	60
D	−20	40	40

What he will have done is to weigh the odds for each of the possible moves by the difference between the payoffs for the other move, to maximize his anticipated gain or minimize his anticipated loss. The solution result is that he should pick C, 60/100 of the time and D, 40/100 of the time. Since he has only one choice, his best bet is to make his choice random, but in such a way that C has a $\frac{3}{5}$ probability of being chosen. One way is to take 5 coins out of his pocket, deciding in advance that if 3 or more are heads he will choose the door to which the princess pointed, but if no more than two are heads he will take the other.

There are many kinds of games studied by game theorists, but the problems become very difficult where there are more than two persons involved or where there is a possibility of mutual gain or cooperation in addition to conflict. In some simple situations, game theory has made it possible to show results mathematically that could not have been anticipated intuitively or proven in any other way before the theory was developed. In most cases at the present time, however, it is little more than a first order approximation to reality. A contestant seeking a guide to successful action against an adversary might do as well to refer to Machiavelli's *The Prince*.

The difficulties between game theory and simulation gaming should be made clear. The playing of games to simulate real situations in war or business or other competitive activities has become a widely used research technique. The poet, William Cowper, once wrote, "War's a game which, were their subjects wise, kings would not play at." Among its practitioners it has, indeed, often been treated as a game. In fact, chess, also called "King's game" in an early modification was used as far back as the 17th Century to represent the movement of soldiers through various maneuvers. War gaming went through many additional changes to reflect more accurately the effect of chance events, but its evolution from a training device for officers to a general research method has fairly well coincided with the use of computers and the introduction of the theory of games. The basic idea is to give the researchers relevant facts with regard to competing strengths of forces, possible attack strategies, etc., to allow them to pick a course of action. The action decision is given to the computer to determine the consequences which are then fed back into the next game. The computer allows contestants to deal with much larger data bases than before, and to calculate theoretical outcomes in a minimum time; and game theory can indicate what the rational solutions to certain problems ought to be. Although simulation gaming can be carried on without either, they have become common accessories to studies carried on for executive training or on military supported research programs. However, as guides to action and problem solution under real conditions, the effectiveness of simulation gaming in a broad sense has yet to be proven.

SYSTEMS DESIGN FOR SOCIAL SYSTEMS – BEGINNINGS

The association of the systems approach with the use of mathematical modeling and computers is natural in a technological society but must never obscure the fact that *ad hoc* reasoning, subjective insight, coercion and impulse play major roles in any attempts to attack problems embedded in a socio-political matrix, and always have. *The Republic* of Plato was a rational, if not a scientific, design for a state and there have been many other tries since that time to construct ideal, or perhaps just workable, social systems.

Methods of mathematical modeling, such as those discussed in the preceding paragraphs, are variants of what Boguslaw has called the formalist approach to systems design[21]. When systems planners use them as representations of social reality they follow the paths (albeit with vastly more impressive tools) which were taken by early counterparts in the 18th and 19th Centuries or even before. The invention of the differential calculus in the 17th Century opened the way to the systematic calculation of optima in problems involving simple linear relationships. The new breed of natural philosophers were then convinced that the material world was rational, and that, given time, it would submit to mathematical description. Furthermore some of them felt that social relationships were not inherently different from material relationships in their susceptibility to quantification and measurement.

Sir William Petty was one of the earliest to try to formulate these ideas. His *Political Arithmetick* published in 1690, three years posthumously, was designed to demonstrate "The art of reasoning by figures upon things related to government" – surely, then, a forerunner of behavioral political science. In his preface, Petty stated "Instead of using only comparative and superlative Words, and intellectual Arguments, I have taken the Course (as a Specimen of the Political Arithmetick I have long aimed at) to express myself in terms of Number, Weight, or Measure; to use only Arguments of Sense, and to consider only such causes as have visible Foundations in Nature . . .".

It was no longer enough merely to speculate on the workings of society. The year Petty died, Isaac Newton published the Principia Mathematica, and with it he had apparently fashioned the key to continuous progress through science. For the next 200 years, and particularly in the 19th

Century, the idea of progress—the determination of its nature and how best to realize it—was a central theme in social thinking.

The stresses and upheavals accompanying political and industrial revolution in the last half of the 18th and the first of the 19th Centuries called forth many proposals for rationalizing views of society along the lines of new moralities and the new science, (the word, science, having a broader meaning then than now). Some of the ideas seem misguided and artless to us, but we cannot ignore that the proponents were fundamentally trying to understand and correct the problems of the world. England and France produced more than their share of social technologists some of whom drew upon their backgrounds in science or industry while others were guided merely by a compelling desire to do good by designing a new social order.

We may note, as Boguslaw has pointed out [22], that the basic idea of linear programming can be found in the works of François Quesnay, a court physician for Louis XV, who used a similar method to try to describe the economic system. In 1758 he produced the *Tableau Économique* which attempted to prove that land is the sole source of wealth of a nation, and only the cultivation of the land (in which he included the seas, the rivers and mines) adds to that wealth. A revised version of the work set up an abstract model of a nation, one-half of whose people were producers who created the nation's riches by farming, one-fourth were the proprietors and the rest were the "sterile" class who were occupied in services other than agriculture. His objective was to follow the profit flow through his model, showing that each aspect of the economy affected every other. He believed the effects would demonstrate that when the system attained equilibrium the optimum economic situation would be realized, but naturally the model required some grossly simplifying assumptions.

A noteworthy modern attempt to use mathematical models for a purpose similar to Quesnay's appeared in the 1930s when Professor Wassily Leontief of Harvard University began to publish a series of articles on input-output relationships in the economic system of the United States. These culminated in his book *The Structure of the American Economy: 1919–1929*. [23] Leontief's work is a composite of highly involved mathematical and statistical analyses, but the intent has been to follow the interdependence of all sectors of the economy by presenting an exhaustively detailed picture of inter-sector transactions [24]. Linear programming is closely related to the input-output approach but differs somewhat in that the latter purports to show conditions as they exist rather than to optimize them according to some criterion. However it is possible to develop an optimizing input-output model[25].

In England, Jeremy Bentham, whose adult life spanned the French Revolution and many years after the Napoleonic Wars had, like Quesnay and Adam Smith, his philosophy shaped in the Age of Reason. He was a lawyer and the founder of a political philosophy which came to be known as Utilitarianism. It was his pleasure to believe there was a science of human behavior based on the fact that man will always act to produce for himself a maximum of pleasure and a minimum of pain, and a community will do the same.

The value of a pleasure or pain would be measured as greater or less according to[26]:

(a) its intensity,
(b) its duration,
(c) its certainty or uncertainty,
(d) its propinquity or remoteness,
(e) its fecundity (meaning the chance it has of being followed by sensations of the same kind),
(f) its purity (meaning the chance it has of *not* being followed by sensations of the opposite kind),

and, in the case of a group,

(g) its extent (meaning the number of persons who are affected by it).

He thought it should thus be possible to construct a "felicific calculus" to attain the logical objective for a progressive society, that is, "the greatest happiness for the greatest number."

Bentham's passion was for efficiency and order and through order, reform. His felicific calculus was thoroughly ambiguous and inadequate, but apart from that he was an able critic of institutions. Although his proposed design of an ethical system had little to recommend it, the design he attempted to work out for a more modest project, a new kind of prison system, was a pioneering effort. This was the famous "panopticon," taken from an idea of his brother, Sir Samuel Bentham[27]. It was a five-storied prison which would also serve as a factory, the expected results being "morals reformed, health preserved, industry invigorated, and instruction diffused." The panopticon did not progress beyond the model stage even though Bentham spent many years working it out in elaborate detail, but later prisons would show the influence of his thinking.

It seems clear that the early natural philosophers and the social

Utopians cherished an optimism about the efficacy of the orderly scientific approach for rectifying the dreadful imbalances in society which was not warranted either by the tools they had available or their imperfect understanding of the scientific method and its limitations. There were many attempts to devise social systems other than those to which we have referred. Some were very loosely structured, and at least one, that of Charles Fourier[28], was planned down to the most incredibly fine detail. Some served as the bases for social experimentation; others were never intended to be anything but idealistic fantasy. The attraction and, presumably, the weaknesses of these constructions usually lay in their evading of existing institutional restraints. We must ask if or how the tactics or strategies loosely bound together as the systems approach offer ways to deal with political and social (as well as technical) restraints and thereby provide a greater promise of creating rational systems for living in the technological society.

A CRITICAL LOOK AT THE SYSTEMS APPROACH

The major effective applications of the systems approach heretofore have been to inanimate systems or to systems in which machines and humans interact, the latter in routine ways determined either by technological demands or by system objectives which can be unambiguously stated in economic or mechanistic terms. The resulting systems have often had social utility. Can the approach then also be successfully called upon to engage problems in new spheres of activity where very large numbers of people presumably exercise their free will to create, cooperate, interfere, defy, breed or destroy? The clear answer seems to be, not too well as yet. It does not appear that the methods of contemporary systems planners have thus far brought us much nearer the point of being able to effect conscious, lasting control of major social processes. Any attempt to recreate the world, or a part of it is beset with incredible pitfalls.

There is a tendency among some to consider the physical sciences the most difficult intellectually. Many of the keenest mathematical minds, (apart from the pure mathematicians), and the most ingenious experimenters have chosen to follow professions in the physical and natural sciences. One of the primary reasons for this must be that the scientist who deals with tangible systems often has a degree of control over his experimental conditions, and a sense of fulfillment from having data correspond with prediction that the social scientist can seldom hope to achieve. Difficult is a relative word. Those social scientists who choose to approach their profession with mathematical methods may have the hardest job of all. The interplay of variables in social systems is of such complexity that the problems of analysis become immense. Not only that, there is a fundamental question whether many of the most important variables are even measurable. Their true significance may lie not in what they mean in any scientific sense, but rather if the message they convey is believed or acted upon at all, or by all in the same fashion.

It is important, however, to emphasize that the systems approach is not synonymous with science if we accept the earlier definition of science which was advanced in Chapter 1. The basic understanding which scientists seek through strict control of experimental conditions in relatively isolated systems is less significant in systems analysis and engineering that what Sackman[29] calls experimental realism. This involves real-time measurement, i.e., the data are gathered and processed as they are generated and then utilized for on-going control of current (or estimates of future) systems performance. The emphasis is on credibility rather than idealization. The systems of interest, in fact all systems of which humans are a part, must be considered open rather than closed, they vary with time in non-deterministic ways, and the models used are necessarily tentative.

All of this is not the same as saying that, because the extemporaneous actions of individual humans defy prediction, no regularity can be detected in the behavior of large groups of people over a period of time. It is not hard to estimate, on the average, how many people will get involved in fatal car crashes on a three-day holiday weekend in the United States. Still to build a practical system for reducing automobile accidents requires the assessment of competing priorities each involving non-linear variables of frustrating complexity. A natural tendency is to try to cope with complexity by dividing the system into subsystems which will yield to independent solutions, and then combining the sub-solutions into a systems synthesis. Sub-optimization has a definite utility for certain situations but inherently it does not conform to the holistic idea that the total system is given meaning by the ways in which its components interact with each other when they are focussing on a given objective. Machol [30] has stated as a principle that optimization of each subsystem independently does not in general lead to system optimization and may actually worsen the overall system. We are, indeed, dealing with a classic question—how to know to what extent a system may be subdivided for examination without destroying the life or meaning of the whole.

Since no theoretical standard for man has yet been devised, as S. P. R. Charter has pointed out[31], mathematical languages cannot propose formulations that apply to man, the individual, in any fundamentally useful way. For those systems which are to be designed to meet large-scale social needs without enmassing individual man, the successes of the systems approach at this stage of development, then, are more likely to be modest than not. This judgment, however, does not deny its merits. There is a real advantage to be gained by fostering a broad, unified outlook, which is at least as important as fragmenting a problem for analytical treatment. Many of those working toward a systems approach to problem resolution have discovered that this unified outlook can spring from the encouragement of more fruitful interpersonal and intergroup communication when individuals from various areas of interest work toward a commonly perceived objective. Then too, there may be additional insights to be gained by developing models such as those for the self-regulating systems to be discussed in the next chapter.

While no one can guarantee that a rationally planned decision will always have a happier outcome than one based on irrationality or whim, not many would choose to accept irrationality as a preferable guide to action. And if the question is asked: who is to decide what is ultimately rational and we cannot give a certain answer, there is surely help to be found in some of the humanistic imperatives whenever they can be separated from traditional pieties.

For all its pragmatic intent, the systems approach may need to be for a long time to come as much a basis for attempts to understand some of the principles obtaining in complex processes, as it is a means of planning for social action. Still, as Churchman states[32], it is not a bad idea.

REFERENCES

1. Gilmore, John S., Ryan, John J. and Gould, William S., *Defense Systems Resources in Civil Sector*, U.S.A.C. & D.A., Denver Research Institute, July 1967.
2. Ellis, David O. and Ludwig, Fred J., *Systems Philosophy*, Prentice-Hall, Inc. Englewood Cliffs, N.J., 1962, p. 4.
3. Hall, Arthur D., *A Methodology for Systems Engineering*, D. Van Nostrand Co., Inc., Princeton, N.J., 1962, pp. 70–71.
4. Boulding, Kenneth E., "General Systems Theory – The Skeleton of Science," *Management Science*, April 1956, p. 197.
5. Kirby, Richard S., Withington, S., Darling, A. B., and Kilgour, F., *Engineering in History*, McGraw-Hill, N.Y., 1956, p. 2.
6. Schlager, K. J., "Systems Engineering – Key to Modern Development," *I.R.E. Trans., Prof. Gr'p. Eng'r. M'gement.*, 3, 1956, pp. 64–65.

7. Hall, A. D., op. cit., pp. 23–58.
8. Page, Thornton, "The Nature of Operations Research and Its Beginnings," in *New Methods of Thought and Procedure*, edited by Zwicky, F. and Wilson, A. G., Springer-Verlag, N.Y., 1967, p. 8.
9. Schuchman, Abe, *Scientific Decision Making in Business*, Holt, Rinehart & Winston, Inc., N.Y., 1963, pp. 3, 4.
10. Hall, A. D., op. cit., p. 149.
11. Sundstrom, J. F., "Simulation—Tool for Solving Materials Handling Problems," *Automation*, January 1969, pp. 90–93.
12. Sundstrom, J. F., ibid., p. 92.
13. Sundstrom, J. F., ibid., p. 93.
14. Ackoff, Russell L. and Rivett, Patrick, *A Manager's Guide to Operations Research*, John Wiley & Sons, N.Y., 1963, pp. 34–56.
15. Brieman, Leo, "The Kinds of Randomness," *Science and Technology*, December 1968, p. 35.
16. Parzen Emanuel, *Modern Probability Theory and Its Applications*, J. Wiley & Sons, N.Y., 1960.
17. Flagle, Charles, "A Decade of Operations Research in Health," *New Methods of Thought and Procedure*, edited by Zwicky, F. and Wilson, A. G., Springer-Verlag, N.Y., 1967, p. 49.
18. Bellman, Richard, "Dynamic Programming," *Science*, Vol. 153, July 1, 1966, p. 134.
19. Von Neumann, J. and Morganstern, O., *Theory of Games and Economic Behavior*, Princeton University Press, Princeton, 1944.
20. David, E. E. and Truxal, J. G. (eds.), *The Man Made World, Part I*, McGraw-Hill, Inc., N.Y., 1968, pp. 52–56.
21. Boguslaw, Robert, *The New Utopians*, Prentice-Hall, Inc., Englewood Cliffs, N.J., Spectrum Edition, 1965, p. 47.
22. Boguslaw, R., ibid., p. 49.
23. Leontief, W. W., "*The Structure of the American Economy, 1919–1929*, Harvard University Press, Cambridge, 1941.
24. Leontief, W. W., "The Structure of the U.S. Economy," *Scientific American*, Vol. 212, No. 4, April 1965, pp. 25–35.
25. Miernyk, William H., *The Elements of Input-Output Analysis*, Random House, N.Y., 1965, p. 62.
26. Bowditch, John and Ramsland, Clement (eds.), *Voices of the Industrial Revolution*, University of Michigan Press, 1961, pp. 43–44.
27. Bell, Daniel, *The End of Ideology*, rev. ed., The Free Press, N.Y., p. 228.
28. Manual, Frank E., *The Prophets of Paris*, Harvard University Press, Cambridge, 1962, pp. 197–248.
29. Sackman, Harold, *Computers, System Science and Evolving Society*, John Wiley & Sons, Inc., N.Y., 1967, p. 517.
30. Machol, Robert E., *System Engineering Handbook*, McGraw-Hill, Inc., N.Y., 1965, pp. 1–8.
31. Charter, S. P. R., *Man on Earth*, Angel Island Publications, Inc., 1st ed., 1962, pp. 138–39.
32. Churchman, C. West, *The Systems Approach*, Dell Publishing Co., 1968, p. 232.

6

Cybernetics — Control and Communication

"But let your communication be Yea, yea; Nay, nay; for whatsoever is more than these cometh of evil."

MATTHEW 5:37

This historical treatment of the man-machine relationship in Chapter 2 introduced the notion that computers, after World War II, promised to transform the face of industry by making possible the development of complex technological processes which require no human supervision. The practical consequences of self-regulating industrial processes are manifest, but many people have also been intrigued by questions about other kinds of self-regulating processes and what common principles relate them all. It has seemed to some that man's design of electronic systems to give them the capability for simulating certain characteristics of animate behavior may foretell ways to develop other brainlike processes and further extend the human capacity for controlling his physical and social environment. By brainlike processes we mean those which acquire and use information adaptively.

Design of inanimate systems so that they are self-regulating does not, of course, need to depend on computers. There are many earlier examples of simple self-regulating systems. One such device was described in 16th Century documents. A profiled shaft attached to a rotating millstone gave a rocking motion to an inclined trough feeding grain to the stones. If the flow of stream water increased, this increased the speed of rotation of the millstone, the rocking of the trough, and also, the amount of grain fed to the stones. However, if too much grain were fed, the motion of the stones would be slowed and the grain supply reduced[1]. A more familiar example is the system which has already been shown in Fig. 2-6 — James Watt's centrifugal governor. It may be seen that as the shaft rotates more rapidly the fly-balls are caused to fly upward and outward. The arms are attached to a fuel-controlling valve, and when they rise the fuel flow is reduced. This in turn reduces the speed of the engine and, thereby, the shaft. The subsequent falling of the arms again increases the fuel supply, and the net result is that the velocity of the shaft remains within mechanically set limits.

Modern control systems can be much more complex than this. What is particularly important is the fact that processes may be maintained in a state of continuous control, even though the input information or energy is not completely predictable. The late Norbert Wiener gave theoretical form to this idea, based in part on his realization that the statistical behavior of systems must be considered when they are too complex to operate according to a precise cause-and-effect relationship. In so doing he established a way of looking at problems which has had influence in many fields. Wiener wrote a book in 1948 called *Cybernetics* [2] which was a distillation of nearly a life-time of insightful explorations into the applications of mathematics to organic and inorganic systems. The word, cybernetics, is rooted in the Greek, *kybernetes*, and the related Latin, *gubernator*, meaning governor or steersman. Analogies are immediately apparent. The steersman controls the course of a ship; and a human governor, the ship of state, if you will. Wiener assumed that he had coined the word, cybernetics, but did not learn until later that it had been used by André Ampère, the French physicist, in writing about the science of government in the early 19th Century.

In the period during and immediately after World War II, Wiener was one of a group of scientists and philosophers in the Boston area who met informally for many discussions touching upon a problem which they commonly perceived, i.e., how to comprehend the nature of organization and control through effective communication for all processes with identifiable goals. In 1947, Warren S. McCulloch and Walter H. Pitts, two members of the group, were working on a device to help the blind to read. This was a most difficult design problem involving an automatic scanning device which would be required to make distinctions between characters of different sizes and shapes. When passed over a line of print it would produce a varying audio signal. One day Dr. von Bonin, an anatomist, saw a schematic diagram of the device, but took it to be a representation of the fourth layer of the visual cortex of the human brain. This appeared, to the members of the group, to be a striking illustration of the underlying unity between control mechanisms in the physical and biological sciences.

FEEDBACK CONTROL

A fundamental goal of cybernetics is to determine why natural systems of control seem to be more versatile and adaptive in certain ways than those which man has designed and constructed. McKay[3] has pointed out that an efficient control system requires a means of *measuring* the degree to which a course of action varies from the path necessary to achieve a specific goal; of *calculating* the amount of correction required to return a straying process to the desired course; and of *selecting* the appropriate response to the calculation which will insure that the process will once again close in on the goal.

To revert to the idea of the steersman again, the way in which a man steers his automobile along a two-lane road is an example of this type of control system. His goal, as long as he keeps moving forward and there are no vehicles in front that he wishes to pass, will be to avoid either crossing the center line or running off the edge of the road. If he is a careful driver he will be continuously measuring by observation the position of the front of his car with respect to the two reference limits. There will also be a process of mental "computation," based on previous experience, as a result of which his nervous system transmits signals to the muscles providing the information that the response of a sharp or ever-so-slight turn of the wheel is required to keep the car on the proper side of the road. Craik[4] suggested that the human operator behaves essentially as an intermittent correction servomechanism, first estimating, then correcting, then re-estimating the residual error, and so on.

The basic mechanism operating here is that of *negative feedback* which was briefly described in the previous chapter. It exists when a signal representing the degree of departure from a norm is fed back into the system, and, by working *against* the initial stimulus, modifies the response of the controller to keep the system operating normally. The controller is never at rest unless the difference signal is zero.

Figure 6-1 represents the characteristics of the closed-loop feedback system of the car and the driver.

The act of driving a car is never, of course, quite this simple. In addition to his concern with the primary goal of keeping on the road, the careful driver will more than likely wish to drive within the speed limit, which may include a lower as well as an upper value along some stretches of

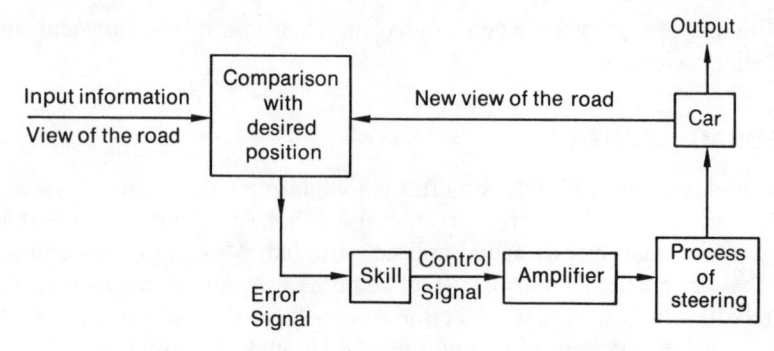

Fig. 6-1. Closed-loop feedback system — car and driver.

road. These two separate goals (and there will be more) are related to each other, because if the speed limit is exceeded it makes that much more difficult the task of staying on a road that curves.

This man-machine system may have basic inadequacies, as attested to by the gruesome annual highway accident statistics in the United States. There may be unexpected inputs such as the darting of a dog across the path of the vehicle. Suddenly there is the new goal of avoiding the dog and this may not be compatible with the first. If the driver's reaction response is dulled by weariness, age or alcohol there is a lag in the system. A characteristic of feedback systems is that lags may cause the output to oscillate widely instead of reducing departures from the norm. Whatever ways can be found to reduce time lags or the number of steps in a feedback loop to the least possible can help maintain control.

An ideal control system would be able to deal with all unpredictable variables. Granting the great versatility of the human organism, a man-machine system will always fall short of the ideal. No matter how akin he feels to his car, man cannot be expected to respond to adverse conditions quickly enough to maintain perfect control over his vehicle at all times. The design of a safer and more efficient way of transporting people, in which man's function as controller is replaced by an electronic system is quite conceivable technologically, but only at a sacrifice in individual freedom of movement, as has already been pointed out. The sociological and political factors involved are not apt to permit this to happen in the near future, at least not as a total transportation system.

There are scientists who insist that from a scientific or technological point of view, astronauts have been unnecessary appendages to the space exploration program — that all of the important observations could be made by instruments and at a much lower cost. Still, the high drama of the first moon landing showed that the vicarious identification of the average citizen with man-in-space is a powerful intangible, and it seems likely that support for the space program would have been markedly less than it has, were only machines involved.

It is the essence of a cybernetic system that each stage of the control process could be made automatic in principle. In the past two decades, the mechanization of the calculation stage by means of the computer has greatly increased the range of utility of man-made systems.

COMMUNICATION

Cybernetics may be described as the study of brainlike processes or equilibrium-seeking processes and in these kinds of terms it is subject to very broad and often disparate interpretations. It overlaps such fields as general systems theory, theory of automata, semantics, information theory, logic and invades important areas of the physical, natural and social sciences. This has led to real controversy. There are those who prefer to use cybernetics narrowly as a synonym for the applied science of computerized engineering control systems, and others who feel cyberneticians have cast their net so broadly that they have forfeited any right to claim for it the status of a separate discipline at all. Apter[5] takes the position that the significance of cybernetics rests in its promise of overcoming the barrier between the systems of physics and chemistry and those of biology. This barrier exists as a result of the inability of the biologist to isolate elements of his system and still retain in them the characteristic of vitality which is his primary concern. Wiener's recognition of an isomorphism between goal-seeking machines and living organisms resulted in a formalization by means of which the behavior of any part of a system could be thought of in terms of the whole system. According to Apter, cybernetics observes systems at the level of organization and information. As we shall see, the consideration of information as a measure of organization is a core concept. Wiener's original definition of cybernetics as control and communication in the animal and the machine embraces these notions since effective control of a complex system can be accomplished with desired accuracy only if communication of information between its elements occurs with a certain amount of fidelity.

Communication has been defined by Dr. R. B. Lindsey as the whole process involving stimulus and discriminatory response in a system. In the man-machine example previously discussed, when the front of the automobile strayed too far from the desired course, the eye communicated an error signal to the brain. The result was a stimulus to the muscles, producing, quickly, a movement of the front wheels to the degree called for by the character of the original signal.

A precise description of factors required for the efficient communication of information is incorporated in the mathematics of information theory much of the development of which can be traced to the works of Wiener and, more especially, Claude Shannon. In 1948 Shannon[6] published a seminal paper which examined the question of how accurately symbols of information can be transmitted in spite of disturbing effects, or "noise." Before this time, earlier workers had also considered measures of information[7] but Shannon's theory, which treated information as a quantity related to the statistical nature of the signal, received particular attention. Not only were communications engineers ready to spend a great deal of time and effort working with the concepts in the hope they would provide practical guides to design of better telecommunications systems, but physicists, psychologists and researchers in many other fields thought they could detect clues to help them achieve a better understanding of their subject[8]. Not all of the early promise of information theory has been fulfilled as far as the application to working systems goes, but cyberneticians still look to it for relevant insights.

If the communication process is analyzed, it is seen that somehow, sensors in the receiving element must be affected. This can be accomplished by certain signals capable of producing some kind of impression. Perhaps the most general understandable statement about it might be: If A is transformed by B, then A contains information about B. Thus an automobile passing along a muddy street leaves a tire mark that provides some information about the tire—its make or perhaps its state of wear. Information as used in this way can be related to a variety of systems. It is possible to say that hormones carry information through the human system telling one part of the body to give a certain response; or that a recently uncovered potsherd contains for the archeologist valuable information about an ancient civilization. In the first case, the information passing through a physical system would seem to be more amenable to some kind of quantitative treatment than the second which is affected by richly subjective factors.

George[9] has described a communication system as being essentially composed of five basic elements: (a) a signal source, (b) a process of coding, (c) a transmission medium, (d) a receiver and (e) a decoder. Perhaps the simplest communication system that comes to mind (although indeed it can be very complex) is a conversation between two persons. The five elements are readily identified as follows. The source of the signal is the speaker. His thoughts are given expression as vocal sounds—a form of code transmitted through the medium of air as mechanically vibrating waves. The ear of the listener receives these vibrations and his mind decodes them into some kind of information. If there is someone else trying to speak to him at the same time or there are other distractions there is a good chance that the audible signal will not be received exactly as intended by the speaker.

Since communication by speaking or writing is a uniquely human skill the thoughts being discussed here will have significance to other topics of interest to us touching on the relation between human and non-human systems—such as artificial intelligence. There is some value in looking briefly at the historical development of human codes of communication to see, from the evidence available, how written language evolved in a historically conceivable period of time into the immensely subtle instrument it is today.

In the early Mediterranean civilizations, recorded communication was in the form of pictorial representation, where each idea was shown in simplified form by a picture or character. This kind of communication was seriously limited and does not qualify at all as true writing. In time the pictures were improved by eliminating some extraneous detail, and sometimes by combining signs. There is, of course, a certain basic efficiency in transmitting information by simple notations. Even today, a nomad of the Arabian desert might inscribe on a rock a circle under a vertical line to represent a bucket hanging from a rope, showing that water may be found nearby. In our own more highly developed civilization the marking ⊢ on a highway sign has an immediate significance to an experienced driver.

The early Sumerians were probably the first to keep records in this fashion. Theirs was a theocracy and to keep track of livestock belonging to the priests, they made impressions on a clay tablet of a picture of a sheep followed by several dots or circles. It was probably sometime about the 4th millennium B.C. that the hugely important step to phonetic writing was taken. This gives a direct relation between aural and written language. We make relatively few sounds that are markedly different from each other when we speak, and consequently only a few symbols are needed to represent them. One cannot be sure why it was the Sumerians who first developed the use of characters to designate specific sounds of speech, but it most likely relates to the fact that theirs was the one agglutinative language in the Middle East, i.e., most of the root words are monosyllabic with one or two audible consonants. As phonetic writing evolved, so did new relations of one sign to another in terms of order and form.

Virtue lies in keeping communication taut and spare by eliminating unneeded words or phrases. On the other hand it is often desirable to use more signs or rules than are absolutely necessary to guard against misinterpretation or the garbling effects of "noise." We call this using *redundancy*. There is considerable redundancy in the English language which makes it possible for a reader to grasp the meaning of a message even though part of the statement is unclear.

If one should change the last letter of every other word in a well known sentence, a reader would probably have little trouble determining how it should have appeared. "The quicd brown fol jumps ovet the lazm dog," is a sentence that will be recognized by most even in this flawed form. A less familiar sentence disguised, say, by replacing all the vowels by z, still contains enough redundancy that it can be made out by the average person. Thus, "Yzz shzzld bz zblz tz mzkz szmz sznsz zf thzs."

Redundancy is one of the remarkable features of the human brain itself, and one of the goals of designing artificial control systems is to simulate this ability. The brain can continue to function rationally in spite of considerable damage, such as occurs in the aging process. Millions of nerve cells in the average brain die each year and are not replaced, but there appear to be numerous alternative networks capable of reacting to the same instructions. Von Neumann[10] developed a mathematical theory to show that this redundancy in the brain produces a degree of reliability not dependent on malfunctioning of the individual components. Artificial control systems similarly designed and constructed would be that much closer to the ideal of maintenance-free self-regulation.

TELECOMMUNICATION AND THE TWO-STATE CODE

The foregoing paragraphs have attempted to show that control engineering and communication engineering are indissolubly related. We shall now consider certain historical developments in the field of telecommunications, because it was from insights provided by the technology of telecommunications systems that the theory of information had important beginnings.

A designer of a telegraphic, telephonic or television network would like to achieve the greatest possible efficiency of transmitting a message along with the least chance of error. To do this it is necessary to know how much information is needed. An approach to this determination goes back to a fundamental principle of communication studies which is that information can be conveyed in what is called a two-state code. A familiar nominal two-state code is the Morse Code, used for many years in telegraphy. In the form in which it was finally perfected by S. F. B. Morse in 1838, letters of the alphabet were represented by combinations of dots and dashes. The dots and dashes were produced by electric current pulses of short and long duration respectively; and separated by a space which was simply an absence of current. Morse must have had efficiency of message transmission in mind, since his assignment of arrangements of dots and dashes to the various letters of the alphabet bore a direct relation to their frequencies of occurrence in the English language. He determined the relative frequencies by the simple but crude method of counting the number of types in each compartment of a printer's box. The most commonly occurring letter in English, *E*, was represented by the shortest symbol, (dot), and so on. Given financial support by the U.S. Congress in 1843 for the purpose of laying an underground telegraph cable between Washington, D.C. and Baltimore, he ran into trouble. He soon found that if the dots and dashes were sent too fast the pulses overlapped and could not properly be distinguished by the receiver. Additional interference was caused by slight, stray electric currents in the circuit. Others who followed Morse experimented with different combinations of intensity or direction of current flow. It was possible to find ways of giving the sender a greater choice of signal and therefore for him to send more letters in a comparable period of time; but at the cost of more insistent noise.

This is of prime practical concern. The number of users of telecommunications circuits grows enormously each year. Even now, engineers and scientists are looking to the time, fast approaching, when telephone cables will be unable to accommodate the expanding demand for services, and they are considering alternative transmission media such as laser beams. The engineering of new systems requires evaluating the relative advantages of many different kinds of communication signals; and determining the information capacity of a system.

INFORMATION AND CHOICE

Shannon's work, anticipated to some extent by R. V. L. Hartley back in 1928 showed that for a receiver, a message carries information only when it contains what is unexpected. In this context information is used in a special sense which should not be confused with "meaning." In the way that we are referring to the word, we may compare two sentences, (one very meaningful and the other containing nothing but gibberish), and determine that both have the same amount of information. This "technical information" relates not so much to what you do say in a message as to what you could possibly say. It is, in fact, a measure of *freedom of choice*. Let us try to clarify this somewhat with an example.

In the first place, as we have discussed, in the two-state codes, (+ —, yes–no, etc.), message symbols are conveniently expressed with the binary digits, 0 and 1. Suppose we have a game whose object is to determine, by asking the least possible number of questions, which of four identical closed cardboard boxes lined up in a row contains a dollar bill. The money has been placed there by a person whom we will call Sender, and only he knows its location. The other player in the game, called Receiver, can ask any question about the location which will require a yes or no response, and Sender must answer truthfully. Let us assume that in the following sketch the money is in the box at the extreme right.

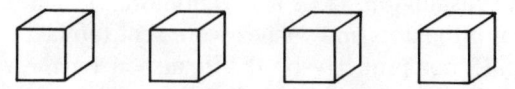

Receiver may be the gambling type and elect to try to identify the correct box with only one guess by asking directly, say: "Is it in the box which is second from the left?" He could be lucky, but the odds are 3 to 1 against it. In fact, he can be sure the odds will be balanced evenly only after he has made two wrong guesses. In other words, approaching the game in this way he has no certainty of making an identification of the correct box with fewer than three questions. On the average, it would be better for him to approach the game more systematically.

He could guarantee a positive identification after two questions by asking, first: "Is the dollar in one of the two boxes to the left?" Answer: no. Then he needs to ask only one more question such as "Is it in the second box from the right?" to determine the location of the dollar. It should then be apparent that it requires only two successive yes or no answers to select one item from four; or, as we could say, the sentence "The dollar is in the box to the extreme right" contains two bits of information where we define *bit* as the number of binary digits required to convey the message.

The same game could be played with a set of eight boxes and, clearly, only three bits of information must be signaled for Receiver to make the identification, i.e., it takes only three yes or no instructions. For sixteen boxes, four bits would be required to select one sign positively. This can be put in the form of a simple mathematical expression

$$2^4 = 16, \text{ or for the first two cases } 2^3 = 8 \text{ and } 2^2 = 4$$

These expressions can also be written in the logarithmic form:

$$\log_2 16 = 4, \ \log_2 8 = 3, \ \log_2 4 = 2$$

Thus it is possible to say that a message can carry information in a quantitative sense. The amount of information, (I), as an indication of the uncertainty of the receiver is given by the logarithm to the base 2 (for a two-state code) of the number of choices:

$$I \text{ (in bits)} = \log_2 N, \text{ where } N = \text{number of choices}$$

Had there been only one box the uncertainty would be zero.

UNLIKE ALTERNATIVES

The foregoing example considers the transmission of signs which have an equally likely chance of occurring, but this is not the prevailing situation with most sources of information. In written English or in other natural languages we can observe that certain letters or combinations of letters occur more frequently than others; or at least have a greater probability of occurring. If we were to count the frequency of occurrence of the letter E in English writing it would be found to constitute about 13.1% of all letters observed. On the other hand, W would be only about 1.5% of all letters. Not only that, but many pairs or longer combinations of letters in English text also produce probabilities greater than zero. *PH* and *TH* are not uncommon. The probability of Q occurring in a text consisting of natural words is small, but it is no different from that for QU. Thus the U in the QU combination adds nothing at all to the information contained in the message.

In his book, *What is Cybernetics*, G. T. Guilbaud[11] has written of an experiment conducted with frequency statistics of Latin letters. He cut letters out of a Latin text, taking into account groups of three. Then pulling them out in a random fashion he obtained the following:

IBUS CENT IPITIA VETIS IPSE CUM VIVIS SE
ACETITI DEDENTUR

To one who has not studied the language, this might easily appear to be an acceptable Latin sentence, and in fact the three underlined words are Latin. No purpose would be served in making too many claims for this kind of experiment, but it does suggest the possibility of developing a statistical arrangement permitting a machine to produce a fair approximation of a text in a given language.

In situations where the probabilities of occurrence of various signs in a message are unequal the equation for the amount of information carried is somewhat more complicated than the one given above. If, for example, there are three choices whose probabilities are, respectively, p_1, p_2, and p_3 then the average amount of information is given by the equation:

$$I = -(p_1 \log_2 p_1 + p_2 \log_2 p_2 + p_3 \log_2 p_3)$$

INFORMATION AND ENTROPY

Wiener and others [12, 13] have written about the potential of concepts based on these ideas for predicting the operation of a system where we can never completely know all of the operative factors, but only something about their statistical distribution.

A factor of uncertainty must always be considered. In sending a message, a human will inevitably exercise a choice as to what he will say next. With choice on the part of the sender there is a corresponding uncertainty on the part of the receiver. We can often deal more effectively with what can be measured, and information theory has provided a means of measuring this uncertainty, in terms of a quality called *entropy*.

The equation for the amount of information given above has the same form as equations representing entropy in statistical mechanics. This fact has led many scholars to probe into the relationship between information theory and thermodynamics. Although it is true that entropy as it is used in information theory applies to problems of a more general nature than the physical problems of thermodynamics, nevertheless some of the questions that have been raised are among the more intriguing in the philosophy of science and are worth examining in a discussion of cybernetics.

The concept of entropy was first defined by a mathematical expression of the Second Law of Thermodynamics, one of the fundamental laws of science. In a physical sense it concerns systems whose infinitesimally small particles are not all at the same temperature. In each fractional increment of time, a very small quantity of heat (related to the average temperature of the system) will change places until the individual temperatures are equalized. Entropy is a measure of this change. Another way of stating the same concept would be that "discrete systems can only proceed to a state of increased disorder," or (equating entropy with disorder) "entropy can never decrease with passing time."

There are ways to try and express these ideas in more familiar terms. One would be to consider the system composed of a hot cup of coffee in a cold room. We know from our experience that in time the cup of coffee will come down to room temperature (which in turn would be raised imperceptibly if the room lost no heat to its surroundings). There will have occurred in the system an increase in entropy which corresponds to a decrease in available energy. That is to say, the energy which was available in the form of heat at the outset for some useful purpose, such as warming a hand that might hold the cup, can no longer be used when the coffee has reached room temperature.

Another statement that can be made about the system, relevant to information, is that an increase in entropy means a decrease in *order*, where order is associated with *knowledge*. We have more knowledge that can be made useful to us about the above system in the first instance where we know the coffee is at a higher temperature than the air, than in the second instance where all matter in the system is at the same temperature.

The more detailed the knowledge of a system the less will be the uncertainty about it, and the entropy will also be less. With more specific references to communications, consider a message source such as a writer, radio transmitter or traffic light which on a given occasion can convey any one of many possible messages. The greater the number of messages available to the sender, assuming like probabilities and independence of sequence among them, the greater will be the uncertainty of the receiver before the message is received and the greater the information, considered as a measure. A message which is one of ten possible messages carries less information than a message which is one out of a possible hundred. Thus a thermometer with 100 divisions can more accurately indicate the correct temperature to an observer than one with ten divisions.

Apparently then, we can put forward the proposition that information tends to bring order out of chaos, i.e., information bears a direct relation to negative entropy.

There is a familiar parlor game in which the participants sit about in a circle, while the initiator whispers a message into the ear of the first person on his right who in turn will whisper his version of the original message to the one on his right and so on around the circle. The message that finally returns to the first person will normally retain only superficial resemblance to his original statement, illustrating the large entropic tendency in human communication.

How far to carry the analogy between information and negative entropy is a matter of disagreement among scientists and philosophers but there are certain implications worth considering.

Many writers have suggested that negative entropy is a measure of organization not only in the inorganic physical systems from which thermodynamic theory was developed, but that it also relates, at the very least as a valuable metaphor, to biological or social systems. In *The Human Use of Human Beings*, Wiener has written: "But while the universe as a whole, if indeed there is a whole universe, tends to run down, there are local enclaves whose direction seems opposed to that of the universe at large and in which there is a limited and temporary tendency for organization to increase. Life finds its home in some of these enclaves."[14] In speaking of the running down of the universe, he is referring to the "heat death" which 19th Century sociologists and scientists foresaw as the eventual gloomy prospect predicted by the Second Law. It is the state of complete randomness wherein all components of the universe will have come to the same temperature after countless millennia of entropy-producing reactions, leaving no further available energy for life-sustaining processes.

Life itself seems to decrease entropy for a time, but only if it is considered as a system closed to the total environment. The organization of molecules that takes place as the fertilized cell grows to be a human body is of profound complexity. At birth the human has a huge available potential but at death maximum entropy has exerted itself. The waxing and waning of civilizations which some historians write about may follow a pattern which has often been linked to the biological life span of an individual. This is an analogy which must be stretched gently to avoid exceeding the breaking point. Nevertheless as human organizations grow in size and complexity with time there does seem to be an optimum point beyond which the normal processes of communication cannot be sustained, leading to an inevitable degeneration of the system.

THE RELEVANCE OF CYBERNETICS

The ideas introduced in this chapter are intended to be more provocative than definitive. A sampling of the literature will demonstrate what we have already suggested, that cybernetics means quite different things to different people. To many engineers and scientists it is a mathematical means of developing the theory and design of computers and other "brainlike" machines; and among them are some who feel that neither cybernetics nor information theory have fulfilled their original promise in attacking specific technical problems. To others it is less a body of facts or equations than a way of thinking to demonstrate the unity existing between disciplines. They see cybernetics pointing to fundamental meanings we already sense, and offering a theoretical justification for describing the limits of—and probing the still-mysterious boundary between—living and non-living systems. This latter point of view is the one favored here.

In concluding this chapter, we shall attempt to suggest something more of the possibilities as well as the limitations of addressing cybernetics concepts to human-related systems. Some of the most fruitful work to date has been in neurological studies. An early concern of the original cybernetics group was with miscarriages of physical function because of physiological anomalies in the nervous system. The development of prosthetic devices for lost limbs has been a direct outcome of this concern. A technical program carried on jointly by the Massachusetts Institute of Technology, the Harvard Medical School, the Massachusetts General Hospital and the American Mutual Insurance Alliance, to which Norbert Wiener contributed his services as consultant shortly before his death, has resulted in the construction of an artificial arm. Designed for the needs of patients with above elbow amputations it differs from the ordinary prosthesis by using electronic control techniques to amplify pulses provided by commands from the amputee's brain.

On a more general level, Broadbent[15] has pointed out that there are three fundamental ways to study the nervous system: (a) physiologically, by detecting electrical and chemical events in an operating system, (b) behavioristically, by observing man or animal from outside, and (c) through mechanical and mathematical models for processes like those carried on by nervous systems. Cybernetics is a rich source of models and offers a unifying framework for these three approaches, each of which employs the common language of information processing.

In terms of human organizations the application of cybernetics has been more tentative, although there appear to be many efforts to construct partial economic or political system models[16, 17, 18]. Beer[19] has created persuasive cybernetic models for management control of industrial processes.

Feedback as a social process was exemplified at least as far back as the

18th Century in a political parable written by William Townsend[20]. Townsend told of a Juan Fernandez who landed a few goats on a remote island. They multiplied greatly and were eventually discovered as a food supply for British privateers preying on Spanish shipping. The Spaniards then landed a male and a female dog on the island and they, too, soon multiplied, feeding on the goats. The natural self-regulating control system thus established kept the number of goats down. Townsend's intent was to show that the poor could only be kept at work through hunger, and that the Poor Laws upset the natural balance between labor and laborer by removing the fear of starvation.

This was an early example of the idea of natural processes as inherently equilibrium-seeking, a concept which has had a powerful influence on important social theorists in the 19th and 20th Centuries. We must conclude that it is also structurally related to cybernetics. Cynthia Eagle Russett has illuminated this notion in her book, *The Concept of Equilibrium in American Social Thought*[21]. She has noted that equilibrium is variously understood as an empirical concept describing closed systems in which entropy is at a maximum or in which the vector sum of the totality of forces acting on an object is zero; as homeostasis; or perhaps as an acceptable but not readily quantifiable definition of normality in any system. Nor are these the only definitions which have been offered.

Living systems exist in a state of dynamic equilibrium wherein conditions are constantly changing but relatively stable, and non-living systems tend to a static equilibrium. When social scientists adopted the equilibrium model they tended to look to either static or dynamic stability as a preferred state depending on their philosophical preconceptions.

Auguste Comte believed there was a grand social unity or equilibrium maintained by a tight interdependence because the division of labor fostered specialization to the extent that the specialized elements could not survive alone. He was, however, less reluctant to accept the possibility of continuous changes in adaptation to new situations than was Herbert Spencer, ever the champion of the *status quo*. Spencer saw a stable equilibrium, despite all social meddling, as the desired and inevitable culmination of human progress, but was nonplussed when reminded by the chemist, Tyndall, that this would be equivalent to the cessation of life[22]. In his static picture this seemed, indeed, to be true.

The American, Lester Frank Ward, attempted a synthesis of social statics and dynamics. In essence he viewed the social structure as being in an imperfect equilibrium, flexible enough to permit the society itself to make controlling adjustments in the name of necessary reform. Vilfredo

Pareto was one of the most influential expositors of a deductive scientific view of society in this century. He framed a mathematical theory which demonstrated conditions necessary for an economic equilibrium, and then attempted to apply the principles to a theory of society as a whole. While his science was on a firmer basis than that of the 19th Century theorists, his assumption of societal equilibrium as a normal state is not now accepted without reservations. Admittedly he made no attempt to establish the theory as anything but a crude approximation to reality. It is interesting to note that each of the men mentioned began their careers in mathematics or a science or engineering-related field. Comte was a mathematician, Spencer, a railway engineer, Ward, a paleobotanist, and Pareto did his early work in applied mechanics.

Can we assume for all dynamic systems that there exists a stable, healthy state from which any departure is abnormal and will therefore be opposed by equilibrating tendencies? Not every social scientist will accept this assumption as readily for social systems as will natural scientists for biological systems. However, the metaphor has a continuing appeal and it is not likely to be submerged by any school of thought which refuses to choose between the relative merits of a stable or unstable society or to search for what constitutes stability. If there is such a thing as a value-free analysis for social systems, of course, there might seem to be no point in setting goals for a goal-seeking process.

Students of government continue to look for understanding of political processes from the standpoint of cybernetics. E. S. Savas has written about an equivalence between government and cybernetics using the municipal government of New York City as an example[23]. He considers that one of the basic attributes of the feedback-control system is the goal-setting which nominally falls under the aegis of the mayor's planning and decision making group. We have already noted that one encounters many and conflicting goals in governmental processes and at least some of them will be hard to articulate. Nevertheless, conditions will exist which seem patently wrong, and the business of the municipal government is to reduce the disparities with what it conceives as the ideal set of conditions. As for the dynamic characteristics of the governing process, there is a poor match between the political time constant (i.e., the length of term of elected officials), and the urban system time constants which do not reduce easily to four year spans. The elected official must remain sensitive to problems posed by initiating programs which would reach culmination when he may no longer be in office or which take so long to complete that the conditions they were directed toward have changed in a basic way. There is of course a more or less permanent civil service structure to provide continuity but there can be serious difficulties coordinating a temporary with an entrenched system. The mayor must rely for feedback on an information system with many channels each with special virtues and defects. If he is active he uses his own eyes and ears as an effective but limited channel. His staff, the Press, and special-interest groups are all part of the feedback process but there are greater or less degrees of distortion or lag from each. Civil disorder and the election process are two sources of strong information signals. The first is a signal of failure in the system and will probably trigger security mechanisms which limit the mayor's options to cope with the problem in a larger sense. The ballot, which produces only one bit of information, is too infrequent to aid the feedback process very much and it may also signal that the mayor has no options left. Savas also suggests that feed-forward (anticipatory) control has definite advantages over feedback control for predictable problems, but the inadequacy of predictive models for social processes limits its effectiveness.

The communications aspect of cybernetics in political systems has also been carefully examined by Karl W. Deutsch in his book, *The Nerves of Government* [24]. He recognizes that the responsiveness of a government or other organization to perceived needs expressed as messages from a subgroup is a cybernetics concept indicating the probability of a "favorable" response in a relevant time. Favorableness or inadequacy would be measured by certain empirical characteristics of the system. Empirical studies of human organizations have sometimes led to new insights which have overturned longstanding articles of faith. Jay Forrester[25] speaks of the "counterintuitive" behavior of complex systems which arises out of the intricate feedback mechanisms involved. His work has led him to believe that many solutions which have been designed to provide job training and low-cost housing for the urban poor have, over the long term, actually caused more rapid decay of neighborhoods and increased relative unemployment. But if intuitive approaches have their serious shortcomings, it is no less true that the difficulties of matching an empirical or mathematical model to a full-scale natural system remain immense. To the student of social processes this is a cross to be borne as well as a challenge.

REFERENCES

1. Klir, Jiri and Valach, Miroslav, *Cybernetic Modeling*, translated by W. A. Ainsworth, Illiffe Books, Ltd., London 1967, p. 18.

2. Wiener, Norbert, *Cybernetics*, M.I.T. Press, Cambridge, 2nd ed., 1961, p. 22.
3. McKay, Donald, "What is Cybernetics?" *Discovery*, Vol. XXIII, No. 10, October 1962.
4. Craik, Kenneth J. W., "Theory of the Human Operator in Control Systems," *British Journal Psychology*, 38 (1947), pp. 56–61. Reprinted in *Cybernetics*, edited by C. R. Evans and A. D. J. Robertson, University Park Press, Baltimore, 1968, pp. 120–21.
5. Apter, Michael J., *Cybernetics and Development*, Pergamon Press, Ltd., London, 1966, pp. 3-6.
6. Shannon, Claude, "The Mathematical Theory of Communication," *Bell System Technical Journal*, 27, 379, 623, 1948.
7. McKay, Donald M., *Information, Mechanism and Meaning*, M.I.T. Press, Cambridge, 1969, pp. 4–6.
8. Gilbert, E. N., "Information Theory After 18 Years," *Science*, Vol. 152, April 15, 1966, p. 320.
9. George, Frank H., *The Brain as a Computer*, Pergamon Press, 1962.
10. Von Neumann, J., *The Computer and the Brain*, New Haven, Yale University Press, 1958.
11. Guilbaud, G. T., *What is Cybernetics?*, Criterion Books (trans.) 1959.
12. Wiener, Norbert, op.cit.
13. Beer, Stafford, *Cybernetics and Management*, John Wiley and Sons, N.Y., 1964.
14. Wiener, Norbert, *The Human Use of Human Beings*, Doubleday & Co., Garden City, N.Y., 1954, p. 12.
15. Broadbent, D. E., "Information Processing in the Nervous System," *Science*, October 22, 1965, p. 457.
16. Bauer, Raymond A., (ed.), *Social Indicators*, M.I.T. Press, Cambridge, 1967.
17. Easton, David, *A Systems Analysis of Political Life*, John Wiley & Sons, N.Y., 1965.
18. Kuhn, Alfred, *The Study of Society*, Irwin & Dorsey Press, Homewood, Ill., 1963.
19. Beer, Stafford, *Decision and Control*, John Wiley & Sons, N.Y., 1966.
20. Polanyi, Karl, *The Great Transformation*, Beacon Press, Boston, 1957, p. 113.
21. Russett, Cynthia E., *The Concept of Equilibrium in American Social Thought*, Yale University Press, New Haven, 1966.
22. Russett, Cynthia E., Ibid., p. 22.
23. Savas, E. S., "Cybernetics in City Hall," *Science*, Vol. 168, May 29, 1970, pp. 1066–1071.
24. Deutsch, Karl W., *The Nerves of Government*, The Free Press, N.Y., 1966, p. ix.
25. Forrester, Jay, *Urban Dynamics*, M.I.T. Press, Cambridge, 1969.

7

Man's Mind and The Computer

"The swiftest thing is mind, for it speeds everywhere."
THALES, *Apothegm*

Man has always had his demons and his angels. More often than not they have been distortions or idealizations of himself. In the folklore these embodiments have helped to give shape to his fear of the unknown; but as science has provided new ways of accounting for the phenomena which we explained by our myths and legends, the shadowy beings have largely receded into impersonality, seldom referred to except in tales for children.

There is, however, one class of folk figure which has persisted, grown out of man's apprehension about the potential for ill in his own creative ingenuity. The modern manifestation is the machine which becomes endowed with an intelligence superior to that of its creator, Man, and eventually turns upon and destroys him. A more optimistic version of the same projection would be that the creation is satisfied to remain man's gifted slave. To use the term, Folk Figure, is not to suggest that only the unschooled retain fears about the powers of the super-intelligent machine. In fact the concept is most often the product of highly trained minds. Their message is commonly a metaphorical warning that Man may be in danger of surrendering a critical measure of his independence to machines which relieve him of the need to perform his physical and mental labor; but the idea of the machine as master is also sometimes proposed and accepted as a literal possibility for some future time.

The objective of this chapter will be to try to interpret a part of the work that is being done toward developing machines, called "intelligent" by some; and to trace the similarities and the differences which appear, at this stage, to exist between these machines and the living brain. Such an attempt must touch upon several fields of scholarship. No one seems to know if the more fruitful programs will be those which try to use the increased understanding of the workings of the brain to devise models for the design of new machines; or those which seek to find clues about the ways people think, using aspects of the science of computers as analogues. Most likely the various research efforts from both directions will continue to reinforce each other as Man seeks to fathom that most elusive of mysteries, "How does a human think?"

THE CREATION OF INTELLIGENT BEINGS—EARLY CONCERNS

The consideration of the human organism as a machine and the endowment of inanimate machines with human characteristics are complementary views. We have already indicated that comparing man with artificial devices became common among scientific philosophers in the period roughly coinciding with that when elaborate mechanical clocks and toy automata were being perfected. Those philosophers, Descartes for example, were not quite ready to deny the presence of the soul in the "human machine," but their ideas were far removed from mystical creations such as the clay Golem of Jewish Folklore which was said, at the end of the 16th Century, to have had the breath of life blown into it by the Cabalist Rabbi Low of Prague. The Golem was still a medieval rather than a Rationalist conception.

In the 19th Century the catholicity of interests which had characterized the earlier natural philosophers had already begun to give way to a widening division between the humanists and the scientists and technologists— a division which sometimes found expression in various literary works in which Man's mechanical creations either overwhelmed him or at least threatened to dehumanize him. Mary Wollstonecraft Shelley's *Frankenstein*, published in 1818, has become the archetype of these stories even though Frankenstein's monster was fashioned of flesh and blood. *Frankenstein* was written on a holiday in Switzerland when members of the holiday party, which included Lord Byron, each agreed to try to write a tale of the supernatural. Whether Mrs. Shelley's work was intended primarily as an entertainment, or as a chilling reminder that Man tinkers with Nature at his own peril is difficult to say, but the latter motive has frequently been ascribed to it and to subsequent stories of the genre written by others.

Another closely related theme favored by certain writers of both the

19th and 20th Centuries is the one which points to the heedless employment of technology as being insidious and inevitably ruinous. Not the first, but one of the more persuasive voices sounding the alarm was that of Samuel Butler, the Victorian novelist. In 1872 he published his satire *Erewhon* as an expansion of an earlier essay, "Darwin Among the Machines." He wrote of an utopian land beyond high mountains where the machine had been abolished. One of his characters says, grimly, "There is no security against the ultimate development of mechanical consciousness in the fact of machines possessing little consciousness now. Reflect upon the extraordinary advances which machines have made during the last few hundred years and note how slowly the animal and vegetable Kingdoms are advancing. The more highly organized machines are creatures not so much of yesterday, as of the last five minutes.... Do not let me be understood as living in fear of any actually existing machine.... The present machines are to the future as the early Saurians to man; what I fear is the extraordinary rapidity with which they are becoming something very different to what they are at present.... Is it not plain that the machines are gaining ground on us when we reflect on the increasing number of those who are bound down to them as slaves, and of those who devote their whole souls to the advancement of the mechanical kingdom?"

Expressions of this kind of sentiment have, if anything, grown in passion and conviction in recent generations. Some are not meant to be judged on the basis of scientific credibility. When, for example, Stephen Vincent Benet, writes in his poem "Nightmare Number Three"[1] of a man-eating concrete mixer and telephones which strangle their users in a maze of wires, one is not inclined to take it very seriously in any literal sense although it is possible to think of the telephone as a communications instrument which has had profound psychological and sociological effects on its users.

There are, however, products of the imagination which are not only impressive from an artistic point of view, but which have such technical virtuosity and verisimilitude that it is easy to accept them on any terms. Such a creation was the prize-winning film *2001—A Space Odyssey* by Arthur C. Clarke and Stanley Kubrick which was one of the most popular motion pictures of the late 1960s, particularly among younger people. The computer versus man conflict has rarely been more dramatically stated than when the "electronic brain" HAL and the last surviving human protagonist on the spaceship bound for Jupiter engage in deadly combat. The human wins, but in the denouement he is obliterated in the vast cosmos.

These dramatic expressions of foreboding have naturally been less common among scientists and engineers than among humanists and artists, but not a few of those who have made contributions to computer science and technology have come forward with cautionary reminders about the yet unrealized potential for evil, as well as good, which may accompany the development of "intelligent machines."

Norbert Wiener, in his book, *Cybernetics*[2] recalled W. W. Jacob's old story "The Monkey's Paw" in which an elderly English couple have just been shown a dried-up monkey's paw by a friend recently returned from India. The friend tells them that the paw has been endowed by a holy man with the power of granting three wishes to demonstrate the stupidity of defying fate. He says that the granting of wishes has brought only disaster to the previous owners and tosses the paw into the fireplace. After he leaves, the old man takes it out of the fire, hesitates a moment, and then asks for two hundred pounds. There is a knock on the door almost immediately, and when it is opened an official of his son's company is there to announce that the son has just been killed in an accident but that the company is prepared to pay two hundred pounds as a death benefit. The official leaves, and the old man, clutching the paw cries out, "I wish my son would come back to life." There is another knock on the door. The old man rushes to it — looks out and staggers back in horror. His third and last wish is that the figure standing there should go away.

Wiener's message here is not so much that the computer is about to become a malevolent, thinking foe of man, nor is it even to be construed as an argument for a fatalistic attitude toward life, but he does wish to warn against the dangers of entrusting crucial decisions to machines, especially before we have learned to ask the right questions.

Is it possible to demonstrate that there are processes in which the computer and the human mind appear to function in similar ways? Statements are sometimes seen which say in effect that computers are just very fast calculators. We do not wish to accept this as a complete statement of fact but it does permit us to begin by examining the fascinating analogies which have been observed between computing machines and human beings with very special mental powers.

MENTAL PRODIGIES AND IDIOT SAVANTS

Numerous historical records exist of individuals who have been capable of performing the most extraordinary feats of mental acuity. Many of these rare persons whose gifts have been especially noticed and displayed have been arithmetical prodigies with the facility for being able to answer correctly and with astonishing speed, problems involving very large numbers. It seems fair to conclude that the phrase "machine-like rapidity" applies very well to the earliest calculating wonders for whom any reliable information is available (that would be in the early part of the 18th Century) because they could surely compete with the few calculating machines of the time. Since then, there have been a number of studies made, of greater or less scholarly value, which allow us to gain a degree of insight into the mental processes of these unusual people. Interesting accounts of some of these cases can be found in books written by W. W. R. Ball[3] and Fred Barlow[4].

There are characteristics which many of the arithmetical prodigies have had in common. For one thing, as Bernstein has noted[5] there need be no relation between the subject's understanding of the abstractions of higher mathematics and the ability to calculate swiftly, although these talents are sometimes combined in one person. A goodly number of the cases studied, have shown the individuals (who were sometimes called "calculating boys") to possess a very limited understanding of mathematical principles.

Zerah Colburn (1804-1840) was one of these[6]. When he was six years old his father, a Vermont farmer, overheard him repeat the product of several numbers. This was surprising, since Zerah had had no such formal instruction. The father shortly afterwards took him to Montpelier, then later to Boston where he gave several exhibitions of his prowess. His early efforts consisted of mental multiplications of moderate difficulty and extraction of square and cube roots. When he was eight, he was taken to England and demonstrated his abilities at many public meetings attended by interested and sometimes academically distinguished examiners. He was asked one time to raise 8 to the 16th power and it took him only a few seconds to give the correct answer, which is 281,474,976,710,656. Five seconds was all he required to provide the cube root of 413,993,348,677; and two seconds to answer 112,640 to the question "How many times would a coach wheel 12 feet in circumference turn in 256 miles?"

Money was raised for his education among those who had become interested in him, and he attended school first in France and then in England. In a very short period of time after he had stopped exhibiting his calculating powers, however, the ability left him and it never returned. For the most part he was unenlightening when asked to explain how he managed to arrive at his answers although in his memoirs, written later, he tried to do so. It is said that when he was a child he twisted his body in a kind of "St. Vitus dance" as he calculated but it is unlikely that this exercise was any more than the lack of inhibition often seen in the young.

Even more difficult to understand are the documented cases of individuals who are to all appearances mentally retarded, yet who are able to calculate in a similar rapid fashion. Barlow quotes Kraitchik of the University of Brussels who was present at an examination of one Oscar Verhaeghe in 1946 when the latter was twenty years old. Verhaeghe was described as having the mental age of a very young child, and yet was able to give the 59th power of 2 in 30 seconds and the square of 888,888,888,888,888 in 40 seconds[7]. The phrase "idiot savants" is sometimes applied to persons of limited mentality who display this special characteristic, although from the standpoint of more accurate psychological classification they would presumably not be able to function in this way if they were much below the intelligence level of a moron.

In contrast to these two cases there are other well-known instances of mental prodigies who retained their gifts throughout their life-times, and several whose brilliance was manifested in other fields as well. The career of George Parker Bidder was very similar to Zerah Colburn's at first[8]. He was born the son of a stonemason in Devonshire in 1806, just two years after the birth of Colburn. Bidder, too, was essentially self-taught and he also was exhibited about the country before he had reached the age of nine; in fact later he engaged in a public contest with Colburn. His father, however, was eventually persuaded to have him enter the University of Edinburgh, where he graduated and then entered the civil engineering profession. He was a highly successful engineer, and seems to have employed his undiminished powers for mental computation usefully throughout his career, particularly as a parliamentary witness. As an illustration of Bidder's talent, a story is cited of an incident that occurred shortly before he died. A friend, speaking of the newly discovered properties of light, noted that 36,918 waves of red light per inch traveling at 190,000 miles per second give the impression of red, and wondered how huge a number of waves must strike the eye each second to have it "see red." Bidder's response was immediate, "You need not work it out, the number will be 444,433,651,200,000."[9] He appears to have been one of the first of the mental prodigies to have made a successful effort to gain some insight into his own calculating processes. He did not work with figures as individual symbols, but envisioned larger numbers in so many groups of smaller numbers. Thus he said one time, "If I am asked the product of, say, 89 by 73, the answer (6497) comes immediately into my mind. I multiply 80 by 70, 80 by 3, 9 by 70 and 9 by 3." Stated in this way, it is apparent that the memorization of multiplication tables was a considerable help. He had also committed to memory various useful facts such as the number of seconds in a year, ounces and pounds in a ton, etc. His auditory sense appears to have been more useful than his visual in recalling numbers from his memory, since he found it more difficult and time-consuming to answer a question submitted on a piece of paper than one that was read to him. Some other calculators have seemed to be possessed of a more strongly developed visual sense.

There appears to be every indication that to reckon swiftly requires an ability to count, an exceptional memory for numbers, and an understanding of their properties. Bidder also felt that the registering of the intermediate results was important. Carl Friedrich Gauss who was himself a mathematical genius as well as a rapid reckoner felt that memory and calculating ability were both requirements for feats of mental arithmetic such as we have been discussing, but that the two characteristics had no necessary relation to each other.

It is surely true that many possessors of prodigious memories have not been particularly interested in numbers, or at least have been able to recall facts of all kinds with equal ease. There are numerous accounts of famous persons who have had this ability. Thomas Babington Macaulay was said to be able to repeat, word for word pages of hundreds of books after having read them only once. General Jan Christian Smuts, the South African stateman, memorized a complete library of 5000 volumes, and Dante, Pascal and Dr. Samuel Johnson were all reputed to have forgotten very little they had read or heard. Rarely, however, have any of these unusual people, famous or otherwise, been studied by trained observers over any extended period of time. One important exception has been discussed in a book about an extraordinary mnemonist by the distinguished physician and psychologist, Professor A. R. Luria[10].

The word "mnemonics" comes from a Greek word meaning, to remember, and refers to various associative techniques for aiding the memory. The mnemonist, S., whom Luria studied over a period of thirty years, had come to Luria's psychology laboratory initially at the behest of his employer, a newspaper editor who had been impressed by his seeming ability to retain everything he had been told without taking notes. Luria found that this was the case—he could find no limits to the memory of S., who could repeat tables of words or numbers directly as given him or in reverse order, and even after months or even years had elapsed. His explanation was that he continued to "see" material that had been shown him. However, there were additional peculiarities which appeared to make his situation more complex than ordinary cases of "photographic memory" which are not so uncommon. If his attention was interrupted by a chance remark of the examiner or any other sudden noise, he saw a blur on the series disrupting his clear mental vision so that he had to shift the table, in his mind, away from the blur. Luria suggests that he had a remarkably developed *synesthetic* sense. Sounds induced in him not only sensations of light and color, but also taste and touch. Each person's voice and each word he had pronounced conveyed to S. a complex mélange of sense impressions. The synesthetic sense, which will cause some people to remark, for instance, that some musical tones are warm, was a fundamental feature of his psychic existence and surely had much to do with his gifts of memory. It was not the only factor. Words, or numbers, caused him to perceive certain images – 3, for some reason, was a gloomy person in his mind, and America would be Uncle Sam. When words or numbers were strung together, he could also distribute his images in a certain order as though along a path on which he was walking. He always asked that words in a series be read slowly and distinctly, for he needed a nominal time to formulate his images. He would occasionally omit a word from a series, but only because he had difficulty in perceiving it. A word he did not know might sometimes appear obscured or poorly lighted along his route.

One example will suggest the degree of S.'s virtuosity [11]. In December of 1937 he was read the first four lines of *The Divine Comedy* in Italian, a language he did not know. Fifteen years later, without warning, he was asked to repeat the stanzas and he did so, using the exact stress and pronunciation.

Most studies of human memory have shown that it deteriorates quickly with time, but with S. the problem was one of needing to forget images which remained with him so strongly that they interfered with the ones he needed at the moment. Strangely, he found that if he wrote facts down he could then forget them, knowing that he had no need to remember.

And yet with this unique, almost overwhelming talent there was a strange, but perhaps inevitable, lack in his psychic makeup. He had limited ability to convert his multitude of images into a coherent body of ideas. To move from the particular to the general, to grasp the beauty of meaning in great poetry, seemed beyond him. His mental pictures, tumbling one after the other in helter-skelter profusion, left no room for profound patterns. He remained a remarkable curiosity.

We have dwelt on the subject of the calculating prodigies because of their peculiar gift, which seems basically to be the ability to perform rapid, orderly sequences of arithmetical operations on numbers taken from memory. It is also the kind of thing that the high-speed digital computer does best. This is not at all the same as saying that the mind of an idiot savant works just as a computer does. For one thing, nobody fundamentally knows how a person, even a simple-minded one, thinks, but since the more subtle thought processes—the mnemonics of S. or the conceptions of an artist—are quite beyond the reach of present understanding it makes some sense to draw preliminary comparisons between the less complex processes.

WHAT IS INTELLIGENCE?

The question "Can a computer think?" is asked again and again and no completely satisfactory answer has ever been given nor is one likely to appear soon except in the unlikely circumstance that a universally accepted definition of the word, think, is found. It is surely possible that thinking is not one process, but many, or at least a process so varied that different manifestations of it little resemble one another.

How can we ever be sure what another person thinks, or if they think at all? We can observe what they do and then try to decide whether their actions were the result of thought, but of course our own conclusions will usually be drawn through an intensely subjective filter of what we ourselves sense and "know."

Still there should be no objection to the statement that a person who has performed a mental calculation and produced a correct answer has been thinking. The purest kind of thought is deductive reasoning, proceeding in logical, finite steps which are susceptible to analysis. We can fairly assume that the mind of someone who has solved a problem has been active in this way. This is not all of mathematical problem solving however. For more complex cases, certainly, guesswork, trial and error and the inductive leap must come into play. Attempts to analyze these kinds of processes, called *heuristic*, become very tortuous.

Computers are machines that can be observed to provide correct answers to problems. They can do so in a progressive, step-by-step fashion according to carefully designed rules; or they can also do some form of hunting out answers through imaginative programming with feedback. In these narrowly restricted functional terms, then, we can say that computers think, but, as indicated at the beginning of the section, saying so does not really mean very much.

The answer to the question of whether we are justified in speaking of "intelligent" machines comes to about the same thing. Intelligence, too, has different meanings. For many years school children have, as the result of taking certain kinds of tests, been quantitatively classified by their I.Q. (Intelligence Quotient). The I.Q. is certainly a valid measure of something, but psychologists and educators who understand the significance of these tests have become increasingly aware that they are so much affected by cultural attitudes, language and reading habits and the like that they should not be regarded as a measure of intelligence, except approximately, and then with a great deal of qualification. Bellman[12] has defined intelligence as "the capacity to solve, in some degree, an adaptive control problem," thus suggesting the variability and flexibility which characterizes human thought processes. He also agrees, however, that the separation between instinct and intelligence is hazy.

Arthur C. Clarke[13] in discussing the various definitions of thinking has proposed that in the final analysis each definer is, in effect, saying, "Thinking is what *I* do." The famous British logician and mathematician, Alan M. Turing[14] attempted a more precise determination of thinking and intelligent behavior in logical terms. His idea was to devise a theoretical logic machine — in fact, a computer — which, in principle, can accomplish any task that any other machine can. It was a concept powerful yet simple. Directing it to the possibility of a machine's imitating man, Turing conceived a game which consists of questions to be directed by an interrogator to a man and a woman in another room. The object of the game is to determine by written answers which respondent is the man. The woman is permitted, as a rule of the game, to assist the interrogator, and she can write "I am the woman," but the man, who is not bound to truth and whose role is to try to cause misidentification can do the same. If, now, the programmed machine takes over the part of the man, the underlying question would be to determine whether the interrogator could correctly distinguish between the two respondents more often in the second case than in the first. This is the question Turing used to replace the original one, "Can machines think?" Turing wrote about this imitation game some two decades ago and he felt that in a relatively few years machines could be built and programmed which would play the game so well that the average person as interrogator could after five minutes make a correct identification no more than seventy percent of the time. There will be some who will be dissatisfied with the Turing test of thinking because it is strictly operational. While it has the advantage of screening out such possible irrelevancies as the greater charm of the woman or the brusqueness of the man, in the long run these characteristics may be most relevant in arriving at the truest assessment of reality. That is to say the "thinking" may still be just a plaster imitation of the actual process of which humans are capable.

The only conclusion to be derived from suggestions in the foregoing paragraphs is that if we carefully and restrictively specify what we mean by intelligence it is acceptable to say that a non-biological system can demonstrate intelligent behavior. The phrase "artificial intelligence" is used to refer to the various kinds of studies being carried on to develop machines (or more often, computer programs to simulate machines) capable of modes of behavior which we have heretofore considered expressly human. Artificial intelligence is not a very satisfactory term, since apparently no unambiguous distinction can be made between what is indeed natural or artificial in the realm of intelligence, but it is generally used for want of a better one. Programs of research follow many avenues in the search for ways of imitating the behavior of the living brain. Some of these approaches appear to have run afoul of choking complexity after promising beginnings, but the primary goals are so challenging that we can expect little diminishing of effort on the part of skilled probers in the fields of biochemistry, neurophysiology, computer sciences, linguistics, psychology, electronics technology and philosophy.

THE BRAIN AND NEURAL NETWORKS

The picture of the brain as a machine for processing data does not appeal to everyone, but it does have scientific value especially if we bear in mind first, that the data may be cast in many forms, and second, that the brain as an adaptive system may react differently at different times to the same inputs. Some of the most interesting neurophysiological research starts from this general premise.

There are obstacles to carrying on research on the human brain. Although it does not always have to be treated as an unapproachable black box, most of the knowledge about it must come from external indications except for post-mortem investigations. The opportunities to observe a healthy living human brain directly are limited. Furthermore, it has become apparent that results from experiments on lower animals do not always shed clear light on human characteristics. Still, much has been learned about the structure of the nervous system as well as about the way it seems to operate.

The nerve cell, or neuron, is its basic unit. The neuron consists of a nucleus with surrounding cytoplasm from which many threadlike connections, called *dendrites* extend. The function of the dendrites is to receive impulses from other cells. There is also a much longer connection, seldom more than one, called the *axon*, which conducts the outgoing impulse away from the neuron. (*See* Fig. 7-1.)

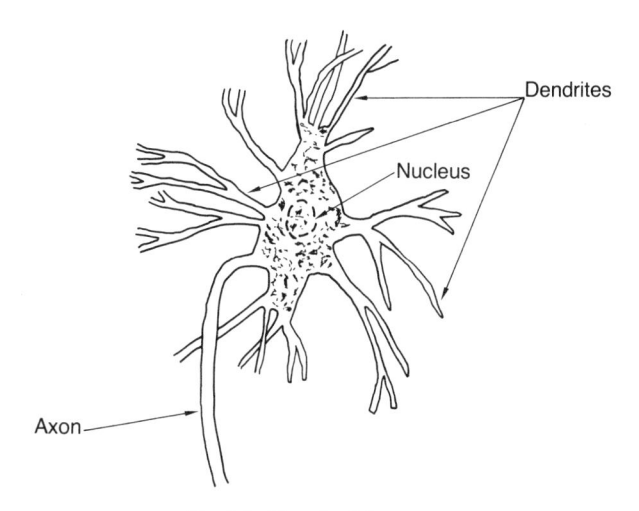

Fig. 7-1. Sketch of the neuron.

For an impulse to be propagated from one neuron to another they must come into a functional relation at a junction called the *synapse*. An axon tends to narrow down and branch repeatedly when approaching a neuron to which it is passing on an impulse. The delicate branches end in knoblike structures either on the surface of the cell body or its dendrites, but in such a way that the junction appears to be structurally discontinuous. All evidence suggests that each neuron is a separate anatomic unit, and that direct transmission of impulses does not continue through the synapses. The theory now most generally accepted proposes that transmission occurs instead by means of an electrochemical mechanism releasing specific chemical substances which regenerate an impulse on the other side of the synaptic junction[15]. It is important to note, however, that not every impulse will stimulate a response in the receiving nerve cell. Indeed some of the axon fibers will tend to inhibit rather than to excite a response. The sum of the exciting voltages of the input axon fibers must exceed the threshold potential of an individual cell before it can "fire" an impulse.

The mechanism of excitation or inhibition has been studied for a long time, most successfully in recent years using the ultrafine microelectrode with which it is possible to observe the electrical activity of a single cell. The compositions of the fluids within and external to the cell, which also affect the transmission of information along the nerve fibers, appear to play a significant role in synaptic transmission. Because of differences in the concentrations of ions in the solutions, there is a voltage drop of about 70 millivolts between the inside and the outside of the nerve cell, the inside being negatively charged. A stimulus from a sense receptor in the "presynaptic" fiber releases a chemical transmitter substance which locally and briefly changes the permeability of the "postsynaptic" membrane. Impelled by the existing negative electric potential, ions can then flow and cause a momentary decrease in the potential. If this depolarization reaches a threshold value, called the "action potential," the postsynaptic neuron will discharge. Inhibitory impulses also arise from a chemical transmitter acting in somewhat the same way except that the membrane appears to be made permeable only to ions below a certain size, and the result is that the voltage of the cell becomes even more negative than it was originally. This effect can oppose excitation and if it is sufficiently strong the neuron will not discharge.

A depolarized membrane is quickly restored to its original state, but during the restoration period additional normal stimuli cannot cause further excitation. However, since the external stimulation of the nerve may be of relatively long duration, longer than the restoration period, there will be a succession of discrete energy pulses whose frequency will increase with the magnitude of the stimulus.

There is nothing simple about the processes described above in such abbreviated detail. Much about the physical mechanisms in single cells is still not thoroughly understood, but beyond that we must remember that the human nervous system has been estimated to contain 10^{10} (ten billion) neurons. Considering that each of these has many linkages through which it affects and is affected by other neurons, the intricacy of neural networks is staggering. In the cerebral cortex of man, an axon of a single neuron may influence as many as five thousand others, and the overall meshwork of the human nervous system may have of the order of one thousand billion synapses[16].

It is no wonder that fundamental neurophysiological investigations have not progressed very far beyond the basic conditions of communication between single nerve units. Still, through research at the neuron level the conditions for further examination of the similarities between brains and computers can be established. The principal analogy relates to the fact that the nerve firing at any instant can be described as an all-or-none impulse; it is either full on or full off. This is not, it turns out, a perfect analogy since the neuron is also sensitive to constantly varying physical and chemical conditions in the body which affect the firing frequency. So, in addition to its digital behavior, the neuron also reports information about continuous changes.

The study of idealized neural networks does not come close to suggesting the versatility of living nervous systems but it is a place to begin. One of the classic investigations in neural network theory was that of Warren S. McCulloch and Walter Pitts, published in 1943[17]. They essayed to treat the logic of situations in any discrete process built of neurons considered as abstract two-state machines. They were forced to make simplifying assumptions but with their model proved that the behavior of any neural net could be described in terms of the symbolic logic of propositions, i.e., there is a simple logic proposition which corresponds formally to each reaction of an ideal neuron. There is theoretical value in showing, as they did, that if a behavior pattern can be described in logical, finite, unambiguous terms, *in principle* a synthetic network can be constructed to achieve it. They felt they had demonstrated that the design of automata capable of copying the functions of the brain is not forever impossible, theoretically, although it may be just as likely, as Von Neumann once suggested, that the diagram of the actual nerve connections may be the simplest possible description of a specific feature of the operation of the brain[18].

PERCEPTRONS

The suggestion of McCulloch and Pitts that neurons could be considered as serving a function similar to that performed by electronic components in artificial circuitry was given more flexibility by the postulation of synapses whose threshold or excitation value was capable of changing under varying conditions, not only momentarily but permanently so that a form of "learning" could be said to have occurred. This was the idea of D. O. Hebb. His theory proposed that a growth process could ensue in one or both interacting cells as a result of repeated stimulation of one cell by a structuring in the overall neural network. Obviously developments along these lines would bring man-made automata one step closer to an identification with natural systems.

In 1958, Dr. Frank Rosenblatt of Cornell University created a random network model of a machine which was capable of adjusting its behavior. This was called a *perceptron* and the term has been applied to subsequent machines designed by others with the same idea. Rosenblatt's first model was a computer simulation but this was followed by a constructed machine which was capable of "perceiving" visual stimuli. A schematic diagram of a perceptron is shown in Fig. 7-2. The perceptron represented by the

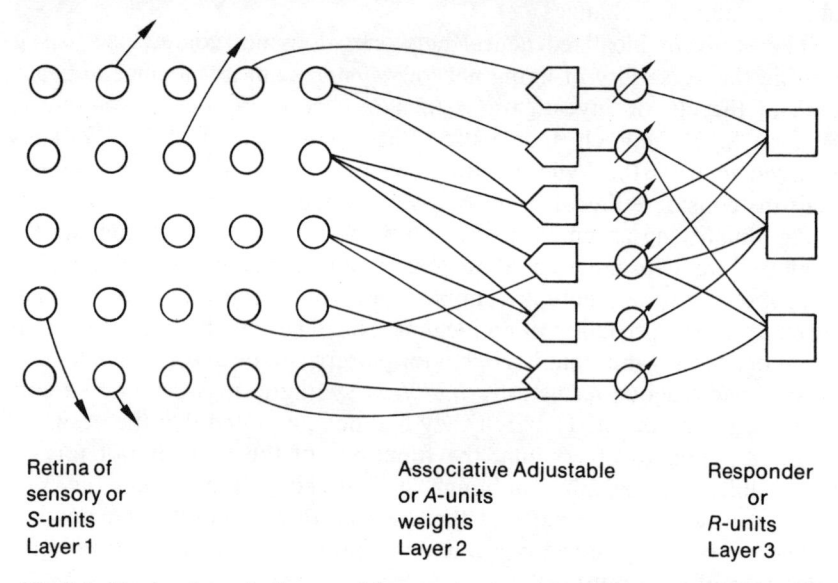

| Retina of sensory or S-units Layer 1 | Associative Adjustable or A-units weights Layer 2 | Responder or R-units Layer 3 |

Fig. 7-2. Simple schematic diagram of the organization of a perceptron (after Cole [10]).

diagram consists of binary (on-off) threshold logic units, designated as TLU, which are arranged in three layers. The first layer, the retina of sensory units or S-units, is an array of separate photocells which will be turned on by a suitable stimulus. The inputs to the sensory units can be supplied in different forms, but an obvious way is to project a picture on the plane of the photocell bank, thus producing a pattern of light and shadow. A photocell which is illuminated by an intensity of light above its threshold value will send an output signal of "one"; or, if the threshold is not exceeded, "zero." Some of the S-units are connected randomly to the A-units in the second layer so that each A-unit may have several inputs from the S-units. If enough of these inputs are "on" at one time the A-unit will fire. Thus for each retinal pattern resolvable by the bank of photocells there will be a specific set of outputs from the A-units.

In order to provide the conditions for the machine to "learn," the binary signals (0 or 1) from the A-units to the R-units are weighted. This adjustable feature in the design allows "training" to take place, and it does so in the following way. First, assume that there are two given visual patterns between which the machine is to be trained to distinguish, for instance the two letters of the alphabet, C and D. We wish to have the R-units, which can give signals identifiable by an observer, respond to inputs characteristic of a C stimulus, but not a D. If we show the C image to the machine, and the perceptron gives a correct output (1), the weights of A-unit signals are to be left unchanged. If, however a correct code is not produced by the R-unit, then the weights of the active A-units need to be changed according to certain rules to give the desired code. As the process continues, with adjustments being made whenever necessary to give a correct response, the required adjustment will be less for each succeeding step. After a finite number of trials the machine will have been "trained" and can be expected to produce the correct output signal each time the letter C, in a standard form, is shown to it. In short, the perceptron gathers data from a number of trials, weights it in a specified manner, and then determines if the collected information is sufficient to conclude that the image is an example of the kind of pattern one wishes to identify.

The simple perceptron will produce only simple results by a form of rote learning. It cannot be considered the kind of pattern-recognizing machine which can recognize and classify the significant feature of, say, visual images which had never been presented to it before. There are such systems, but they do not exist in an advanced state of development, ordinarily being designed to meet a very specific application. Perceptron machines, a good deal more complicated and versatile than the original one, have been designed and built; but it is not surprising, when we consider that their active elements number some few hundred or perhaps a few thousand artificial neurons, that none are more than feeble imitations of the human brain. The term *self-organizing system* has sometimes been applied to these concepts. One interesting example of a perceptron-like machine, called Madaline (a multiple of an earlier version, Adaline) was built at Stanford University by Professor B. Widrow. A unique feature of this machine is a threshold logic section the units of which automatically correct their weight settings during a training period, through electrochemical action.

Studies on perceptron-type conceptions have followed three main channels of investigation: mathematical analysis, simulation on digital computers and the construction of actual machines[20]. The advantages of building machines are first, that precision in construction or reliability of components turns out to be not important and second, that results can be achieved more rapidly than with computer simulation in which the processes must be carried out serially. Nevertheless, the obstacles to constructing machines capable of approaching the adaptive performance of even the lower animals are truly formidable. The visual perception process is imposingly more complex than it was once thought to be. At one time the theory was fairly well accepted that an image impressed itself on the brain by a point-by-point television-like mapping. It appears now that there is a substantial amount of pre-processing of information that occurs in the eye itself before the signal ever reaches the brain. The retina uses certain qualitative criteria about boundaries, contrasts, movement, etc., for its pre-analysis and sends this information to the colliculi for a separation of functions. Then, too, more than one section of the brain is involved in the reception and utilizing of the information.

There are those who believe that the mathematical approach to the theory of perceptrons offers the best hope of arriving at any significant improvement in the understanding of information processes. Some important contributions in this direction are contained in the book, *Perceptrons* by Minsky and Papert[21].

In general, we can summarize this section by saying that programs designed to try to devise physical imitations of the human nervous system appear to face inherent barriers to progress which are not likely to be overcome in the reasonably near future. Some computer scientists are disinclined to include this kind of work under the heading of artificial intelligence research, preferring instead to reserve that term for the simulation of human thought processes on the computer. Some examples of these latter programs will be discussed in the sections immediately following.

GAME-PLAYING MACHINES

Can a machine learn?

This, of course, is the same type of question as the one we examined earlier, i.e., "Can a machine think?," and the answer would have to be the same type of answer. It depends upon how one defines the word, learn. One could agree we have already given an acceptable answer to the question of machine learning in the instance of the perceptron, but at the same time that must be recognized as a severely limited case. An approach which offers a wider range of possibilities occurs in the development of game-playing computer programs.

Norbert Wiener offered the following operational definition of learning, "A learning machine is an organized system which can transfer a certain incoming message into an outgoing message according to some principle of transformation. If the transformation has a criterion of merit of performance and the performance improves, then the machine has learned."[22] Should we accept this definition then we can turn to game-playing where a valid criterion of performance is a victory in a game which is played according to stated rules. This suggests a primary reason why the development of game-playing machines is one of the favored areas of artificial intelligence research. The winning of the game is an unmistakable measure of success, and other secondary criteria related to winning can also be recognized. If a computer program is developed to play a game against a human opponent, and the computer continues to improve its performance on the basis of an examination of how a previous match was played, the definition of learning is met in an interesting way. Two other requirements for a machine-learning program are also satisfied by game-playing. The rules of common games can be expressed without ambiguity, and most of the games do not take long to complete, thus there can be many trials of the capabilities of the program.

Still, not all games are equally useful for these studies. Some are trivial in the sense that their complete theory is known, and any player familiar with the theory need never settle for less than a tie. Others are so difficult that any appreciable progress in adapting them for significant artificial intelligence research seems beyond likelihood for present-day computer science. Tic-tac-toe is an example of the first kind. The reader may recall (Chapter 2) that Charles Babbage had considered building a machine to play tic-tac-toe and to exhibit it in contests throughout England in order to raise funds for the completion of his Analytical Engine. His reasons for not doing so were related to his being convinced of the impracticality of the scheme as a money raiser, for we can be sure that the theory of construction of the machine, at least, would not have offered him any great difficulty.

The relative simplicity of writing a program for playing tic-tac-toe, or noughts and crosses as it is sometimes called, is not hard to demonstrate. Most people are familiar with the principle of the game which is to place three marks in a line — horizontally, vertically or diagonally — on adjoining squares of a 3×3 matrix before the opponent, taking alternate turns, can do so. In a computer game, of course, there must be some means of indicating the choice of play of the computer either through a visual display of the matrix or through some other output which will permit the selection to be marked on a board. Although there are ostensibly nine possible plays to make initially, because of the symmetry of the matrix the player having the first move has really only three unique plays — either the center square, a corner square, or one of the four squares in the center of each border column or row. Assuming that the computer has the first choice of moves it would begin by marking an "X" randomly in one of the squares. The human opponent then can mark his "O" in any of the eight remaining positions, and this continues until the game has been won or tied. Although a game can be won in only five total moves, usually if the play is random it takes six or seven moves to bring it to a close. This means there may be some 100,000 or so possible games. This number of possibilities could be stored in the memory of a computer in such a way that it can be instructed, say, to search out the future steps most likely to produce a win according to the game theory.

A second alternative method of attack, that of a rote memory scheme, is a more useful demonstration of the learning ability of a computer. Over a sequence of a large number of games the computer can store in memory the picture of the board as it was in the next-to-last move each time the opponent won a game. When the same situation is encountered again the computer will not make that move but instead try others. It has begun to learn to avoid undesirable situations, and it can continue to improve until it reaches a point of development where it need not lose any game. A rote learning program is depicted on the flow diagram shown in Fig. 7-3.

A move to a much higher order of complexity can be taken by going to the game of five-in-a-row where the objective is similar to tic-tac-toe except that five consecutive squares lined up in any direction are to be filled with the same mark, instead of three. There need be no limit to the playing area, but a 20 × 30 square array makes a practical game board.

Cole[23] has described how one might proceed in a general way to write a program to play this game. Assuming a general understanding of the principles one must define in detail certain promising short cuts or rules-of-thumb to a winning scheme. We call this a *heuristic* method of approach, as compared with an *algorithmic* procedure wherein each finite step, taken in order, is formally determined by pre-assigned rules. Obviously, a player, A, cannot permit his opponent, B, to mark four consecutive squares in any direction with both ends of the row unobstructed because he will then have lost the game. For his own part, A would like to have such a situation in his favor but there are also other positions toward which it would be very desirable to aim. If, for example, he can manage to set up two three's in a row, B will not be able to stop him, and that would

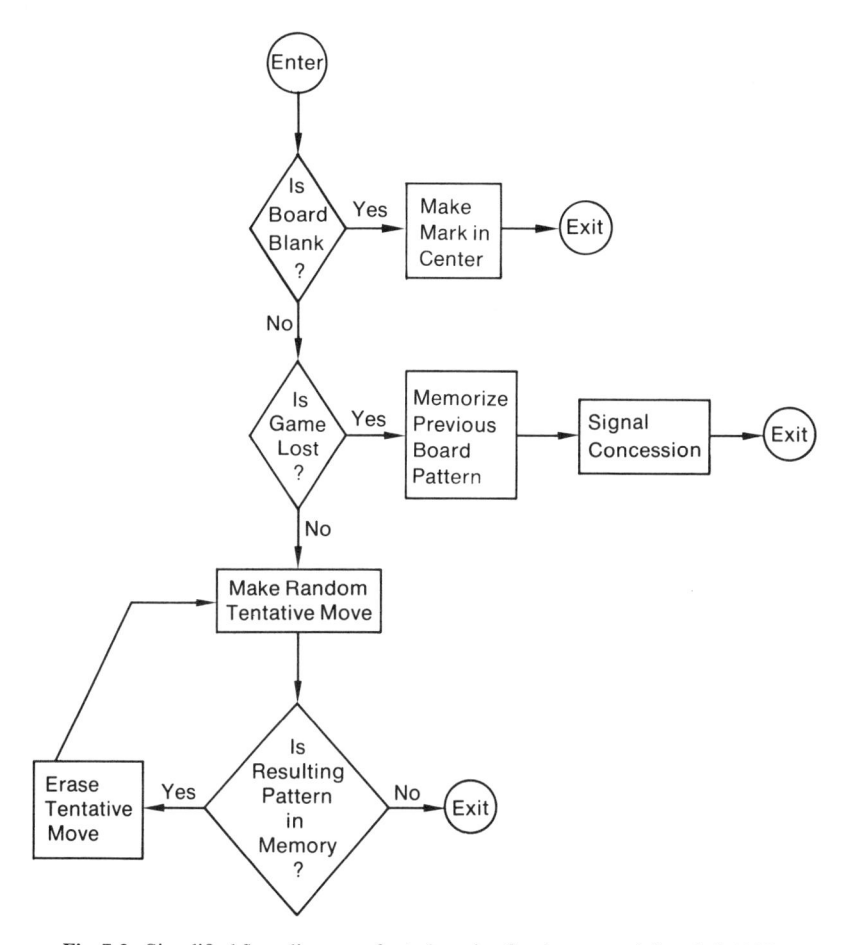

Fig. 7-3. Simplified flow diagram of rote learning for tic-tac-toe (after Cole[23]).

also be true for three two's in a row. The program can then arrange and classify certain patterns in order of priority, with four-in-a-row to receive first attention, three-in-a-row next, and so on. The pattern dictionary would have to be brought up to date after each move, and then the move made in terms of the highest priority pattern appearing. Learning can be induced by raising priorities on patterns shown to be successful, and reducing priorities for the unsuccessful ones.

A more desirable generalized picture of the game could be provided by computing a single figure of merit for each allowable choice for a given turn. This composite score is more universal than the scoring of individual squares and patterns, and permits recognizing the potential interrelation of influences from patterns widely separated on the board. A definite advantage is realized by computing scores for each possible move from a given position, and also for each possible response by the opponent. The advantage is enhanced by selecting a move which promises to lead to the highest score several moves ahead. This process of exploring imaginary moves and countermoves several steps in advance while retaining the original board position from which to make the most promising actual move can be represented by a decision "tree."

Perhaps the most significant programs which have been written up to the present time for a game that has no complete theory are those developed by Dr. A. L. Samuel, formerly of I.B.M. Corporation and recently of Stanford University. Dr. Samuel's work has been described many times [24, 25], but some aspects of it are so important in the history of artificial intelligence research that they bear summarizing here. No complete theory exists for checkers for reasons that are not at all difficult to understand. Samuel estimated that to explore every possible path in a checker game would involve analyzing of the order of 10^{40} (ten thousand billions of trillions of quadrillions) choices of moves which at three choices per millimicrosecond would take about 10^{21} centuries to consider. The possibilities of choice of movement in the game of chess are even greater and the rules for chess are much less simple than for checkers, so that even though admirable chess-playing programs have been written, the computer is still more competent at playing checkers. How competent it is can best be examined by going through some of the steps taken when a match is played, and then seeing what kind of outcomes have been reported.

First, naturally, the rules of the game are programmed into the computer, and some way arranged for reporting the moves of the computer, as well as for recording the data concerning moves made and the current board situation. There are 32 playing squares on a checker board, and these are represented in the computer by a 32-bit word, the occupied squares being represented by 1's. When a piece is moved from a square, the bit for that position becomes 0. Thus, computer words, representing the board position after each move, can be formed and stored. The computer can also look ahead several plays in advance by computing all possible next moves and their significance in the light of the new board position. The results of looking ahead in this way are represented by a decision tree such as that seen in Fig. 7-4. [26]

We observe that each horizontal set of branch points on the diagram is

Game-Playing Machines 241

Ply 1 – Machine chooses branch with largest score.
Ply 2 – Opponent expected to choose branch with smallest score.
Ply 3 – Machine chooses branch with most positive score.

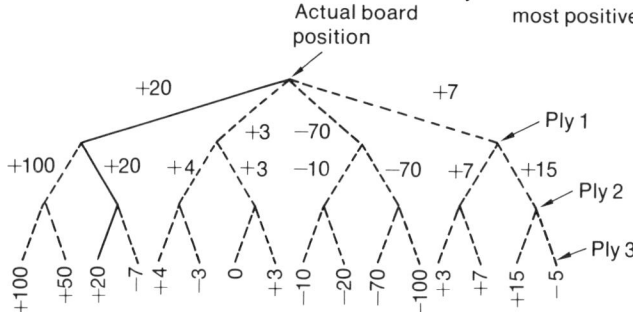

Fig. 7-4. Simplified fragment of decision tree for checkers playing program (Courtesy of A. Samuel and IBM.)

called a "ply." Each ply represents one projected move. When the machine was beginning to learn, the program was set to look ahead to the third ply but not beyond that unless a jump was imminent or had just occurred, or there was a possibility of an exchange offer. If any of these conditions existed, continued exploration would take place along that branch until a ply was reached that contained no situation considered interesting according to previously defined criteria. In no case would the exploration go beyond the 20-ply level. That would be too costly and complicated.

The computer with its great speed can and will, however, compare several paths of the tree. If it has the first play at the start of the game it has a choice of 7 moves open to it, but then the number of possibilities quickly opens up. It can anticipate the results of carrying on down a path after a given move on the assumption that after each move it makes which is favorable to its perceived objective, the opponent will make a countermove to frustrate that objective. Presuming that the computer has evaluated the board positions down to the 3-ply level, it can then imagine what would happen if it had replaced the latest move with a different one. This will permit a decision about the next move. Now the process is reversed and the computer backs up the tree, always remembering that the opponent has the initiative at each alternate level. Back at the start the computer can then assess how successful the initial assumed move would have been.

Referring again to the diagram, Fig. 7-4 we can see that the computer has examined several possible paths in this way and has concluded that the moves most likely to give the best results at the third ply are those following the solid path down to the +20 score. Of course, in an actual situation there would be many more paths to consider.

The problem of assigning quantitative values to the various steps has been a difficult matter. The board positions are scored relative to their value to the machine. The overall board position for each of the two branch points at the end of a line segment in the diagram is analyzed in smaller groups, rated according to certain parameters. These parameters, abstracted from the experience of the researchers and expert checker players, relate to such factors as the relative piece advantage, where it is accepted that the machine's chances of winning will be favored by a move reducing the number of the opponent's pieces relative to its own. The parameter values of the board positions at the branch points are subtracted one from the other and multiplied by a weighting factor derived from experience, whereupon all the weighted terms are added to give the value for the line segment. It has turned out that the numbers given to the parameters, while not arbitrary are often imprecise, and moreover, many of the inter-parameter effects have been found to be not expressible in any useful way. Dr. Samuel has indicated that some of the concepts used by master checker players elude formal definition. The hope of producing a program that can develop its own terms is still not realized.

Two primary learning procedures were studied in detail. "Rote learning" which was mentioned previously, required keeping a record of the board positions as well as the machine analysis of each situation; and then using these records when similar situations appeared. The number of words required for this method is very high, however. The programs based on rote learning have improved slowly, but have not reached championship caliber. The "generalization" learning procedure provided a means for the program to maintain a continuous readjustment of the weighting coefficients for the numbers evaluating the board positions at the important points. The tendency was for this type of program to learn more rapidly but then level off, as the ability of the programmers to make their parameters more sensitive and accurate ran into limitations.

It has been suggested that one of the critical failings of checker playing programs has been their apparent inability to sustain a continuing strategy[27]. An expert player will observe certain indications of a favorable position which will then suggest to him a strategy to be maintained for at least several moves in advance. He may consider only a few dozens of positions but these he analyzes shrewdly, as much in a descriptive as a quantitative way. The computer will remain at a disadvantage as long as it treats each situation as a totally new problem.

SIMULATION OF HUMAN THOUGHT

There has been another aspect of heuristic programming research to which A. Newell and H. A. Simon have made major contributions. It concerns the study of ways in which people actually go about solving problems in a logical fashion. This is not quite the same thing as computer game-playing, where the first aim of a program was to get the machine to play a competitive game by any useful means. The program which Newell, Simon and J. C. Shaw[28] have described is called the General Problem Solver (GPS). While it could conceivably be used to play games or discover the proofs of a theorem, the intent is rather to have it search for, discover and administer a hierarchy of actions leading from a given to a desired situation without regard to the subject matter of the problem.

Very simply stated, in the original studies the experimenters had a human subject describe aloud the thought process he felt himself to be carrying on as he worked out a logic problem new to him according to certain given rules. His remarks and those of the experimenter were included in a "protocol," and the protocol when sufficiently analyzed and defined was incorporated in a computer simulation model. A program was then run comparing choices made by the individual and by GPS when solving the same logic problem. It was found that both often follow the same path to a solution at least for limited periods of time.

In solving any non-trivial problem, the human mind seems somehow able to select certain sub-goals which promise to be more readily achievable than the main goal. The number of possible paths to be taken to meet the successive linked sub-goals is enormous. A measure of human intelligence is the ability to select the most promising possibilities and to do so sparingly. It recognizes the given situation, postulates a desired situation which is somehow different from the given, and chooses a process which ought to affect a change. As a goal-seeking system the mind must receive information about the state of the environment through its afferent, or sensory, channels; act on the environment through its efferent, or motor, channels; and store necessary information about the original state and the effect of its actions on the environment. The young child gradually learns how the afferent and efferent worlds are related, and GPS is programmed to try to build similar relations[29].

On the afferent side, GPS attempts to characterize given and desired situations and their differences by symbolic objects; and on the efferent side, to represent, by *operators*, actions which will affect changes, (also symbolically). The operators selected by GPS are the ones deemed most likely to eliminate the difference between the given and the desired situation. If they do not, others are applied when available.

In spite of its nominal general purpose capability there is ordinarily no obvious way to denote difference types in a GPS program for a problem less specific than one, say, proving a theorem, nor is the information relating each operator to the relevant difference type commonly available, except in certain artificial problems. Even if it is available, the unknown effects on other differences may be the more important. The versatility of the General Problem Solver has been expanded continuously since it was first introduced, but it remains a limited means of simulating the human reasoning processes.

THE LANGUAGE MACHINES

Earlier in this chapter we referred to a conversation program devised by Turing to determine whether or not a machine was capable of demonstrating intelligent behavior. Since the time of his article the exploration of linguistic relationships between man and the computer has come to represent one of the more active branches of artificial intelligence research. The research also has its hostile critics, some of whom suggest that the studies are fruitless, that language is the unique expression of humanness, and that it, above all else, separates man from the other animals and from the machine.

Nevertheless, there is both practical and theoretical interest in the problems raised. One of the most tiresome and costly phases of computer usage exists at the man-computer communication interface, and derives from the fact that man has been unable to communicate with the computer in his own natural language. This is a difficult enough problem between two humans who do not speak a common language, but at least they may be able to achieve a degree of understanding through non-verbal means. When man communicates with the computer, it is the language of the computer that he must use, and unless he is familiar with that language he needs the services of a translator, who is the programmer. So it is man who must adapt to the computer rather than the other way around.

If we were to achieve a desired goal of, say, direct and unambiguous verbal communication with a computer, what conditions would have to obtain? At least five distinct requirements have been identified[30]. In the first place the computer must be able to accept the vocal signal of the human. We might note that verbal communication between humans is often aided by the sense of sight, but we can assume here that the voice sounds are sufficient to carry the message, and in that case the first requirement can be handled very well with a microphone. Second, the machine needs to be able, in the entire acoustical output it receives, to distinguish between the sounds which are needed to transmit information, and those others—lisps, wheezes or other audible, individual idiosyncracies—which are essentially irrelevant to the basic message. This capability for voice recognition is not highly developed in any existing machines, but should substantial additional progress be made in this direction then the next requirement would have to be a means of reconstituting the message from the meaningful part of the speech pattern. This is the process of linguistic analysis and it is the point at which many programs in the technology of language begin. Should we be unconcerned about the way the message is offered to the machine, e.g., if it is entered by punched cards or by keyboard, then it is the understanding of the syntax and semantics we are striving for and the changes in air pressure produced by the human voice would not be of consequence. Fourth, once the message has been distinguished, presumably there must be a means of reporting or otherwise making use of it. This presupposes that the machine has stored within its memory the necessary information in a readily retrievable form, and that it possesses the capability of somehow ascertaining the relevance of this information to the input message, and of composing a suitable response. Finally, the response must be expressed in natural language, phonetically formed for understandable audible delivery.

The problem, then, is multi-faceted and each part is of forbidding complexity. We will take the position that if fundamental advances were to come, they would come through a deeper understanding of linguistic analysis, and under those circumstances the problems of automatic speech recognition would then be largely technical.

MACHINE TRANSLATION

The following paragraphs will concentrate on summarizing some of the features of reported work on machine translation in programs in which the input message is in the form of a written text in natural language.

The original impetus for mechanical translation research was as much political as scientific. By the early 1950s competition with Russia had reached such a stage of intensity that every aspect of Soviet research and development concerned the United States government. Rapid improvements in computer technology provided grounds for hoping that a partial solution to the need for translating into English the many Russian documents appearing in the technical literature might be found through perfecting mechanical translation programs. The scope of the problem is suggested by the statistic that nearly one-fourth of the articles appearing in the journals *Chemical Abstracts* and *Nuclear Science Abstracts* are from Russian periodicals.

An experimental program, jointly conducted by IBM Corporation and Georgetown University was completed in 1953. It was a modest endeavor involving the translation of a brief, selected text with a few hundred words but it was well-publicized, and contributed to a widely held assumption that a large-scale capability for mechanical translation of Russian text to English was imminent[31]. The problems of bringing this about, however, proved more intractable than many had thought they would be. Research studies were funded and soon took two general directions.

The first method, sometimes called the "brute-force" approach, did not have much relation to artificial intelligence. The assumption was made that there is an equivalence between words, or at least between phrases, in different languages, and that a large machine-readable dictionary, the inherent speed of the computer and a few rules of syntax would permit translations of good quality to be produced. This was not the case. For such a process, the dictionary has to be not just large, but impractically, if not impossibly large. A few operational systems grew out of this approach but none in which the quality was very good, at least not without extensive pre-editing or post-editing.

The polar opposite of this method of attack, the one which has received the most sustained attention, is really an admission that no fully satisfactory machine translation programs can be achieved or anticipated unless the fundamentals of languages can be described in terms of their structure and meaning. This work has sometimes taken the form of mathematical modeling of linguistic structures[32], but these formalisms have not resulted in any realistic syntactic description of natural languages. Other research efforts have dealt with more approximate models in which the machine is programmed to simulate the procedure by which an intelligent person would attempt to translate a text written in a language he does not know into his own native tongue with the help of a dictionary and a set of rules. The programs attempt to resolve ambiguities in both word meanings and structural groupings by heuristic methods of improving the problem-solving capability of the machine.

Let us examine the general process by which the translation of language by a computer might take place. The text of the source language can be entered by means of a teletypewriter into a machine especially designed for translation. If the language to be translated is Russian, then the keyboard must have the symbols of that alphabet. It could, for instance, have the Russian Cyrillic alphabet in upper case, and the Roman alphabet with standard numerals in lower case. The text is of course typed in letter by letter and word by word, starting from the left.

At the outset there is a problem of recognizing words. The word "state" will appear in English dictionaries but not "stating" or "states," nor, generally, will most of the words with endings that signify number, tense, etc. To find a way around this difficulty one would have to consider two possibilities [33]. One is an editing routine which chops off word endings on prefixes; searching in its store after each letter or combination of letters has been eliminated for the basic word stem. This could have complications. If, for example, the machine were to search out and eliminate "-ing" endings, words like bring or King would not survive. The alternative of listing words in all their forms requires a much greater store of words but with the large storage capacities of contemporary computers, this is preferable.

The translation problem itself is crucially complex. The machine program must not only recognize every word, but it must do so in the context of groups of words. Words by themselves can have many meanings. The simple word "state," irrespective of endings, has as Fink points out [34] at least nineteen definitions as a noun, four as an adjective and three as a verb. Only by analyzing the sentence in which the word is included (sometimes even the sentence cannot be understood except as part of the whole paragraph) is it possible to grasp the intended meaning. One question then is, how should the size of the groups of words to be stored be determined? Since there are, for example, some 10^{18} possible three-word combinations in the English language, of which millions certainly have meaning, the problem of storing combinations of words is seen to be very great.

One helpful approach is to store phrases which appear commonly in textual material of a certain kind. Many scientific words and phrases occur again and again in the literature, and would appear as such in the machine dictionary. When the text, in the form of machine code, enters the input register the first input characters give the indication, from a directory, where to start the comparison with the stored entries. This is similar to the procedure followed in looking up a word in an ordinary dictionary, wherein the efficiency of the search is improved by glancing first at the key words at the top of the page.

The computer search then continues from the bottom of the located dictionary "page" to the top, seeking to match to the input characters the longest stored phrase or word beginning with the appropriate letters. The longest is presumably the most significant and if a match occurs, that part of the search is successful. Each dictionary entry, in addition to the words of interest in the source language and the target language, contains coded grammatical information about the source language words. This grammatical information is designed to help resolve semantic and syntactic ambiguities before the translated phrase appears at the output. Unfortunately, the rules are not always sufficient to cope with the complexities of the translation, since no adequate explanation of the way that human language is organized has yet been proposed. This is not surprising if we try to imagine how languages must have evolved.

Bowers and Fisk[35] have given an example indicating the kinds of instructions which had to be contained in the entries for translating a simple two-word phrase from Russian to English. In the case cited, which is from a program demonstrated on a machine translator exhibited at the New York World's Fair, the normal left-to-right pattern of the program had to be altered and a total of eighteen dictionary searches needed to be made. The phrase С ГРУППАМИ (substituting English letters, the approximate equivalent would be seh groupamee), means "with groups." However, the preposition С, cannot be translated directly, because its translation (which in another context might be "from") depends on the case of the noun in the prepositional phrase. Thus the program must look up С and find out through the coded rules in the entry what the irregularities are. It next looks up the stem ГРУПП and the ending АМИ to learn that ГРУППАМИ is the instrumental plural, and then goes back again to С to choose the translation "with." If the noun had been in the genitive, the correct translation of С would have been "from." The last step is to translate ГРУППАМИ as "groups." One can imagine the magnitude of the problem facing anyone who wishes to devise a program to analyze the syntax of a long involved sentence from the order of the words, their endings, etc. A fundamental difficulty with translation programs is that it may very well be necessary first to have a clear idea of the meaning of a sentence before its syntactic structure can be resolved.

We can quickly summarize the present state-of-the-art by saying that the goal of producing high quality machine translation of language has not been met. A report issued in 1966[36] by the Automatic Language Processing Advisory Committee of the National Academy of Sciences – National Research Council concluded that no machine translation of general scientific text was in immediate prospect, and that machine output which had not received editing from human translators was "sometimes misleading" and inclined to be "slow and painful reading."

While it is true that the work on translating machines has stimulated research in theoretical linguistics, it is also true that these latter studies have made rather slow progress. We are still a long way from knowing the rules by which our languages are structured. Indeed it seems likely that a full set of rules may never be enunciated, so, although humans can learn to speak and to understand language without paying attention to formal rules, the machines of the present show little promise of being able to do so.

The problems of machine translation are to an important degree similar, while not identical, to those encountered in man-machine communication. One difference is that man-machine communication systems designed for question-and-answer dialogue do not have to deal with disparities between two natural languages. They must also have an information processing component which identifies and retrieves information for answers to questions directed to the computer[37]. The need for "understanding" still exists, however, to the extent that man-machine communication and also information retrieval might be considered special, though hopefully somewhat less difficult, cases of machine translation.

The field of automatic search and information retrieval has received a lot of attention, for the obvious reason that no one who is looking for relevant documents on almost any given topic can hope to find more than a fraction of the published information in a reasonably short period of time. If the speed and large memories of computers could be coupled with an ability to specify and recognize without error the object of the search, there would be a large practical gain in having automatic systems to scan the literature.

The two kinds of systems which are normally envisioned are for the retrieval of either documents or facts[38]. Cataloguers will pre-identify documents with "descriptors," which are key words significant in each scholarly field. All publications relevant to that field are placed in the memory with the list of descriptors. The person doing the searching makes up a list of the descriptors which he feels would relate to the topic or field in which he has an interest, and the machine then performs the search.

Unfortunately, he cannot always be sure of the criteria the cataloguer used. If facts are to be retrieved, both the information and the requests will be prepared in a more rigid formulation.

As might be expected some categories which spill over into many subject areas pose additional problems for the developer of a system for automatic information retrieval. If, for example, a systems engineer would wish to conduct a literature survey for a project on the design of an advanced food processing plant[39], some of the major categories of document topics to be searched might be *Cryogenic Freezing, Radiation Preservation, Soil Properties, Food Production, Refrigeration, Packaging, Processing Equipment, Synthesized Foods*, etc. Literally hundreds, if not thousands, of titles such as "Effects of Ionizing Radiation on Lipids of Fish," "Application of Radiochemistry Techniques in Food Processing," "Chemical Synthesis of Protenoids," "Studies of Algae in Hungary," etc. could appear under the various headings. Surely the original cataloguer would have to know a good deal about library science and computer science as well as the significance of the subject topics even to make a start on a creditable information system research. Although information retrieval studies are not, strictly speaking, a part of artificial intelligence there are programs which attempt to have a machine draw the kind of inference about requests for information that a competent librarian might. Lindsay[40] developed a basic English program from the observation that there is a natural model for family relationships such that a machine could conclude that "Jane is Bobby's aunt," from the two input statements "Jane and Mary are sisters," and "Mary is Bobby's mother." Presumably an extension of this line of thinking might produce programs which would facilitate a search for genealogical references, but there may be legitimate doubts whether the results would justify the effort and the costs to a sponsor of the project.

Salton[41] made a study to assess the effectiveness of automatic document indexing and classification methods. He pointed out that the measure of effectiveness in satisfying the user's need can be met in experiments by two criteria—"recall," meaning the proportion of material actually retrieved, and "precision," meaning the proportion of retrieved material actually relevant. The criteria are on the average mutually antagonistic, i.e., a broader search formulation leads to higher recall but reduced precision. Studies on the Medlars (National Library of Medicine) retrieval system using manually indexed documents[42] showed an average recall of 0.577 and an average precision of 0.504 for a document file of over 600,000 items. This is considered a reasonably good performance. A surprising result of certain studies was that, on the average, the simplest indexing procedures of single-term content words manually chosen from the documents gave better results than more complex indexing methods. It was also noted that retrieval performance was improved by multiple searches based on user feedback information. In any event, although the proportion of relevant information retrieved by *any* procedure is not high, the average performance of automatic document analysis when it can be implemented seems about as good as but not really better than that of those manual methods presently being used.

Most predictions about the facility with which information may be retrieved automatically have been over optimistic. Reimars and Avram [43] have reported that as a result of a study completed in 1963, a target date for implementing a system of automatic organization, storage and retrieval of information for the U.S. Library of Congress was set for 1972. The target date will not be met. There is a tendency in such studies to underestimate problems in information control. In large libraries, for example, reference location and retrieval is only a minor part of the necessary total activities which include selecting, acquiring, cataloguing and shelving of holdings. All of these are closely interrelated. Quite possibly, an information system as complicated and as large as the Library of Congress with its sixty million items will never be completely automated.

SELF-REPRODUCING MACHINES

We can conclude that automata which can perceive or learn, even in a restricted sense of the words, seem little nearer being a reality than they were a decade ago. The problems encountered in simulating the activities of the brain or analyzing them, which may well be prerequisites to producing such automata, are intimidatingly difficult. Some theoreticians have suggested that a feature of living systems or of automata perhaps holding greater promise of succumbing to logical analysis is the ability for self-reproduction. The point is that even the simplest organisms, some without nervous systems, are capable of reproducing themselves. Can it then also be demonstrated that machines can do the same?

John Von Neumann, who was responsible for so much original thinking in questions relating the computer and the brain, offered the first organized description of how such a process might occur [44]. His concept hypothesized an automaton equipped with a set of instructions, and placed in an environment containing, in ample supply, all of the components needed for the construction of other similar machines. He concluded there was no theoretical reason to believe that the automaton could not assemble the necessary combination of parts to duplicate itself. Moreover, the newly produced automaton could do the same. The logic of self-reproduction in complex systems can be developed on the basis of their being described as states or processes, i.e., blueprints or recipes, as Simon expresses it[45]. Von Neumann's conception had to deal with the problem that a simple set of instructions to tell the machine how to build itself is insufficient, since the set would need to include, along with the description of the automaton the directions themselves. The solution to the problem was to envision two machines. The first one (A) is similar to a punched-card reproducer—it can copy the blueprint as a set of instructions. The second one (O) is able to produce another automaton when furnished its description, and the product, or offspring, will itself contain the ability to receive a similar program and to act in the same way. If now, these two automata are combined with a sequence-control device (C) for turning each on at the proper time, the result is one large automaton which functions as follows. O is furnished the instructions for fabricating the new machine and the controller then provides the command for carrying on the fabrication. The program is next duplicated by A on orders from C which include instructions to furnish it to the automaton just produced. Thus a complete self-reproduction process can commence as soon as a program for creating the complex aggregate of automata (considered as one automaton) is fed into O.

Moore[46] has drawn a parallel between Von Neumann's theoretical machine and what goes on in a living cell. He suggests that a function similar to that provided by the automaton, A, is performed when the enzyme DNA polymerase copies the set of genes composed of DNA in which hereditary characteristics are encoded. The fabricating automaton, O, on the other hand may be thought to behave like the system of messenger ribonucleic acid (RNA), enzymes and ribosomes which assemble amino acids following the program of the DNA to produce a new cell.

Von Neumann's machines have never been built—in fact he chose later to turn to strictly mathematical models. The geneticist, L. S. Penrose, has however constructed actual hardware models of self-reproducing systems which illustrate how a template can influence coupling of inanimate objects[47]. They are admittedly simple. One such model has wooden blocks in two different shapes so that a block of shape A could adhere only to one of shape B in one of two ways and not to another A.

Neither could *B* adhere to another *B*. Furthermore either an *A* or a *B* would adhere only to a pair already connected. When a number of such blocks are shaken together in a box with *AB* "seeds" present, the slant of the combined *AB* shape forces the individual pieces to assume a similar position. This induces any *A* to attach to a *B* next to it and more *AB* units are nucleated. If a seed having the alternate form *BA* is present it will act to nucleate more *BA*'s since the slant will be in the opposite direction.

There will be differences of opinion about the relevance these concepts have to biological reproduction but it is useful to note Ashby's reminder that no organism, save the legendary phoenix, reproduces *itself* [48]. There must be a complex interaction between a nurturing environment and a form such as a fertilized egg from which is generated a new form similar to the original. In Von Neumann's model, one machine makes another like itself but in the case of animal reproduction the progeny is originally a much simpler organism than the parent and only reaches maturity like the parent after an extended and complicated period of development [49]. So in this case we have really not succeeded in avoiding the difficulties of trying to understand how an evolving system achieves its full development.

ARE THE ROBOTS COMING?

We have come this far without mentioning robots, although in the popular press and the less imaginative science fiction the robot, usually conceived as a huge, forbidding-looking metal figure, is perhaps the most common 20th Century symbol of man's ability to create a tangible image, albeit a crude one, of himself. The computer is sometimes referred to as a robot, but more often the term connotes the more titillating idea of a mobile creation free to perceive, to pursue and, in general to interact physically with its environment.

By this time it is fairly well known that the word was first used in approximately the sense it now conveys by the Czech writer, Karel Čapek in the title of his play *R.U.R.* (standing for Rossum's Universal Robots), produced in 1921. Robot comes from the Czech *robota* meaning "forced labor." The original developers of the robots, Old Rossum and Young Rossum, (the Czech word rozum means "reason"), never appear in the play. The latter had started a factory to mass-produce soulless humanoid slaves from vats of protoplasm. Now Domin, the archetypical profit-oriented manager, is running the factory which keeps turning out thousands of robots. They will be sold to perform without complaint or emotion man's labors; and even to fight his wars. Helena, the daughter of the president, and wife of Domin, depressed by the lack of feeling in the robots convinces one of the scientists to change the formula to make them more sensitive and understanding. Instead, like Frankenstein's monster, they turn on their creators and at the end have destroyed all of humankind. The reader is left with the feeling that even before the robots began to evolve as conquerors, their makers had sewn the seeds of their own dehumanization and destruction by playing God.

For many years now, engineers have been working on the development of mechanical robots. A part of what they have been doing coincides with research on various aspects of artificial intelligence which have already been referred to, but some of the robots now commercially available have been produced with strictly practical reasons in mind. They are essentially substitutes for man in situations where the work to be done is of a repetitive nature but difficult or dangerous by virtue of certain peculiarities of the environment. The first such robot went on the market in 1963, and by the spring of 1970, 270 had been sold[50]. These are relatively simple looking devices with a single, hydraulically-powered arm extending from a circular metal trunk. The trunk does not move from a given position, but the arm, controlled by a solid-state memory system, can precisely accomplish sequences of five separate kinds of movements: up-down, in-out, rotary motion from the swivel point, and a bending and a swivel motion at the outer end of the arm[51]. This is all, but it is sufficient to do jobs like transferring white-hot metal billets from a furnace to a forging press, dipping pails into a molten zinc bath for galvanizing, or holding a welding gun for spot-welding automobile body panels.

These robots have obvious advantages. They are impervious and tireless, they do not get bored, and they can easily be retooled to do other jobs of the same general type, which basically means "putting and taking." On the other hand, they are quite expensive, especially if they are not kept working almost continuously, and so they cannot really be considered for other than mass production operations. For the first few years of the 1970s at least, the outlook is for a relatively slow increase in the rate at which these industrial robots are put to work.

For jobs which are less repetitive than those discussed above or in Chapter 3 — where manipulations to be performed are sensitive or complex — automatic machines have not been able to eliminate the need for human control. The human ability to integrate continuous sensory perception and manipulative dexterity with low power consumption is not to be underestimated. Many industrial activities require that the unique human capabilities be extended or amplified by mechanical means wherever the job requirements call for greater speed, strength or exposure to potentially dangerous conditions. The steam shovel is a familiar example of such a symbiotic relationship between man and machine, but far more versatile and sensitive devices are now being developed.

Mosher[52] has described a system with feedback controls which has a demonstrable kinesthetic sense. Its ability to maneuver with the proper delicacy and degree of force is comparable to that of a human in some situations. Often developments of new technologies have been delayed by attempts to copy from nature. The flying machine is an example. However, the system Mosher called the Cybernetic Anthropomorphous Machine (CAM) actually couples a human operator to the device by a "follower rack" which senses and measures the positions of his relevant joints, acting together as a kind of template, as he goes through the desired manipulative motions. Signals are thereby generated to operate the powered joints of the remotely situated machine in the same fashion but without some of the human limitations. The human can observe the movements of the machinery directly or indirectly. Either way there is a problem of orientation for which a training period may be required.

This type of machine is still very far removed from a robot with a computer brain which can observe, make independent decisions and function freely and effectively in a variety of environments. Models leading to the latter kind of capability are at about the stage of "hand-eye" experiments, the eye being a television camera and the hand, a flexible vise-like contrivance. A coupled computer can read into its memory what has been "seen" and can direct the hand to perform simple functions such as stacking blocks the camera has located, although the programming is not at all simple.

MIND AND MATTER

> "Numberless are the world's wonders, but none more wonderful than man."
> SOPHOCLES, *Antigone*

The answer to the question heading the previous section seems to be: "No, the robots are not coming soon," at least not the humanoids, fearsome or benign, which have populated a certain class of our fiction for so many years. Some computer scientists are willing to assert flatly that the mind of man is unique and unknowable, and that the bulk of artificial intelligence research which attempts to transcend certain modest limits is doomed to frustration. They foresee complexity compounding complexity in an infinite regression instead of an eventual unravelling of the brain's secrets.

History, of course, has a cautionary message for those who would insist on man's uniqueness. Mazlish[53] reminds us that the concept Western man has retained of himself as above and forever distinct from all other creatures or inanimate things has received some powerful buffetings. The challenging, by Galileo, Copernicus and Giordano Bruno, of the official dogma that the Earth is the center of the Universe was bitterly resisted by the Christian Church in the Renaissance. Also repressed and forgotten for many centuries only to be revived in the time of the New Science was the theory of the pre-Socratic Greek philosophers—Thales, Leucippus and Democritus—that all matter was the same and could be reduced to absurdly small particles surrounded by space. To know that man, at the most fundamental level, is composed of the same stuff as the lowliest plant or rock was grist for the materialists to grind, and a source of heresy. Can man, indeed, be described as *nothing more than* an intricate assemblage of atoms?

Charles Darwin and Thomas Huxley forged another link—between man and the beasts—and shocked followers of fundamentalist creeds into an angry confrontation with science which has not subsided. Strong forces, particularly in the southern and western parts of the United States, still actively resist the teaching of the Darwinian concept of evolution as an acceptable scientific theory.

Sigmund Freud chose to include himself among the list of those who had divested man of his separate and central status in the Universe. Freud felt he had shown that man's vaunted free will was nothing of the sort—that human behavior is shaped and driven by the shadowy subconscious of which the individual has no awareness and over which he has no control. Freud's contentions are more tenuous than Darwin's and less influential now than at one time, but unquestionably the powerful light which they threw on the sources of man's actions has helped lead to a more organized picture of the mental processes.

Not a few subscribe to the belief we now must learn to accept the idea that just as there is no basic discontinuity between man and the other animals, neither is there between man and his machines. The man-machine dichotomy (with God tending the soul) was, as has been stated before, a 17th Century concept. Mazlish quotes from Descartes' reply to the question: How would a mechanical man operating entirely according to natural laws be different from a living human? The reply [54]

> ... if there were a machine which had such a resemblance to our bodies, and imitated our actions as far as possible, there would always be two absolutely certain methods of recognizing that it was still not truly a man. The first is that it could never use words or other signs for the purpose of communicating its thoughts to others, as we do. It indeed is conceivable that a machine could be so made that it would utter words, and even words appropriate to physical acts ... but it could never modify its phrases to reply to the sense of whatever was said in its presence, as even the most stupid man can do. The second method of recognition is although such machines could do many things as well as, or perhaps even better than men, they would infallibly fail in certain others, by which we would discover that they did not act by understanding, but only by the disposition of their organs.†

These questions which have puzzled men for centuries are obviously not going to be answered here. All we can do is assert a few points in something less than a philosophically rigorously framework, and tentatively advance a few more.

First, is man unique? In some respects he is surely unique. Certain observable characteristics, among which one of the most remarkable is the use of language, are not possessed by other creatures nor by machines in any ordinarily acceptable sense. The concept of language is expressly relevant to an examination of thought processes. Indeed, Sapir[55] and Whorf[56] have contended that the relationship is central because a person's view of reality is confined within the language he speaks. If this should be so, the construction of a machine which could carry on, indefinitely, an intelligent conversation in natural language would be a demonstration of thinking of an order that would meet any reasonable criterion.

Assuming that man has unique characteristics, will they be forever unique? This is quite another question and it bears on the possibility of developing intelligent machines. Theorists have suggested it cannot be proven impossible to design (this presupposes the ability to construct) machines to carry on certain processes which any impartial observer would call "intelligent." Even if true, this is not the same thing as saying that such artefacts *will* emerge, although fundamental scientific advances have sprung from no more encouraging bases than this.

If scientists could develop an artificial brain, would it also function as

---

†From René Descartes: *Discourse on Method*, translated by Laurence J. Lafleur, Copyright © 1950, 1956, by The Liberal Arts Press, Inc., reprinted by permission of The Liberal Arts Press Division of the Bobbs-Merrill Company, Inc.

a mind? This forces one to consider the classical physico-philosophical question of mind-body dualism of which the mind-brain problem is a special case. The mind-body question has been called a "non-question" not only because of the ambiguity of the words needed to discuss it, but also because few scientists or philosophers would any longer accept that the mind and the body are two separate entities. Materialists would have it that matter is all that exists, (accepting that matter is interchangeable with energy); that the world makes itself apparent to the material brain through the senses; and that every state of individual being is ultimately describable in physical terms. To the average individual this view is less congenial than his intuitive feeling that the events which come into his consciousness from the external world and those—such as his intuitive feeling—which occur as seemingly spontaneous thoughts independent of outside stimuli are of two different kinds.

A position alternative and preferable to the materialist's would be that both kinds of pictures, either externally originated or internally composed, are constructs of the mind which is the only reality. That is to say, there is no unbridgeable discontinuity between conceptualization and perception. Consciousness, which has been considered proof of the existence of mind, turns out to be a small part of the psychic events which include the larger realm of the unconscious, and thus mind and brain are seen to merge[57]. It is not surprising that physical illnesses will stem from psychic traumas, just as physical or chemical treatments will affect the mental state.

We begin to find a possible key to the philosophical dilemma as we shed the "thingness" of our thinking. The dematerialization of matter by modern physics makes it necessary to refocus our concept of reality in the direction of events rather than objects as the characteristic phenomena of human existence. An individual does not have to perceive given sense data the same as any other individual. The brain does not afford descriptions in any absolute quantitative measure, but because of its capacity for simultaneous parallel activities it is singularly well-suited to deal with comparisons and relationships between patterns. Those events having properties which can be described in terms of such relationships are said to have meaning. Doran has termed the mind a social phenomenon [58]. Each person, physically unique at birth, learns by stumbling and making mistakes, but they are mistakes made, recorded and corrected within the milieu of social tradition and the spoken word. The concept of the outside world only develops with time and it is conceived as temporally changing. The comparison the individual mind makes with all subsequent inputs is against the structure of previously experienced formal and informal acts and their consequences, somehow rearranged, grouped and coded in the nervous system.

Man is what he is not only because he is descended from other men, but also because he lives, works and plays among them. That is why the development of machines endowed with the whole range of human intelligent responses seems unlikely. However ingeniously men are able to design mechanisms capable of performing some human-like functions, human-level intelligence would be denied them without their having the total human experience.

Should we then expect that at some future time, as machines proliferate, they would interact with each other (and with man in special contexts) to evolve their own culture and their own kind of intelligence? One might say that if it were to happen we would hope they would be good neighbors; but in fact, it seems fanciful at this time to expect any artificial intelligence systems to develop which do not have a substantial human component. Fein[59] suggests that some scientists might well spend a part of their time searching for "postulates of impotence" which would assert the futility of trying to develop automata to copy all human intelligence as surely as the Second Law of Thermodynamics asserts the impossibility of building a perpetual motion machine. Perhaps, too, man should spend more time studying his own behavior in the company of machines, to determine if the nature of his responses is becoming restricted by limits set by those machines.

REFERENCES

1. Benet, Stephen Vincent, "Nightmare Number Three," *Selected Works of Stephen Vincent Benet*, Holt, Rinehart & Winston, Inc., N.Y., 1935.
2. Weiner, Norbert, *Cybernetics*, M.I.T. Press, Cambridge, Mass., 1948, (Paperback Edition, 1965), p. 177.
3. Ball, W. W. R., *Mathematical Recreations and Essays*, (revised by H. S. M. Coxeter), the Macmillan Company, 1962, 418 pp.
4. Barlow, Fred, *Mental Prodigies*, Philosophical Library, N.Y., 1952, 356 pp.
5. Bernstein, Jeremy, *The Analytical Engine*, Random House, Vintage Books, N.Y., 1966, pp. 19, 20.
6. Barlow, Fred, op. cit., pp. 21–27.
7. Barlow, Fred, op. cit., pp. 66–67.
8. Ball, W. W. R., op. cit., pp. 357–62.
9. Barlow, Fred, op. cit., p. 34.
10. Luria, A. R., *The Mind of a Mnemonist*, (translated from the Russian by L. Soloturoff), Basic Books, Inc., N.Y., 1968, 160 pp.

11. Luria, A. R., ibid., p. 45.
12. Bellman, Richard, "Control Theory," *Scientific American*, Vol. 211, No. 3, September 1964, p. 200.
13. Clarke, Arthur C., "Are You Thinking Machines?" *Industrial Research*, March 1969, p. 50.
14. Turing, Alan M., "Can a Machine Think?" *The World of Mathematics*, edited by Newman, Vol. 3, Simon & Schuster, N.Y., p. 2099.
15. Eccles, Sir John, "The Synapse," *Scientific American*, Vol. 212, No. 1, January 1965, p. 56.
16. Singh, Jagjit, *Great Ideas in Information Theory, Language and Cybernetics*, Dover Publications, Inc., N.Y., 1966, p. 132.
17. McCulloch, Warren S. and Pitts, Walter H., "A Logical Calculus of the Ideas Immanent in Nervous Activity," *Bulletin of Mathematical Biophysics*, Vol. 5, University of Chicago Press, Chicago, 1943, pp. 115–33.
18. Moore, Edward F., "Mathematics in the Biological Sciences," *Scientific American*, Vol. 211, No. 3, September 1964, p. 154.
19. Cole, R. Wade, "Artificial Intelligence," unpublished manuscript, December 15, 1966.
20. Arbib, Michael A., *Brains, Machines and Mathematics*, McGraw-Hill, Paperback Edition, N.Y., 1965, p. 45.
21. Minsky, Marvin and Papert, Seymour, *Perceptrons*, M.I.T. Press, Cambridge, Mass., 1969, 258 pp.
22. Wiener, Norbert, *God and Golem, Inc.*, M.I.T. Press, Cambridge, Mass., 1964, chapter 2.
23. Cole, R. Wade, op. cit.
24. Samuel, A. L., "Some Studies in Machine Learning Using the Game of Checkers," in *Computers and Thought*, edited by Feigenbaum, Edward A. and Feldman, Julian, McGraw-Hill Book Co., Inc., 1963, pp. 71–105.
25. Samuel, A. L., "Some Studies in Machine Learning Using the Game of Checkers. II – Recent Progress," *IBM Journal of Research and Development*, Vol. II, No. 6, November 1967, pp. 601–17.
26. Samuel, A. L., op. cit., (Computers & Thought).
27. Samuel, A. L., op. cit. (2nd ref.) p. 616.
28. Newell, A., Shaw, J. C. and Simon, H. A., "A Variety of Intelligent Learning in a General Problem Solver" in Yovits, M. and Cameron, S. (eds.), *Self-Organizing Systems*, Pergamon, N.Y., 1960.
29. Simon, H. A., *The Sciences of the Artificial*, M.I.T. Press, Cambridge, Mass., 1969, pp. 66–67.
30. Garvin, Paul L., "Language and Machines," *International Science and Technology*, May 1967, pp. 63–64.
31. Peschke, Sergei, "Machine Translation – The Second Phase of Development," *Endeavor*, Vol. XXVII, No. 101, May 1968, p. 97.
32. Chomsky, N., *Syntactic Structures*, Mouton & Co., The Hague, 1957.
33. Singh, J., op. cit., p. 285.
34. Fink, Donald G., *Computers and the Human Mind*, Doubleday-Anchor, Garden City, N.Y., 1966, p. 229.
35. Bowers, Dan M. and Fisk, Miles B., "The World's Fair Machine Translator," *Computer Design*, April 1965, p. 20.

36. "Language and Machines: Computers in Translation and Linguistics," Publ. No. 1416, National Academy of Sciences, Washington, D.C., 1966.
37. Garvin, Paul L., op. cit., p. 65.
38. Raphael, Bertram, "SIR: Semantic Information Retrieval," in *Semantic Information Processing*, edited by Marvin Minsky, M.I.T. Press, Cambridge, Mass., 1968, p. 35.
39. Glaser, Peter E., Jolkovski, Robert M., Margueron, Claudio and Noel, Walter M., *Space Technology Transfer and Developing Nations*, NASA CR-1222, October 1968, pp. 124-25.
40. Lindsay, Robert K., "Inferential Memory as the Basis of Machines Which Understand Natural Language," from *Computers and Thought*, op. cit., pp. 217-33.
41. Salton, G., "Automatic Text Analysis," *Science*, Vol. 168, April 17, 1970, pp. 335-43.
42. Lancaster, F. W., "Evaluation of the Operating Efficiency of Medlars," *Nat. Libr. Med. Final Report*, 1968.
43. Reimars, Paul R. and Avram, Henriette, D., "Automation and the Library of Congress —1970," *Datamation*, Vol. 16, No. 6, June 1970, pp. 138-43.
44. Von Neumann, John, "The General and Logical Theory of Automata," in *Cerebral Mechanisms and Behavior*, edited by Lloyd A. Jeffries, John Wiley and Sons, Inc., N.Y., 1951.
45. Simon, Herbert A., op. cit., p. 113.
46. Moore, Edward F., "Mathematics in the Biological Sciences," *Scientific American*, Vol. 211, No. 3, September 1964, pp. 154-55.
47. Penrose, L. S., "Mechanics of Self-Reproduction," *Annals of Human Genetics*, 23, Part I, November 1958, pp. 59-72.
48. Ashby, W. Ross, "The Self-Reproducing System," in *Aspects of the Theory of Artificial Intelligence*, edited by C. A. Muses, Plenum Press, N.Y., 1962, pp. 9-10.
49. Apter, Michael J., *Cybernetics and Development*, Pergamon Press, Oxford, 1966, pp. 64-65.
50. Stephan, Jane, "Engineer's World," *Engineer*, May-June 1970, p. 4.
51. Harrah, Jule, "What Automation Can Do With Industrial Robots," *Metal Progress*, Vol. 97, No. 2, p. 127.
52. Mosher, Ralph S., "Handyman to Hardiman," *S.A.E. Transactions*, Vol. 76, 1968, pp. 588-97.
53. Mazlish, Bruce, The Fourth Discontinuity, *Technology and Culture*, Vol. 8, No. 1., January 1967, pp. 1-15.
54. Descartes, René, *Discourse on Method*, (translated by Laurence J. Lafleur, Bobbs-Merrill, Indianapolis, 1960, pp. 36-37.
55. Sapir, Edward, *Culture, Language and Personality*, University of California Press, Berkeley, Calif., 1958.
56. Whorf, Benjamin L., *Language, Thought & Reality*, edited by John B. Carroll, N.Y., J. Wiley & Sons, Inc., 1956.
57. Von Bertalanffy, Ludwig, *Robots, Men and Minds*, George Braziller, N.Y., 1967, p. 96.
58. Doran, F. S. A., *Mind, A Social Phenomenon*, William Sloan Associates, N.Y., 1952.
59. Fein, Louis, "Impotence Principles for Machine Intelligence," in *Pattern Recognition* edited by Laveen N. Kanal, Thompson Book Company, Washington, D.C., 1968, p. 443.

8
Technology, Science and the Arts

"When nations grow old the Arts grow cold,
And Commerce settles on every tree."
WILLIAM BLAKE (1757–1827)

The rebellious, mystical Blake was often difficult to understand, his moods were changeable and his statements inconsistent; but still he saw as clearly as any other artist early in the 19th Century that the wedding of commerce and the machine would produce a different, fragmented kind of world in which artists would have more trouble finding their way. Of course since his time many artists have succeeded in creating exciting new forms of expression but as the pace of life has continued to quicken, and the freedom to "enlarge the boundaries of the possible," (to use the words of Debussy), has become virtually unlimited, one can wonder if creativity has not become overshadowed by random experimentation.

There is a body of opinion holding that technology and the affluence it has produced in the highly evolved Western nations have so accelerated the rate of artistic innovation that the arts are losing inspiration and closing in on themselves in a sterile cycle. Roy McMullen has examined this thesis in his book, *Art, Affluence and Alienation*[1]. He concludes that it seems possible some of the modern arts may run out of interesting new ideas to explore by, say, the 21st Century, but he doubts that they can retain their vitality by resorting to attempts to recreate the past. Is there, then, any alternative to plowing old furrows? Perhaps, he suggests. By then the modern cultural dynamism which demands continual stylistic innovation may have been displaced by a new acceptance of the traditional, not as models to be copied but as part of an eclectic, atemporal intertexture of the arts of all periods and places. This could be closer to art as it was in the older times—a fusion of craftsmanship, religion or magic, and folk customs; all color, action, and participation—instead of what it has become.

Many modernists reject the assumption that the arts are moribund in any sense. They look for endless possibilities for novelty and fresh experience in the waves of contemporary technological and scientific achievements. As long as science and technology progress it would appear that art should be able to progress with them. Why not engage the computer, electronic synthesizers, plastic-forming machines, the laser beam and inert gas-shielded welders in the creative process? Why not, indeed? The materials of the painter, the tools of the sculptor, and the instruments of the musician are all products of technology, albeit sometimes a crude technology developed by the artist himself. Art has always relied on it to some degree. However, the relationships between art and the newer technologies have become more subtle and complex. Composers can use the electronic media to produce dimensions of sound never heard before, well beyond the constraints imposed by ordinary instruments. The invention of the camera was not the only reason that painters stopped concentrating their art on the reproduction of subjects in their visible surroundings but undoubtedly it was one. Now the electron microscope has unveiled a hitherto hidden world, so that the art of abstraction finds it is duplicating other patterns of nature. The programmed computer produces pictures or musical compositions which appear to have the same qualities as other works judged by critics to be art. These kinds of developments are raising questions about the arts in the technological society which are not easy to answer, but which we shall wish to examine in this chapter.

THE "TWO CULTURES"

There are many uncommon interfaces between technology and art or science and art, but also many that turn out to be familiar. One purpose of exploring these interfaces is to examine implications of the proposition that in a period in which the nature of work seems to be changing, the arts may be due for a renaissance. Another is that in a technology-dominated culture, no less than in any other kind, the arts of the time are organically related to other prevailing customs and are among the most sensitive indicators of social process and change. This hypothesis we can now accept with little question, but as Raymond Williams contends[2], prior to the 19th Century, men who wrote about art were generally inclined to refer to it as something entirely apart from and unaffected by trade, or political economy or war.

By the middle of the century, art critics had also become social critics. They were moved by the Victorian doctrine of progress, but to them the progress that counted was that toward beauty and truth expressed in art. John Ruskin is perhaps the foremost example, although his conception of the goal of art was almost a retreat toward a Gothic ideal, at least as a solid rock from which to embark toward a new tradition. For him, Renaissance art relied too much on the systematic methods of the new science, Baroque art was grotesquely contrived, and the art of his own era, corrupted by the deadening force of the new machines.

Ruskin's rejection of industrial vulgarity was only one manifestation of a continuing controversy which had its origins with the Industrial Revolution. Its most recent re-elevation to the status of cause célèbre came as a result of the sharp response of the literary critic, Dr. F. R. Leavis to the 1959 Rede Lecture of Sir Charles Snow on "The Two Cultures and the Scientific Revolution." It would appear by now that almost too much has been written and said about the debate because to some extent from the beginning it was based on ill-defined premises and misunderstandings. The reader may recall that Lord Snow made what was essentially a reasoned argument in favor of the need for new attempts to span the gulf between humanists and scientists — a gulf that had all but frozen communication between the two disciplines. Snow later acknowledged he had oversimplified the issues, especially by assuming a simple bi-polarity which did not include the social sciences and by not sufficiently recognizing the communications gaps within the disciplines. Perhaps none of the points drew the lightning from Leavis and others more than the assertion that an understanding of the Second Law of Thermodynamics was as valid a mark of culture as a familiarity with Shakespearean drama. It was possible to question this parallel without going so far — some did — as to hurl the charge of philistinism at a man who merits attention. His advocating a measure of science training for everyone who really hopes to achieve a relatively undistorted picture of the world was unexceptionable.

A number of writers [3, 4, 5] have observed that basic features of the arguments were raised before, notably in another Rede Lecture by Matthew Arnold in 1882, directed as a reply to an address entitled "Science and Culture" given by Thomas Henry Huxley. Huxley had contended that the monopolization of the curricula in the universities by the classics to the virtual exclusion of any worthwhile science training was no longer acceptable. He made clear that he was not solely concerned with the status of science as a necessary ingredient of culture. The application of science to manufacturing, replacing rule-of-thumb procedures, was indispensable to the prosperity of the nation. Nevertheless, a definition of culture and a consideration of the means to acquire it were very much on his mind. He was willing to agree to the assertion contained in Arnold's writings that the essence of culture was the criticism of life, but not to the additional assumption that all of the elements needed for a searching criticism could be found in literature, nor (as the consensus of the time would have it) that culture should be equated with morality. To Huxley, the most distinctive difference between the modern world, (of 1880), and earlier times was the new understanding of science. It was not too difficult to agree with this view of his taken alone, but certain excessive claims for the efficacy of the scientific method were less well received. His declaration that "for the purpose of attaining real culture an exclusively scientific education is at least as effective as an exclusively literary education" was hard for classicists to swallow.

Matthew Arnold was a religious liberal who inherited from his father, Dr. Thomas Arnold, the famous Rugby schoolmaster, a deep concern about the effects of industrialization, particularly with regard to the widening chasm between rich and poor caused by the application of the *laissez-faire* doctrine. He wrote, in *Culture and Anarchy*, that culture, the "pursuit of total perfection," must be the key to a new orientation in society in which "reason and the will of God" would prevail. The middle class—the Philistines as he called them—were primarily responsible for the grossly mechanical and materialistic style of living which existed in Victorian England. As the most energetic and influential group in society, they would have to be weaned from their coarse habits and pleasures and moved toward a state of perfection by a system of humane education in the "best that has been thought and written." This typical 19th Century elevation of culture almost to an aerie for an elite still has an appeal for many humanists who hold that the humane intellectual tradition provides the thread of continuity sustaining standards of taste and sensitivity.

It is unlikely that the idea of permanent, transmittable, intellectual standards of culture has ever been challenged so strongly—by the young and by a growing number of their teachers—prior to the present period. Roush[6] contends that as technology pervades society the resulting democratization of education and power forces a massive change in value systems, and the past becomes a museum to many of the young rather than a temple. Value, or "goodness," then, (if it is accepted as a valid concept at all) derives from the perceiver and does not cling to any specific work of art. From this commitment to impermanence and change—this dismissal of the past as almost irrelevant—come some clues about the state and direction of modern art.

Every culture in each era must look at art in its own way, even though there are some similarities in aesthetic responses that span the centuries. In more primitive cultures and earlier times art has been sensed and accepted by all within the community as a living function. There was no need to search for its meaning. However, the division of labor in highly industrialized societies has fragmented human functions. The artist, the academician and the lay critic have all sought to apply their own standards to art, with the result that no generally accepted standards can be found. To some, this is intolerable. Others are willing to accept this absence of definition, believing that the concept of art must be fluid and changeable; and that with the pressures and opportunities of technology accelerating those changes, past standards can only become arbitrary dogmas for the present.

WHAT IS ART?

In our times when science and technology provide the means for terminating human existence or altering society beyond recognition, there was an understandable urgency in Lord Snow's plea to "breach the gulf of mutual incomprehension" between men and women of science and of the humane arts. Yet there has always been a relation between science and the arts which is more homologous than antagonistic. To demonstrate that this is so, one ought to begin by specifying as clearly as he is able what he means by art. (A definition of science has already been essayed in Chapter 1.)

Philosophers and artists have tried for centuries to develop a theory of art but no matter how wisely or ingeniously they have proposed their definitions they have invariably ended with paradoxes and contradictions. By now we have learned to accept this imprecision. One of the most difficult relations to see clearly is that between art and beauty. Aesthetics is the name given to the branch of philosophy which studies this relationship. Apparently, Alexander Baumgarten coined the name in his book, *Aesthetics*, in 1735, although aesthetic thought really goes back as far as Plato. Plato had fixed ideas of the arts as specific realms of being. Works that we now consider artistic masterpieces—the Iliad, for example—failed to meet his standards of perfection calling for an ultimate vision of truth and beauty reached through abstract contemplation rather than sensuous physical experience. Nevertheless visual beauty or erotic experience could help guide the beholder. The Greeks had no word for what we might now call art in its aesthetic sense. They did draw a distinction between the useful and the fine arts and did not deny the virtue of craftsmanship in either.

Aristotle saw nature and art as the same. He believed that a human enterprise, whether it happened to be the creation of a statue or a state, works inherently in the direction of a fully developed form which is its ideal nature. Unlike Plato, he was capable of finding artistic virtue in poetry since it could purge impure emotions and allow man to reach the highest state of intellectual joy. Aristotle had a penchant and a gift for classifying. Following his lead the medieval scholastics set up categories of human activities, classifying the humanities (arithmetic, music, astronomy, geometry, grammar, rhetoric and logic) as liberal and therefore superior, while assigning manual or physical activities, including architecture, sculpture, painting and music to a lower category of mechanical arts.

In the Christian era, the Platonic idea of beauty had been further refined by St. Augustine and then St. Thomas Aquinas into an abstract concept of unity, proportion and clarity, the manifestation of which should be pleasing to the beholder.

It was in the time of the Renaissance that the artist came into his own, respected as a masterful creator and as an individual, conscious of his own significance, sensing the quickening spirit of science about him, and sometimes seeking inspiration from nature to complement the ancient traditions. If there was any thoughtful reflection on the nature of art and its aesthetic value it had not cohered into a mature philosophical system, yet creative spirits in all of the arts, (speaking of art in the modern sense), were becoming aware of a commonality of vision and a feeling which formed vital bonds between them.

By the 17th Century and into the 18th, reason prevailed, and rational thought was bent on perceiving and following a structure of rules extracted from the patterns of ancient wisdom, but fortified with empirical knowledge. Following the rationality inherent in nature, the poet or the musician would not fail to give pleasure by his art, and thus create beauty. Alexander Pope wrote:

All Nature is but Art, Unknown to thee
All Chance, Direction which thou canst not see,
All Discord, Harmony, not understood...

The beauty in poetry was conceived to be a matter of form and proportion rather than passion or imaginative metaphor. The ideal form was ultimately mathematical. In fact a committee of the Royal Society in the 17th Century set up for the purpose of studying the English language recommended a "return to primitive purity... bringing all things as near the mathematical plainness as they can."[7] This was all a part of the Cartesian view of the world which held that the best approach to reality is through mathematics. As Harries suggests, such an approach could only result in a down grading of the sensuous in the arts[8].

By the middle of the 18th Century there was an accepted discipline of aesthetics soon to have its own schools of thought applying the rigors of philosophical analysis to ancient questions. German philosophers and writers were influential, and among them, Immanuel Kant (1724–1804) became pre-eminent. Kant would not accept the use of scientific or metaphysical standards of reason for trying to understand concepts beyond the reach of the senses. Science could explain the visible world only. Beauty was a key phenomenon capable of giving sensuous pleasure. It could intuitively, but never rationally, be recognized as contributing to a humanized spirituality. Beauty was the bridge between the knowable and the unknowable.

No one could have been more typical of the contemplative scholar, steeped in reverie, than Kant. But as Sypher[9] has phrased it, the 18th Century reverie was about to become an illumination with the emergence of romanticism. The new, creative romantic tradition was informed and sanctioned by Kant's rejection of the adequacy of pure reason but he, on the other hand, would not have been very comfortable with the basic romantic emphases. Romanticism was a philosophy which exalted the real beauty, and therefore value, of nature in its every aspect. It was an emotional tradition rather than ratiocinative, mystically aware of artistic inspiration as a kind of sanctified experience.

The romantics were revolutionaries—their emotions were revolting against the coldness of the intellect, as Hauser puts it[10]. Essentially, too, the movement was a revolt of the rising middle class against the sophistication and the decadence of the aristocracy, and in its reaching for individual freedom it was linked to the *laissez-faire* economic spirit. However romanticism could not help but become, in the industralized nations of Western Europe, also a crying out for some way to escape from the excesses of industrialism.

On the Continent, Rousseau's naturalism provided the model for the romantic wave, but it was in Germany, rather than in France, that the spirit took hold with a peculiar intensity. There the middle class intelligentsia had been kept in the service of the petty princes in subordinate careers as schoolmasters or minor officials. Their exclusion from any position of influence had bred in them a tendency to create for themselves a picture of an idyllic, hazily-conceived inner life, quite divorced from sober practicality. So although their idealism was rooted in Kantian philosophy, it turned toward a pronounced anti-rationalism, constantly searching for original experience. The aesthetics of art were expressed in terms of a cosmic spirituality which denied the primacy of a reflective intelligence.

The philosopher who perhaps best expressed the spirit of romanticism in relation to art was Schopenhauer. Like Kant he felt beauty was a key to the unknowable world. He defined art as "the way of contemplating things, independent of the principle of understanding, in opposition to the way of contemplating things which proceeds in accordance with that principle, and which is the method of experience and science." To see truth, man must rise above his will, which can be done effortlessly in the presence of art. He wrote "Every work of art... really aims at showing us life and things as they are in truth but cannot be directly discerned by everyone through the mist of objective and subjective contingencies. Art takes away this mist.... Music expresses in a perfectly universal language... the inner structure, the in-itself of the world." As for the poet, "... all that human nature in any situation has ever produced from itself, all that dwells and broods in any human breast, all this is his theme and his material.... He is the mirror of mankind, and brings to its consciousness what it feels and does." So at last the philosopher, who since Plato, had been the arbiter of truth, had conceded his place to the artist.

A term like romanticism is a tempting convenience but if it is used to confine artists and a style within a limited historical period it gives a false picture of uniformity where none exists. In a broad sense, some manifestations of romanticism are evident from the time of the French Revolution throughout the 19th Century, and, in music at least, up until World War I. In fact, according to Hauser, all the exuberance and anarchy of modern art are related to romanticism [11]. Within the period generally described as romantic, moreover, there were many different styles and artists whose art showed more classical than romantic influences. In fact, the whole romantic movement had many ambivalences, but its primary significance seems to be that it represents one side of an abiding polarity between a subjective, full-blooded, individualistic sensibility on the one hand, and a more conservative, intellectual acceptance of the universality of certain rational standards on the other.

Keats expressed the anti-rationalism of the romantic movement this way:

> Do not all charms fly
> At the mere touch of cold philosophy?
> There was an awful rainbow once in heaven:
> We know her woof, her textures; she is given
> In the dull catalogue of common things.
> Philosophy will clip an Angel's wings.

In some periods, the predominant style of the arts tends to be more rational than irrational, more abstract than sensual, and then it changes back again. While these attributes can serve as descriptions of art they do not satisfy as a definition. To narrow the search toward a definition, some philosophers of art have tried to break it down into subclassifications in the manner of Aristotle and the medieval scholastics. Hegel was one who tried. But the problem of categories is that particularization robs a sweeping abstraction of its essence about as surely as a dissection will rob an organism of its life. Benedetto Croce has said there can be no classification of an activity for which limits cannot be described, and the polar limits mentioned above are hardly more definable than art itself. In terms of its Latin antecedent (*ars*), art was thought of as a specific kind of skill or ability. More often now we think of it as the product of that skill, which may stimulate a response in a listener or beholder—perhaps a quickening of interest, a wave of nostalgia, or a sense of sublimity or fulfillment. Can we be sure that art needs a vessel at all? Can art exist without an audience? As for its aesthetic value, aestheticism is not so pure a concept as it seemed to many art critics in the later 19th Century. The term smacks of a false elitism in the minds of moderns, and the moralizing of that earlier period is rejected almost out of hand.

It is no help either to try to draw a line between fine arts and applied arts. Is architecture any the less a fine art for being useful? An ancient urn, hand made by a primitive technology, might at first have had whatever beauty it possessed subordinated to its utility. Now shorn of serviceability, it can be appreciated by the modern day beholder for the grace of its lines (for which its maker may not consciously have striven) as well as for the entire imagery evoked by its envisioned history.

It no longer seems right, either, to deny entirely the fact of serviceability to a work of art. Can we then relate its value to purpose in any way? Probably only in a negative sense, if at all. It does seem that a good case could be made for witholding the seal of artistic merit from any creation whose purpose is totally commercial in that it is designed to appeal to the largest possible audience. Such a proscription would then dictate that most products of industrial design could not qualify as art (a standard which some creative people would find unacceptable) but would still allow the artistic value of one-of-a-kind works created with the aid of modern technologies.

What it comes to is that Croce and all of the others who have agreed that no fully satisfactory definition of art is possible, are of course, right. But at least art is what artists do, and those who are not artists can observe the artist at his work, listen to him speak of it, then reflect on their individual reactions and draw their own conclusions about the nature of their experience. We are not even sure a consensus is a valid criterion for judging the worth of a work until time has accorded its sanction. Yet great and minor art does appear and is recognized, and so are the artists. Art may be much else besides the creations of a Rembrandt, a Mozart or a Shakespeare, but these, though created in and for the culture of other periods, we can unequivocally accept as art in our own time.

THE RELATIONSHIP OF SCIENCE TO THE ARTS

In comparing the arts and the sciences, one finds the matter of critical standards to be a major point of difference at the outset. The scientist generally has the means of empirically testing the validity of the theories of another scientist. Sometimes there are experimental problems preventing an immediate evaluation, but if these can be overcome, results can be obtained which can be reproduced at any time within certain limits of accuracy. This constancy attests to the "truth" of a theory. There can be no such empirical test of truth in art, and yet there are some subjective truths that appear enduring. The empirically validated truths of science, moreover, as we have begun to learn in this century, are indisputably true only as part of the picture of the universe accepted at the time. As T. S. Kuhn has pointed out in his book, *The Structure of Scientific Revolutions* [12], the nature of the universe changes whenever certain paradigms, against which all theories are tested, are overthrown. Thus, not all of the scientific truths in Newton's world were exactly the same in Einstein's.

Ludwig Heydenreich[13] has proposed that the differences and similarities between science and art can be understood better if we consider them complementary sides of the creative ability of man, with intuition generally predominating in the creation of art, and reason, in the creation of science. In either, sometimes the mode is more abstract and sometimes more sensory; but always imagination leads to the crucial point of inspiration. Russell[14] suggests that the difference between specific arts or sciences may be as great as those between what we call art and science; and that important creations of both scientists and artists seem to result from much the same kind of mental processes. There is, at first, a time of preparation during which the individual soaks up knowledge of many things. It is a period of information gathering, of observation, and of concentrated reflection on the problem and its antecedents. Then there is likely to be a quiescent period followed by a sudden flash of insight, and at last an utter certainty that the solution is the correct one. Many artists, and scientists too, have written about the long, wearing effort required before their creation leaped into form in their mind's eye. While our concern of the moment is with comparisons of processes of creativity in art and science, there is no reason to think that the psychological processes resulting in the conception of significant mechanical inventions are different. James Watt's moment of inspiration came suddenly, the reader will recall, on a Sunday afternoon walk after just the kind of period of preparation described. Although this description of the stages of creation will probably suffice for most situations, some of the greatest works of art have been produced on order, to keep food in the artist's mouth. We cannot capture the genius of Mozart with simple formulas. Anyone who wishes to probe deeper into the subject of creativity in science and art should make a special effort to read Arthur Koestler's splendid book, *The Act of Creation*[15].

The polarity between science and art expressed as reason versus intuition does not uniquely distinguish them. As was discussed earlier, a comparable polarity, represented at one time in history by the transformation from classical to romantic modes of thought and expression, can also occur within the arts. In contemporary science it is unquestionably the case that intuition is not so highly esteemed as it once was; but few can deny it a measure of importance in conceptualization, or a distinctive place in discovery. It also seems apparent that among many younger people who have been attracted to science and who may have an undeniable talent for it, there are signs that criteria which from the point of view of traditional science might be considered intuitive, emotional, and out of place are having an increasing influence on their view of purpose.

Roger Fry[16] has observed that the opposite yet supportive characteristics of particularizing and generalizing in science, (represented on the one hand by curiosity for and seeking out of facts, and on the other by constructing general systems of relationships), have recognizable counterparts in art. The curiosity of the artist is expressed in his constant search for new ways to encounter nature in all of its variety; and his need to generalize, in his attempts to gather emotion or experience into universal form.

It is sometimes said that the primary objective of artists and scientists alike is to communicate reality in some form, but with the difference that, while the symbols of communication of the scientist are austere and precise, the artist-humanist feels no similar need for economy of expression. At least, although it is true that the best artists choose to avoid overstatement, they are able and willing to use any connotative means at their command. The modern physical scientist often deals with abstractions that have no counterpart in human experience. Mathematics is the only language rigorous enough to express his ideas, (or, perhaps more precisely, to give form to a procedure which leads to similar results), with the clarity required to make them useful to other scientists. In the natural sciences, although the tendency now is to employ mathematics for fundamental communication, there are many topics whose interrelationships are so rich that they cannot meaningfully be reduced to purely mathematical expressions. Natural language is then an indispensable adjunct to, or even a substitute for, the mathematical discourse perhaps preferred among the scientists. Physical and natural scientists are both concerned with some concepts that have a universal significance, and when it seems important to make these concepts understandable to the general public the scientists are encouraged to use more colorfully expressive language. Harold G. Cassidy[17] has noted that the 19th Century physicist, Clerk Maxwell, saw this to be a useful function of the scientist. Maxwell wrote that "for the sake of persons of different types, scientific truth should be presented in different forms and should be regarded as equally scientific whether it appears in the robust form and vivid coloring of a physical illustration, or in the tenuity and paleness of a symbolic expression."

But poets have also been able to convey scientific ideas with a wealth of allusion so that the basic concepts are vividly communicated to the lay reader. Cassidy[18], a scientist himself, has provided an illustrative example of this in comparing two statements of Sir Isaac Newton's inverse-square law of attraction — the mathematical:

$$F \propto \frac{m_1 m_2}{d^2}$$

which says there is a force of attraction, F, between two bodies, directly proportional to the product of their respective masses, designated by m_1 and m_2, and inversely proportional to the square of the distance, d, between them, and the poetical:

> When to the eyes of thee
> All things by immortal power
>   Near or far
>   Hiddenly
> To each other linked are
> That thou canst not stir a flower
>   Without troubling of a star.

The poem, written by Francis Thompson, is in the rather florid "aesthetic" style of the 1890s. Its central idea, the same as that of Newton's formula, is presented in a form that a non-scientist would not ordinarily find exceptionally difficult to understand. The scientist could comprehend both statements but he could use only the first in his work, and would not consider the second as an aid to understanding the theory. The layman would need to have the symbols explained before the formula had any meaning for him. Much poetry is neither very clear, nor meant to be, of course. When modern poets feel that the freshness of ordinary words or traditional patterns of phrasing has been exhausted they may choose to write poetry, by design, as obscure to the non-scientist, or the scientist, as some of the most involved mathematical expressions.

One should not assume that because the formula is predictive, (i.e., it can be used with confidence to predict the motion of two measurable bodies with respect to each other, or to design a system) that it somehow offers a *higher level* of understanding than that provided by the poem. This is not to deny the beauty and elegance of the formula, nor its demonstrated success in dealing with natural phenomena, but merely to reiterate that art can seek understanding in ways that are different but not entirely foreign to the ways of science, nor necessarily inferior. If scientists are justified in making a claim for special status as interpreters of the world, it is because of their willingness to accept the risk of failing to prove their point after following a generally accepted set of rules. The artist can choose his own rules and does not have to feel under an obligation to prove anything. However, if he is a good poet, novelist or dramatist he will not write serious nonsense in his work about factual matters. Alex Comfort[19] proposes that the mathematician occupies a position between that of scientists and artists in that his chosen form of communication, unlike the artist's is unambiguous; but he can select his own system of rules, of the form "if ..., then ...." At the same time, however, he does not have to care whether the relationships ever check out against a physical situation.

Artists and scientists seek to satisfy their curiosity about the world in different ways. In each case their quest is guided by a special sensitivity, and by their talent for developing creative means to help them look for solutions. Even though their distinct modes of expression offer no aid to communication between them, somehow they still seem to act in parallel as both products and precursors of change, mirroring the spirit of their times. At certain periods in history, artists and scientists alike have become aware of unbearable inconsistencies in the contemporary view of nature. For men and women of special gifts these contradictions are chafing bonds from which they must cut themselves free. The Renaissance was such a time, and so were the years which led out the 19th Century and ushered in the 20th. In the Renaissance the problem of communication hardly existed as science and art came together to serve each other. In order to see nature as it was, undistorted by encrusted superstitions of the preceding centuries, scientists and artists (who were sometimes one and the same), probed the visible world. The artist looked to mathematics for aid in representing perspective, to chemistry for improving his raw materials, to physiology for explanation of visual images, and, as in the case of Leonardo da Vinci, to the mechanical sciences perhaps simply out of fascination for the way things worked. Vesalius, the great anatomist of the 16th Century, was one of the foremost examples of a scientist who used artists to help further understanding through illustration[20].

But it is the second period—the latest turn of the century—which is of more interest to us now, because the view of the world which began to be formed then is the one that is still with us. It was a foreign, upsetting concept which rejected the objective reality of the world, judging it to be mere illusion. The year 1900 makes a convenient reference point marking the beginning of the new age in science because it was in that year that Max Planck published his famous paper demonstrating the quantum nature of radiation. It hypothesized the startling idea that the energy of electromagnetic waves is propagated, not continuously as intuition and classical scientific thought would have it, but discontinuously, in discrete jumps. This called for a revolutionary picture of the universe as one in which very small things behave in quite different ways from large things. Just five years later, Albert Einstein's paper presenting his Special Theory of Relativity was published. However, puzzling contradictions in chemistry and physics which led to the discoveries of Planck and Einstein had already appeared in the latter part of the 19th Century. The observation that the visible spectra emitted by electrically excited gases were unexpectedly complex had strongly suggested that the simple picture of the atom emerging from various experiments following John Dalton's original work at the beginning of the century was not correct. Also the Michelson-Morley experiment intending to show the relationship between light and the hypothetical ether through which radiation was presumed to propagate had given no evidence of the existence of the ether. Another paradoxical discovery by Thomson and Kaufman had shown that electrons increased in mass the faster they traveled. So the experimental foundations for the staggering break with the classical Newtonian system were already laid before the present century began.

As for the visual arts, it seems that the fundamental movement away from the older forms had also become evident before 1900, first with the Impressionists in the 1860s. Humanist scholars and artists find it hard to decide among themselves whether the Impressionists, (insofar as the individual artists characterized as such can be considered as having a collective style), represent the beginning or the end of an era. Very probably it was both. In any event it was an important transitional movement in which a group of young painters including Monet, Degas, Manet and Renoir rejected the statically elegant art of the academies, not by denying reality totally but by fragmenting it and catching it on the wing. Impressionists elevated technique above the subject, and sunlit, evanescent moods over permanence. Characteristically they broke up the familiar pictorial continuity into blobs or dots of pure color—in short, the subject was reduced to its elements, and the observer had to make his own integration from a distance. Peter Blanc[21] suggests that there is a clear similarity between the fragmented view of the world adopted by the Impressionists and the 19th Century atomic scientists. It is problematic whether the Impressionists were strongly affected by or fully cognizant of contemporary scientific progress but apparently there was an influence, especially with regard to the physics of light. The neo-impressionists, Seurat and Signac, attempted to reduce their work with scientific detachment to a carefully calculated arrangement of "molecular" dots, and Blanc quotes Paul Gauguin as writing in 1903, "Artists have hopelessly lost their way in recent years due to physics, chemistry, mechanics and the study of nature."[22]

Musicians also worked in the same style to a less marked degree, emphasizing mood above structure and blurring the sharp outlines of their images by using hazy snatches of melody or phrasing. The "vague impressionism" of Claude Debussey's "Printemps" was disparaged by the *Journal Officiel* in 1887. It is less easy to draw a clear connection between literature and impressionism although Sypher[23] notes that Emile Zola, for instance, in *Une Page d'Amour*, develops a hazy, moment-by-moment impression of Paris which gives the same effect as a canvas by Monet. There is also a close tie between impressionism and the symbolist French poets. Men such as Rimbaud, Verlaine and Mallarmé sought to project a sense of the subtle and ineffable through calculation.

The impressionist movement in painting, in its active sense, was not long-lived. There was actually a preparatory phase preceding by a decade or so, the first collective exhibition in 1874 but other trends had begun to appear even before their last exhibition in 1886. Post-impressionism, expressionism and l'art nouveau are some of the other terms used to identify certain characteristic styles of art practiced near the turn of the century, but they are more convenient than definitive. Sometimes the stylistic affinities among artists commonly considered to be part of a given movement are not very prominent. Very often an individual artist – Paul Cézanne is an example – will experiment with many styles during his career, and even a single work may show various influences. Cézanne began as early as 1870 to paint in a manner that reflected his admiration for Manet, but by the end of the decade he was ready to go beyond impressionism to another stage. The post-impressionist paintings of Cézanne, for all their liberated brushwork and freshness, show an intelligent, disciplined conception which is at once more solid and ordered than a typical impressionist work. But the order is his, and where his sense of composition requires that nature be redesigned or distorted, this is the solution he chooses.

This willingness to distort – to break away from pictorial verity – really represents the beginning of modern painting. It was to be an extraordinary artistic revolution to match the remarkable conceptual changes that were taking place in the scientists' view of the physical world. From the time of the Cubists in the first decade of the 20th Century, the important trends in art have been those denying the supremacy of a reality which had to be confirmed by the senses. It has been an art to which the beholder looks in vain for the beauty of the older arts, for indeed the artist has had no wish to cater to the senses either by sentimental or sensual means.

There is perhaps a danger in carrying the parallels between the 20th Century revolutions in art and science too far, but one can ask how might the artist have reacted otherwise to the apparent overthrowing of the orderly logic of cause and effect. It was not just that the science of the 19th Century had pictured the so-called material world to be mostly empty space. After all, if one did not look too deeply, the view of matter as little worlds of atoms consisting of a positively charged nucleus circled by electron satellites was not beyond reason. But then in 1905 Albert Einstein was to show that Newton's majestic definition of space as absolute—"eternally similar and immovable;" and time as absolute—"flowing equably without relation to anything external" was not correct. Einstein had discovered the only way to dispose of the paradoxes, which were troubling his contemporaries in physics, was to conclude that neither space nor time were absolute at all—they depended on the instruments of the observer. The reality of space was contingent upon the observer's perception of the order in which objects are placed in it; and the reality of time upon the sequence of events by which it is measured. There could be no absolute past nor absolute future. From one standpoint space could be considered to exist as a function of the way in which matter and energy were distributed, but it had been shown that matter was converted to energy by motion. Matter and energy were interchangeable and perhaps fictions, neither definable except in terms of something else no more certain. Thus tangibility and permanence were blown away. The arch of certainty under which man had constructed his concepts of reality was shattered.

Most of the younger scientists whose systems of reference had not hardened in the Newtonian mold were prepared to accept the new world on its own terms. Laymen, on the other hand, were hardly aware of the radical overthrowing of the old physical order. They might make little jokes about the dimly comprehended concept of relativity; perhaps a few might deny the conclusions of the new physics as a challenge to faith and reason; but most probably they would continue to ignore them as being of no practical consequence in their daily lives. To a large extent this last assumption remained correct for more than a generation until technology showed that the convertibility of matter to energy, represented by the equation, $E = mc^2$, was indeed a physically demonstrable phenomenon.

Early in the 20th Century, painters and musicians, ever alive to current views of reality and with the way well prepared for them, now moved into another medium. In painting it was the cubists who first began to bring their art into consonance with the new scientific thought. At the focus of their attention was the theoretical manipulation of space. They may have begun with an object, it is true, but only to dissect it—to observe how its form changed when observed from various points of view while abandoning the perspective base. No longer would they limit themselves to the physical dimensions, but they would try to explore the dimension of time. They abstracted geometrical elements and then reassembled them to give a sense of movement through both time and space from contrasting but characteristic reference points.

Pablo Picasso, probably the most renowned and influential painter of the 20th Century, was, along with Georges Braque, the originator of cubism. He was born in Spain in 1881 but his early work appeared in Paris in the post-impressionist period. Of all painters, it would be hard to find any who have had a longer or more productive career, embracing many styles and new directions. At first, he undoubtedly admired such painters as Van Gogh, Gauguin, and Toulouse-Lautrec who are sometimes called expressionists because they seemed to have a powerful need to express their emotions in their art. These emotions were often so passionate, particularly in a painter like Van Gogh, that they could only find realization in exaggerated distortion and vivid colors. However, Picasso's painting in the period from about 1903 to 1905 was not notable for vivid coloring—running rather to sober bluish tones, and was illustrative in character. The early cubist work shows some of the disturbance of expressionism, but the more direct influence seems to be that of Cézanne who had said, "You must see in nature the cylinder, the sphere and the cone." It must also be mentioned that there is a suggestion of the primitivism of African sculpture in the cubists' work as well.

The cubist influence has been very powerful in 20th Century arts not just in painting but also in sculpture and architecture, and even, it appears, in the drama and the novel. Sypher[24] writes of Pirandello's play, *Six Characters in Search of an Author*, (1921) and the narratives of André Gide as essaying to portray a compound image of dramatic or literary subjects given up to dissection and analysis. Past, present and future are drawn into a complex intersection where the boundaries between art and life have been rubbed out.

From cubism it was just a step to abstractionism, which meant a total divorcement from the object. The Russian-born Wassily Kandinsky has sometimes been called the father of non-objective painting. He was one of a group of artists who worked in Munich at the start of the second decade of the century and were referred to as Der blaue Reiter (The Blue Horseman). They were expressionists, too, imbued with some of the emotionalism and sense of deep disillusionment that had characterized certain of the earlier German romantics. Kandinsky, however, unlike some others in the group believed that the emotions he wished to express were contained in the basic elements of the painting—line, color, form and shape—from which spiritual "vibrations" would emanate. It was a versatile company in the sense that the members had a lively concern for the relationships between the arts. Kandinsky felt that abstraction brought painting closer to music, and Arnold Schoenberg, a composer in the group, occasionally attempted painting in the abstract fashion[25].

The distinction between the sensually objective and the abstract in the visual arts is easier to comprehend than in music, although in neither case does it yield to unambiguous definition. An observer viewing a recognizable image in a picture will have certain mental associations which his intelligence derives more or less directly from the objects he sees. These associations are not necessarily the same as those which other observers make but the chances are fairly good that there will be some relation between them. When music affects the individual there can be no direct connection to the object. The associations are more emotional and personal than those from the visual arts and they depend to a large extent upon relations among familiar sounds. Meaning is indeed a very vague word when applied to music but to the extent that it signifies, it is essentially the succession of notes and chords which supplies it. This should be kept in mind if we are to speak of Schoenberg as one of the important figures in the 20th Century movement toward "abstraction" in music, comparing that movement to similar directions taken in science and the visual arts. It might be preferable to refer, instead, to the asensual character and logical design of his music.

The fundamental 20th Century innovation was the change in the manner in which music was composed. For about 300 years prior to the start of the 20th Century, Western music with few exceptions was based on conventions of functional harmony or tonality, using a seven-tone diatonic scale. Actually, the very early beginnings of music for which we have written records (beginning with a few fragmentary examples from the Greeks) had been at first a development of scales and melody, and of written notation. Rhythm, of course, was an integral part of it. A rudimentary rhythm is inherent in many natural processes as well as in some of the earliest human rituals. We can gather from early records that fairly complex rules of rhythm and melody were followed in the production of music. We also know that melody, defined as the organization of consecutive musical notes according to pitch, was sustained in both Greek and other primitive music as a single unharmonic line, with the occasional accompaniment of an instrument. The formal music of the Western world from the early Christian era was that of the Church. Melody usually meant song chanted without accompaniment. By the 9th Century, however, rudimentary efforts at combining sounds into harmony were being attempted. The single line of chant was supported by a rigidly concurrent accompaniment of fifths, i.e., when the voices sang G, this was paralleled by other voices singing C, five notes below. This was called organum, at the time. The octave and the fourth interval were also considered to be consonant.

The development of music for the next several hundred years consisted mainly of advancing the art of combining melodies, known as counterpoint. After the organum was freed, by varying the intervals at which the doubling of a chant occurred, many experiments in performing different melodies at the same time were attempted. One example was the familiar round, the earliest surviving example being the English song "Sumer is icumen in" from the 13th Century. With the recognition that voices could progress even more independently in direction and rhythm, composers began to follow a generalized series of rules in order to avoid harsh sounding combinations of notes from the interweaving of melodies. This ordering of melodies — modal music — became an increasingly formal exercise in the same way that mathematics and science were becoming formalized. Although there was some breaking of the rules by troubadors and minnesingers, the conventions of modality were paramount until the beginning of the 17th Century when composers began to achieve extraordinary innovations. The fundamental idea of the relationship of the melody to the whole composition was changed as various progressions of chords were tried, the bass line became more important, and some of what had been considered dissonance was found to add an emotional impact to music. Unlike the case with counterpoint, the separate parts of the music in harmony did not all have to have a distinct melodic life. A great area of exploration into symmetries of harmony was opened up. Tonality had almost completely replaced modality.

It was this tradition of tonality that seemed to have reached the end of the road at the beginning of the 20th Century. Many composers, of course, continued to write admirable works loosely based on the fundamentals of tonality, often by resorting to nationalistic or neo-classical themes. This essentially conservative path, however, did not appeal to those who were convinced they had exhausted possibilities for new forms of expression within the framework of the existing system. (It is interesting to note that one famous scientist in the 1890s declared that the important discoveries in physics had been made, too, and as far as he was concerned all that needed to be done was to fill in some of the existing data gaps.) The younger composers saw little hope of capturing the power of the masterpieces with new works created in the existing musical vocabulary. Richard Wagner, an unquestioned romantic, had anticipated the crisis and had carried the challenge to the tonal system as far as it would be carried in the 19th Century. Wagner, in his search for dramatic expression, had chosen to develop his themes by means of an intense chromatic style which for substantial periods of time during the composition deferred the classical principles of harmonic logic. His chromatic harmonies were a kind of dissonance—a blurring of tonality—resulting from the constant changing from one key to another during a passage of music. There were numerous examples of isolated dissonance in earlier compositions, but Wagner emphasized it to such a degree that many listeners never accepted his music. With familiarity, however, many other listeners found the harshness of dissonance, when used as an amendment to consonance, making a pleasing contrast.

The early work of Schoenberg had developed in the tradition of Wagner, but from about 1906 on he began to move toward a new idiom increasingly blurring the relationship of tones and chords to a basic keynote. In 1908 he wrote a composition that was not in any key, and thus for Western composers broke the long domination of the principle of tonality. This revolution against tonality carried with it the danger of musical anarchy, but Schoenberg and his followers were intellectuals for whom a degree of order was instinctive. Eventually he perfected the atonal system, (he preferred the term, pantonal), based on a twelve-tone scale using every chromatic tone in an octave. Rather than relying on a central key-tone, each of the tonal centers is separately related to the others, and all are of equivalent importance. In the strictest form of twelve-tone compositions none of the notes can be heard for the second time before the rest of the notes in the series have been sounded. Moreover, the chosen series can be inverted, reversed, and the reversed can also be inverted, i.e., with the intervals ascending rather than descending, or vice versa. All of these possibilities offered modern composers a stimulating intellectual challenge. They have been strongly exploited by men like Berg and Webern. For the listener, the works are intensely demanding. They make no effort to appeal to the memory, (Schoenberg expressly wished to avoid reminiscences of tonality because of the false expectations of consequences they would create), and are therefore ordinarily unappreciated by ears that are not highly trained. The average listener waits in vain, and often with mounting frustration, for resolutions which are never meant to occur.

To return to the analogy with modern science, the composer of atonal music and the scientist have chosen, each in his own way, to fragment their worlds into the most basic elements where the older macrostructures of organization no longer prevail. They each work with abstractions based on a powerful internal logic to produce results which are intellectually satisfying but which require the use of an increasingly private language. To the layman, the languages are foreign and will remain that way unless he is willing to make an exceptional effort to learn them. Thus the results have no rational meaning or emotional gratification for him, and may even be profoundly disturbing. Of course the artist is unlikely to know the language of the scientist at all, and only the exceptional scientist will take the time to try to fathom what the modern artist is trying to say. Still when old systems wear out and change is in the air, both artist and scientist are able to feel it early. With their special sensitivity they find other forms, as we have already noted, which, in parallel ways somehow, seem to meet the need for new systems.

TECHNOLOGY AND THE ARTS

The fundamental analogy between art and science need not be taken any further than it has been here to demonstrate its validity. It is enough, in any case, to suggest that the theme of the Two Cultures — the antagonism between art and science — is partially based on a misconception of their nature. The actual schism, such as it is, lies between the humanities on the one hand; and on the other, the technologies which for understandable reasons are so often confused with science. Science and technology have too much in common — rely too much on each other — to be discussed as completely separate entities. This is also true of the "pure" and the "applied" arts. Nevertheless, certain aspects of technology such as its commitment to efficiency, its close identification with commercial and political purpose, and the direct, ubiquitous role it plays in the daily lives of the people set it apart from basic science. The important relationships between technology and the arts are of a different nature from those which were discussed in the preceding pages. This section will be devoted to exploring them.

At the beginning of the chapter, we noted that modern artists are fully alive to the potentialities for enhancing their art offered by technology. If this is so, then one can fairly ask how can it be said that they are estranged from it? Part of the answer, of course, is that not all of them are. There are as many different temperaments and reactions toward technology as there are artists. There are artists who will have as little to do with technology as possible, and those who have expressed their distaste for the whole technological civilization in savage or bitterly satirical works. Art as an enterprise tends to be life-affirming, and many artists have chosen to judge technology according to that standard — rejecting it when its purpose seems to be destructive but embracing it with enthusiasm when it has offered means for expanding their creative horizons.

As for the technologists and engineers, generalizations about them as a group will be neither more nor less valid than those about artists. Lord

Snow, in his lecture on the Two Cultures suggested that engineers tend to be more "conservative" than scientists. At the time he wrote, this was no doubt true—in fact there have been serious sociological studies attesting to this conclusion—and it may still be true as a relative statement. The word, conservative, can have either a pejorative or complimentary connotation depending upon the orientation of the person using it; not quite apart from the political significance, which need not be considered at this point, there would be reasonable agreement that a person so identified would tend to subscribe to the older verities and approach change with more than the average amount of caution. What has this to do with the attitude of an engineer or a technologist toward art? According to Snow, technical people as he knew them displayed relatively little interest in the arts, except possibly for music. Although he did not go on to suggest the kinds of music they listened to (nor the kinds of literature, painting or architecture they preferred when they chose to pay attention to it), one could reasonably assume that their preference in serious art ran to the standard classics, and that they were not favorably inclined toward the discordant and the unfamiliar.

Actually the view of the engineering profession as being divorced from the arts seems to overlook a rather important consideration. Engineers have sometimes created structures and products as the end result of their work which, in the grace of their form or in their dynamism, come remarkably close to art. The qualifications that art must be one-of-a-kind, or non-commercial, or conceived by a single intelligence would—if adhered to—eliminate most of these creations from being accepted as such, but there are some structures that could meet rather demanding criteria. This being stated, it is only realistic to concede that they are not so common as they ought to be. Engineers and those who employ them have certain constraints that they must operate under. Of these, the question of cost and the need for efficiency (which may indeed become a constraint), are among the most limiting. Those in the practice of engineering have ordinarily accepted these constraints as a normal part of doing business. However, in the last several years there have been some rather remarkable transformations in the attitudes of younger professional people toward the social responsibilities of their profession. This includes a growing reluctance to contribute their services to projects whose end results will be out of harmony with the environment. How far-reaching this commendable attitude will become is not easy to predict, but its influence on the relation between technology and art could be significant.

One interesting conjunction has occurred in the form of an organization called Experiments in Art and Technology, Inc. This group was set up in 1966 under the leadership of Dr. Billy Kluver, an electronics engineer at the Bell Laboratories, and the artist, Robert Rauschenberg. The purpose was to stimulate cooperation between artists who have ideas for employing advanced technologies in their work (but neither the required technical knowledge nor entree to the necessary equipment), and engineers willing to apply their professional talents in a collaborative effort to create new art. The organization has attracted more artists than engineers, as might be expected, (the technical people involved are mainly electrical engineers), but the group as a whole has had several major exhibitions featuring many ingeniously mobile and sometimes audible works. The average engineer would most likely choose to call this kind of activity purposeless and the general response might well be typified by the judgment of one critic who, after attending an exhibition, stated that it was fun but it was not art. These kinds of reactions are understandable but the exhibitors would also find them debatable. Two points reflecting on this difference of opinion might be made briefly before passing on. The first is contained in the respectable theory, going back as far as the 18th Century, that the enjoyment of art is related to the normal impulse to play. The second has to do with changing views of purpose, not only among the young but also among people with many years of experience in the technologies. In the aftermath of the most recent downturn in the national economy, beginning in 1969 and 1970, which resulted in massive layoffs of highly trained engineers and scientists, there has been an observable trend toward re-evaluating the traditional objectives of professionals in technological fields. Questions about the roles of the engineer and the scientist in times of war and peace have become more insistent, often taking forms we have touched upon in previous chapters. One which is commonly posed, as we have just suggested, concerns the role of the technologist in helping to transform the whole human environment in the so-called leisure society. This will be discussed at greater length later on.

EARLY INTERACTIONS BETWEEN TECHNOLOGY AND ART IN THE UNITED STATES AND EUROPE

Much of the innovative activity resulting in the production of electronically-supplemented art has been centered in the United States. This is unremarkable when we think of the pre-eminent position this country has long held as a technologically oriented nation, not to mention an affluent one. More than any other people, those who have grown up in the United

States have had technology "in their bones." They have been accustomed to expensive mechanical toys, appliances, conveniences of communication and at a relatively early age, personal powered vehicles in their daily routine, as well as to numerous mechanical and electronic aids in their work. There is no reason to think that young artists in North America have been unaffected by these circumstances. In the 1970s this has also become quite true of young Western Europeans and Japanese but in those countries the expensive technologies, such as computers, are not yet so common nor so available for experimental uses. It turns out that the great variety of post-World War II technologies, together with the expansion of opportunities to use them for novel purposes is what is new — not the idea of advancing art through the application of electrical or mechanical methods. That goes back much earlier in this century.

Some critics, and artists too, suggest that the use of engineering and scientific experiments can produce at best a minor kind of art. At this juncture it is not easy to tell if these judgments are premature or correct, for there are serious questions being raised about the validity of contemporary art in general. The old certitudes against which the majority tested the worth of a new work of art were often arbitrary and vague, but they did offer an effectual standard. This has now disappeared, with nothing satisfying to take its place. Not only is it presumptuous to classify art unequivocally as minor (or major), it is also possible to wonder whether anything like a major fine art, in its earlier sense, is being produced today. This changed situation has its traumatic elements, but in one way the trauma may be less pronounced in the United States than in Western Europe. That is because in this country there has been no long-standing tradition of great art as a vital force helping to shape the culture of the nation.

Prior to this century, surely, the average American had very little feeling for painting, music and sculpture as a part of everyday life growing naturally out of the domestic environment. European literature, painting, architecture and music largely set the standards for refined tastes. Herman Melville said one time that Americans felt somehow their native literary genius would arrive in the costume of Queen Elizabeth I's day[26]. Art, in fact, was something for which a busy nation, bound to make its way with Puritan single-mindedness, had little time. The major artists were an isolated few. Fewer still were those who worked successfully with American themes. We are less impressed today than their contemporaries were with some of the popular poets of the time who tried to write in the vernacular. Among the major poets who are still respected, Emily Dickinson and Emerson looked backward introspectively, and Poe's visionary work could have been written anywhere. His poetry was more highly praised in France than in his own country. Walt Whitman worked to portray the contemporary America, but often by creating atmosphere through indirection. The nation's taste in prose literature ran to the sentimental novel, especially if it were English. Of the novelists, no one writing in a completely American idiom before Mark Twain was accorded full international stature, and Twain's genius lay in his ability to evoke a nostalgia for a time that had already slipped away.

Nowhere in the few classics of 19th Century American literature do we find a serious attempt to get at the roots of the dynamism of the emerging culture which were to be found in the coupling of an expanding technological economy with a democratic system of government. Perhaps under the circumstances this theme could only be dealt with through reportage and the essay. The romance of the period for young and old still lay westward with the winning of the land; or, if eastward, it was chiefly back across the ocean toward the aesthetic landmarks.

The situation with regard to painting was similar. There were few painters who had won serious acceptance as fine artists. Painting, in this country, was characterized by careful, almost photographic, draftsmanship. Some of the artists and sculptors had received early training in the mechanical arts. John Ruskin, in 1856, criticized some American paintings he had recently seen as being obviously "true studies" and therefore, he assumed, "the ugliness of the country must be Unfathomable." [27] The emphasis on reality comes across most often as an explicit, journalistic accounting of natural scenes. Still, we can readily appreciate the fresh exuberance of Winslow Homer's seascapes, and can acknowledge a triumph of realism in the dramatic integrity of Thomas Eakins' work. Lewis Mumford considers that Eakins was able to carry out the themes of Walt Whitman in his art[28].

As for music, composers chose to probe little further than the fertile soil of the folk song.

The condition of the arts in the United States could not help but affect the attitude of those Americans who were earnestly trying to pursue culture. Feelings of inferiority were common. The tendency, then, was to react by aping European ways the more closely; or, more rarely, by rejecting them as effete and decadent. Some artists could not live with that sort of ambivalence. James McNeil Whistler, Mary Casatt and Henry James, for instance, went back to Europe, to the roots of their culture, for their inspiration.

Basically, to find an expression of the irresistible movement toward industrial dynamism in the United States in the 19th Century one cannot look to the formal arts. The development has to be traced in what we might call the folk arts, which, broadly interpreted, could include the beginnings of simple forms of industrial design.

The ugliness of industrial civilization had not, in the last half of the 19th Century, become so apparent — certainly not so concentrated — in the United States as it had in England. It was not that American technology was more primitive, because in many ways it was more advanced. Some of the reasons why it had surpassed British methods were discussed earlier. American manufacturers had to make the most of their labor supply, and so they moved very quickly to develop and adopt ingenious new methods of mass production. There were big jobs to be done with modest resources and this meant that function was more important than form. The result was that tools — machine tools as well as hand tools — were usually characterized by simple, clean lines with a minimum of wasted bulk. When the mighty Corliss steam engine, a giant structure weighing 1,700,000 pounds, was exhibited in the Philadelphia Centennial Exhibition in 1876, its classic design prompted the French sculptor, Bartholdi, to write that it had "the beauty and almost the grace of the human form;" and the *Atlantic Monthly* concluded "surely here, and not in literature, science or art, is the true evidence of man's creative power." [29] The last statement is really too extravagant to be swallowed whole, but it no doubt represents an honest contemporary admiration for the imposing technical virtuosity which American industry was able to muster. This characteristic grace and restraint was to be found in the designs of other kinds of structures as well. The elegant clipper ships and the classic Brooklyn Bridge of John and Washington Roebling established beyond question that frugal but imaginative engineering sensibilities could produce works of real beauty. In the end, though, one must conclude, on balance, that examples of clean, flawless functionality in the mechanical arts were less typical of the age than the out-of-scale and the tasteless.

This was no less true in Great Britain; in fact there were probably fewer redeeming examples of a sensible machine aesthetic there than in the United States. Mumford has written about the attempts of British industrialists and engineers to atone (whether consciously or unconsciously) in some small way for their reckless ravaging of the environment[30]. Unlike their American counterparts, the British could turn on all sides to models from a long cultural heritage. What could be more natural than to adorn machines and manufactured articles with ornaments of the past?

Never mind that rococo flourishes and Gothic traceries did not contribute in any way to the function of the product; the obeisance to art would be made. But it was spurious art.

The response of the artists themselves to this kind of deception comes down to us mainly from the writers of the period. For the most part they considered it to be one more count against the machine civilization and the entrepreneurs who were profiting from its grossness. Yet the overall reaction of creative people toward technology and the industrial society was ambivalent. There was usually a tendency to separate the social effects of industrialization from the technology which made it possible. As Sussman observes [31], most of the major Victorian writers chose not to treat the physical images of technology or the workings of the industrial system as the center of their interest. Rather, the machine was a symbol—either of progress and order, or of ugliness and human degradation. The ambivalence is often found in the work of a single writer. Much earlier, William Blake had expressed his rejection of machine technology in the two often quoted phrases, "dark Satanic mills" and "cogs tyrannic," but he also sensed that some forms of technology would have to be used to deal with the new environment. Charles Dicken's novels contain many colorful descriptions of the operations of machines. He often used metaphors involving obstreperous beasts or remorseless efficient monsters, but these are isolated pictures. He was not interested in exploring the details of technological processes, nor in questioning the inevitability of mechanization. His sympathies for the oppressed were boundless, and he clearly recognized the unhealthy emotional effects which could result from wearisome indentureship to the machine, but he would have rejected any attempt to relate the effects to basic flaws in the economic system. Dickens's art was as natural as life. One cannot imagine him regarding himself as an artist; in fact he viewed artists and intellectuals as objects of ridicule [32]. Thus, we would look without success in his work for any direct commentary on the connection between technology and the arts.

It is necessary to turn to the writings of John Ruskin and William Morris for the most searching discussions of the influence of engineering and industrial methodologies upon art and the general quality of life. Ruskin had followed Thomas Carlyle in denouncing the essential immorality of the utilitarian culture, but Ruskin's especial concern was to construct a Victorian aesthetic which would reinstate a union between spiritual beauty and simple natural forms fashioned in undefiled, pristine surroundings. Of all the writers of his period, Ruskin was most clearly able to uncover the intrinsic poverty of Victorian bourgeois art. True his penchant for fuzzy moralizing about the divine attributes of art often gets in the way of modern readers. However, despite his later tragic descent into madness, his best contributions to a theory of culture still have an originality and vision that commends them to any readers of the present generation who are concerned about the debasement of their natural or contrived environments. The machine, he believed, by displacing men from healthy, productive work within their capabilities had fractured the organic fulfillment of function which served as the natural condition for every human being. Without this fulfillment, life was bound to become less satisfying as the mechanical society forged ahead. Ruskin deplored the corrupting of public taste for commercial reasons as we see in this quotation from a speech to a group of manufacturers about the need for integrity in industrial design:

> You must always remember that your business, as manufacturers, is to form the market as much as to supply it. If, in short-sighted and reckless eagerness for wealth, you catch at every humor of the populace ... no good design will ever be possible to you ... Every preference you have won by gaudiness must have been based on the purchaser's vanity; every demand you have created by novelty has fostered in the consumer a habit of discontent; and when you retire into inactive life, you may, as a subject for consolation for your declining years, reflect that precisely according to the extent of your past operations, your life has been successful in retarding the arts, tarnishing the virtues, and confusing the manners of your country.

This was not at all the sort of thing Victorian entrepreneurs were accustomed to hearing.

Morris was an equally penetrating critic of Victorian ways. Like Ruskin, he abhorred bad art, not only for its own sake, but also because it must connote a pervading degeneration in society. He was also well-educated, although he said he learned more by self-study than through formal instruction either at Marlborough College or at Oxford. The study of history, especially of medieval times, always intrigued him, but he was too much a man of the present to lose himself entirely in the past. His wish, rather, was to see a fusion of the past and present through a rebirth of the handicraft arts. A superb craftsman himself, Morris knew firsthand the emotional satisfaction to be gained from exercising the hand and the eye to make a deep-felt personal statement. The industrial system had eliminated this tactual element which had served as a catharsis in previous societies. He believed that the placing of art on a pedestal where it could be cultivated and admired only by the wealthy had been alien to its true spirit, and had refined it to the point of exhaustion. It was not really the machine, *per se*, that he resisted, (except for steam power which he disliked), but rather the spirit of capitalism and the division of labor which helped make the system profitable. He saw no justification at all for heavy, repetitious toil if it could be lightened by the machine, but where the machine interfered with man's instinctive impulse to create beauty in his work, the result was a blatant social evil.

Morris, along with a group of enthusiastic friends, including architects and professional as well as amateur artists, started a decorating business with the intention of putting these principles into practice[33]. The idea was that each would share in every aspect of the work irrespective of his own particular background. They would all help design, paint, weave, fabricate, experiment with new materials or do whatever else needed to be done. Artistically, the results were praiseworthy. Their tapestries, furniture and wallpapers were charming and very well received. In fact the business was so successful that it proved to be the undoing of Morris's ideal system because when it grew larger some specialization became almost inevitable. The alternative would have been utter confusion. As time went on the designers and the managers found less and less time to take part in the actual execution of the jobs and other skilled workmen had to be hired for the purpose. Ironically, the prevalence of handcrafting in the products of the firm raised the cost of its services to the point where only the well-to-do could afford them. Thus, his hope of starting a new direction for society to follow by creating an "art of the people" was unsuccessful.

Nevertheless, Morris's influence should not be judged on that basis alone. This vibrant, talented man was not only an exceptional artist and designer, he was also a writer, a poet and one of the foremost spokesmen for socialist thought in the 19th Century. The relevance of his work can be found in his fervent championing of an aesthetic for the industrial age to counter the corrupting waste and vulgarity. He was not a doctrinaire socialist — he blamed all of the classes for the poor state of the environment — but his wrath fell most heavily on business and science (by which he meant applied science) as we can see from this quotation from his first public lecture delivered in 1878[34].

> "Is money to be gathered? cut down the pleasant trees among the houses, pull down ancient and venerable buildings for the money that a few square yards of London dirt will fetch; blacken rivers, hide the sun and poison the air with smoke and worse, and it's nobody's business to see to it or mend it: that is all that modern commerce, the counting-house forgetful of the workshop, will do for us herein.
>
> And Science — we have loved her well, and followed her diligently, what will she do? I fear she is so much in the pay of the counting-house, the counting-house and the drill-sergeant, that she is too busy and will for the present do nothing. Yet there are matters which I should have thought easy for her; say for example teaching Manchester

how to consume its own smoke, or Leeds how to get rid of its superfluous black dye without turning it into the river, which would be as much worth her attention as the production of the heaviest of heavy black silks, or the biggest of useless guns. Anyhow, however it be done, unless people care about carrying on their business without making the world hideous, how can they care about Art?"

The present-day significance of these sentiments cannot hide the fact that, while Morris was a forward looking visionary in some respects, he was never able entirely to cut himself off from his 19th Century middle class roots. Sussman proposes that he was a transitional figure between Victorian medievalism and morality and 20th Century primitivism[35].

TECHNOLOGY AND ART IN THE 20th CENTURY

The Victorian ethos was very tenacious. In the second half of the 20th Century, strong traces still linger in Western concepts of property and morality, although it is true that standards of sexual morality appear to be crumbling. However, as we have observed, within a decade after Morris had died (in 1900), some artists, convinced that prevailing traditions had outlived their value, were working toward a complete liberation from the metaphysics and sentiment of their predecessors by insisting on the irrelevance of any cultural or moral precepts. This of course was similar to the amoral, acultural basis upon which the foundations of science were presumed to be constructed.

But technology has been effective in influencing the direction of art too, and not always in the same way as science. When we speak of new discoveries in science, we imply that inherent characteristics of nature which can be described by verifiable "laws" have been uncovered. These can make a great difference in the way people think and live, or very little. Sooner or later, though, if the scientific laws are transformed into technology, some kinds of change in styles of living can be expected. Basic new laws of science are not found very often, while innovative technologies, many of which can evolve from just a few scientific principles, have multiplied at an unprecedented rate in this century. The resulting quickening of the pace of life has been the hallmark of the post-World War II years.

It has been noted earlier that discoveries in science gave fresh meaning to concepts of space and time. So did the new technologies. Mechanization of transportation on the ground and in the sky made high-speed travel available to all in the Western nations. Space could be explored to the fullest, and time would be considered in other terms. The seasons no longer had the same meaning, and the accepted permanence of ancestral roots and familiar scenes had little meaning at all.

In the first few years of the century the effect of these trends upon art were not so strikingly apparent. The flight into abstraction has been portrayed as almost a philosophical or historical necessity, but it was still barely noticeable in literature and less so at first in music than in the visual arts. The artists were ready for change, (some more wary than eager), but as yet there were no styles or schools which confronted the fact of technology head on and sought either to challenge it or to seek a symbiotic accommodation with it. The Futurists, in the time shortly before World War I, became the first to exalt the dynamism of the machine as a model for art. Taking their cue from the Italian poet, Filippo Marinetti, who published the manifesto for the movement in 1909, they insisted that art was to be found in form as movement with the spectator at the center. All of the aspects of the mechanical world — the gleam of steel and the motion and sound of the industrial process and the machine — were vigorously embraced as the means by which the old forms would have to be exorcised. The movement was also called "antipassatismo," meaning, "down with the past." The pronouncements of the Futurists about their concepts were generally more enlightening than the works themselves, but that has now become a not uncommon situation in modern art. They often seemed to take cubism as a point of departure for their work. Sometimes experiments in photography helped them to reach toward the image of motion for which they were striving. Futurism faded out during World War I but as Poggioli says[36], elements of it are present in all *avant-garde* art. Some of its attitudes toward the mechanical aesthetic can surely be sensed, for example, in the *Ballet Mechanique* (1926) of George Antheil, which featured two whirling airplane propellers among the percussion effects.

THE BAUHAUS

The Futurists had no clear, practical plan for making man at home in the technological environment despite their progressive stance and the appreciation for the spirit of the 20th Century which some of their best work represents. In 1919, however, a school was founded in Weimar, Germany which had just that kind of ambitious program as its aim. This was the Bauhaus.

The Bauhaus, (Das Staatliche Bauhaus Weimar), was basically a school of design which sought to bring the arts and crafts together in a new guild of craftsmen; yet in a way that would meet the technical and commercial demands of mass production industries. Statements of purpose made at various times by associates of the Bauhaus show a strong link to the ideals of William Morris, but since it was a state-supported institution, they were bound to take a more realistic attitude toward their country's need to compete in the international markets.

During the first nine years of its fourteen year existence the Bauhaus was directed by Walter Gropius. Gropius, an eminent architect, had been a part of the artist–socialist movement in Berlin which sought to show the way to a rebirth of man's essential humanity, in the tumultuous period after the World War I armistice[37]. This intensely idealistic spirit stands out in the Bauhaus manifesto of April 1919. During his tenure as director, Gropius brought together an accomplished group of artists and technicians to serve as teachers. They were actually called "masters," using the medieval terminology, and the students were apprentices. Wassily Kandinsky and Paul Klee—and Marcel Breuer and Laszlo Maholy-Nagy of the newer generation—were among the imposing talents who worked at the Bauhaus.

From the beginning the enterprise was plagued with artistic, economic and political problems. There was a constant shortage of funds, (most of the students were impoverished veterans), and continuous tension between the more mystical expressionist artists and those among the staff who believed that art was simply the creation of natural harmony through the constructive application of rational, objective laws. The impartiality of Gropius was constantly tested, but he did not lose faith in the validity of the idea of a productive synthesis of the studio and the workshop. Externally, the school was being harassed by a hostile bureaucracy in Weimar, which thought it to be a center for subversive activities; so in 1925 the decision was made to re-establish at Dessau. In that more friendly community atmosphere the activities took on new life for a few years, and the coordinated efforts resulted in a complex of buildings embodying the original founding spirit[38]. Gropius resigned in 1928 but the Bauhaus continued on until 1933 when it was closed down for good by the Nazis. Many of the staff found posts in the United States, where their dominating effect on styles of architecture and industrial design is still apparent.

Structures and product forms which we now accept as natural, such as glass and steel "functional" architecture, clean-lined tubular and cantilevered furniture, and unadorned, geometric lighting fixtures came to be thought of as the Bauhaus style although Gropius denied the existence of a style, since "form must follow function." As for the works of art they produced, none is more interesting than Moholy-Nagy's "Light-Space Modulator." This kinetic light-play sculpture which was completed in 1930 was one of the first such works powered by electricity and a very early example of cooperation between modern engineering and art[39].

Some of the more idealistic aspirations of Gropius and his colleagues were thwarted by social realities. Industrial technology moved too quickly to submit to a full course of humanization. Nevertheless there has been no more influential attempt to link art and technology in this century than the experiment of the Bauhaus.

DADAISM AND SURREALISM

One can find none of the idealism of the Bauhaus in dadaism. The dadaists, while rejecting the past, refused to concede that there would be a future worth thinking about. It was a movement conceived in 1916 by a group of young artists and poets who had taken sanctuary in Switzerland from the war. Their only creed was nihilism. Questioning the validity of all rationalism, they tried to compose into their work a rejection of the whole technological culture which could mount such a war. The works of the dadaists were impudent, self-mocking and anarchistic in tone, and have been called the essence of anti-art. Even the name, dada, which is a French baby-word for horse, was apparently selected by a random thumbing through a dictionary and stopping at a word which seemed best to convey a spirit of nonsense. After the war was over, the political, anti-war emphasis of the movement shifted toward an attack against formal values in art and the presumed pomposities of the art establishment. Dadaists were fond of saying that life is more interesting than art; so that everything in life from fur-lined teacups to broken sewing machines to cancelled theater tickets—either assembled into a composition or left as found—were considered fitting subjects for their attention.

Marcel Duchamp is often called the archetypical dadaist even though he never really belonged to the group. His participation for the most part consisted in collaborating in some multi-media performances with his friend, Francis Picabia, a robust dada rebel. From the time Duchamp was asked to withdraw his most famous painting, "Nude Descending a Staircase," from a cubist exhibition in 1912 on the grounds that it was unrepresentative of cubist theory, his work was marked by an ironical, humorous irreverence toward sober tradition or authority. In 1913, several years before the dadaists appeared, he had created the idea of the "ready-made," which was defined by André Breton as a "manufactured object promoted to the dignity of object of art through the choice of the artist." Duchamp did this by mounting a bicycle wheel on a kitchen stool, and exhibiting it. Later, for his own pleasure he took an ordinary bottle-drying rack and merely by signing his name to it presumably changed the context in which it would be recognized. At any rate, this ready-made and others were widely exhibited, and replicas have been sold as art objects[40]. He was also interested in chance happenings as a possible manifestation of the subconscious. He and his sister produced a musical composition in 1913 by randomly drawing the notes of the musical scale out of a hat.

But the dadaist movement by the very nature of its abandonment and lack of direction was bound to exhaust its vitality and to suffer the reaction of tedium. Not all of its members were capable of maintaining the amused detachment of Duchamp, and some of the exhibitions (or performances) in Europe were marked by violent political confrontations. Several years after World War I, the movement petered out, to be succeeded by surrealism.

The surrealists were at first primarily a literary group strongly influenced by the work of Sigmund Freud. They believed in the superiority of a reality seen through the subconscious mind. André Breton, a French poet, stated the definition of the aims in a manifesto published in 1924. It was: "pure psychic automatism by which one intends to express, whether verbally or in writing or in any other way the real purpose of thought. It is the dictation of thought, free from any control by the reason and of any aesthetic or moral preoccupation." Surrealists were as concerned as were the dadaists with the absurdity of the human condition and the pretensions of artistic formalism, but with the difference that surrealism had an object. To the dadaists, chaos was the only reality and the end; whereas the surrealists hoped to find form in the irrational by retrieving from the unconscious a new language of art. Since surrealism was primarily a literary activity, given to searching for the springs of inspiration through the method of "automatic writing" it is not surprising to find the paintings developed in an illustrative way. Recognizable objects often predominate, albeit in unnatural, schizophrenic juxtaposition with unrelated objects or backgrounds. Just so, Breton wrote, "the apparent antagonism between dream and reality will be resolved in a kind of absolute reality—in surreality."

There are many well-known artists who have been exponents of surrealism or whose work relates closely to it, including Joan Miró, Salvador Dali, Jean Arp, Max Ernst and Giorgio de Chirico. The surrealists also regarded Duchamp as one of themselves although he had completely withdrawn from any formal art activity in 1923. These artists have chosen to use a great variety of subject matter, from the sprightly, elfin, abstract forms of Miró to the anxiety-ridden nightmares of Chirico, but the typical work couples spontaneous inspiration with careful draftsmanship.

ART IN THE UNITED STATES

Although both dadaism and surrealism had their origins in Europe, (and the most famous names associated with each movement were Europeans), after World War I, American artists had so many unanswered doubts about the presumed benefits of the industrial society that they were often receptive to the ironic ambivalences of those two art movements. We must note that before this time artists in the United States had remained only marginally affected by the torrents of innovation which had characterized the European art scene for more than a generation. In the first decade of the century while art circles in Europe were being rocked by change, North American artists virtually ignored the excitement and concentrated on trying to develop a distinct native style. In doing so, many of them turned back to pre-impressionist work as a model hoping to infuse it with new life by using indigenous themes. The dominant characteristic of their work was realism. One group, who referred to themselves as The Eight, reacted against the stuffy academicism of the Gilded Age by painting scenes from the run-down areas of large Eastern cities, and their seedy inhabitants. For this reason they were later called the Ash Can School[41]. But their style in no sense represented an aesthetic revolt and indeed there was little indication yet that the United States was ready for the *avant-garde*. Small exhibitions of the works of modern painters were being given, starting about 1908, primarily in the tiny New York gallery of the photographer, Alfred Stieglitz, but only to the unconventional few[42].

The first large show of European moderns in the United States, the International Exhibition of Modern Art, was held in the New York Armory in 1913. President Theodore Roosevelt, who visited the show found himself unable to accept the "European extremists." A typical, shocked critical reaction was one which referred to Marcel Duchamp's "Nude Descending a Staircase" as "an explosion in a shingle factory." Not too many years before, a French critic, in speaking of a violently colorful painting of one of the young expressionists who were called Les Fauves, (the Wild Beasts), had referred to it as "an explosion in a paint factory." It appears that American criticism may have been no more original than its art.

Only two years after the Armory Exhibition, however, when Duchamp came back to the United States, he was lionized. From that time on, he spent a good part of his time in this country, maintaining a studio in New York. Along with other visitors or expatriates from Europe, he became the center of much of the dadaist activity in the early 1920s, and was also involved in surrealist circles later. American isolationism in art now seemed to be dead or dying. Not only did large numbers of expatriate and refugee artists bring European ideas to the United States during the 1920s and 1930s but there was also a reverse migration. Many young American's had flocked to Europe after the war. Although some stayed on, many of them later returned. They were now full of new ideas for expressing themselves and ready to regard the American scene with a fresh eye. The dada movement was important because it created a climate for outlandish experimentation which had hardly existed before. The total abandonment of any rules or discipline was later renounced by many artists, but in recent years it has again become a major trend.

One could also see the influence of cubism, and futurism too, in the geometric style of a group of painters, including Charles Sheeler, Charles Demuth and Joseph Stella, who have been called precisionists. Unlike the dadaists who used the machine as a basis for satire, these men saw in the structures and machines of America bold, interesting images. Other painters in the 1930s turned back to conservative ways in their art if not always in their political outlook. The Great Depression cried out for social commentary, and there were artists who were ready to provide it, sometimes harshly and movingly, and sometimes to the point of cliché. However, there are few paintings of the genre which surpass the sensitivity of the best photography in this period. In the work of the regionalists, typified by Grant Wood's *American Gothic* (1930), conservatism in art was matched by a conservative appreciation for sobriety, hard work and the older, rural virtues as a reaction to the bohemianism of the 1920s [43].

Barbara Rose contends that the only major international art movements between the two world wars were dadaism and surrealism; and that by the 1940s, with the center of the art world having shifted to New York, American artists had taken up the challenge of trying to synthesize the two[44]. Since then, during the years between the end of World War II and the present, so many different terms for art styles have appeared that it seems unrewarding to try to name them. The central point to be observed is that none of them represents an entirely new direction. In each development it is possible to see elements of experiments which have been tried before.

There would appear to be no serious difficulty in accepting McMullen's judgment that, since just before World War I, the basic "energy centers" of visual art activity can be described as abstractionism, primitivism and quidditism[45]. Quiddity refers to the philosophical fuzziness arising when attempts are made to penetrate the barrier of obscurity separating a viewer's presumptions from an artist's activity by asking the question, "What is it?" (*Quid est?*) The word, activity, is used here rather than intentions because not all artists will admit to having specific intentions. For a 1970 example, what does one say about a well-attended exhibition where one exhibitor standing on a ladder held a transistor radio suspended from a fishing line in a bucket of water until it stopped playing? Dadaism appears to be the obvious illustration of quidditism. Duchamp contended that an artist was nothing more than a medium whose principal mission was to create a connection between a work and the spectator. All of the questions about "art as truth," fine art, decorative art, or art as business acquire a special edge when dadaism or neodadaism are being considered. Of the three primary forces, abstractionism has undoubtedly been the most significant, but since the terms are not precise no sharp distinction can really be made between them. McMullen notes that abstract expressionism with its chance drips and dabs raises some of the same kind of questions as dadaism, and surrealism can be thought of as coming close to primitivism on a psychological level, although the normal understanding of primitivism is something different.

Whether one uses these or other terms to describe the main directions in visual art, it is worth reiterating that the basic denominator has to take into account the artist's character and purpose. He may approach the world in its many aspects as an intellectual problem to be solved, he may seek to find in it sublimity or sensuousness, or he may see it only as an accidental collection of objects and events in which it is absurd to look for any significance. In this third instance, then, his only purpose can be to point out the irony of existence.

Artists in the 1970s, attempting to carry out their activities on any of these levels or somewhere in between, seem to find it increasingly difficult to come up with something original by simply relying on traditional resources. Edgar Allen Poe said the very essence of an artist's insight lies in a refusal to do what has been done before. In much of modern art this has come to mean a search for the interesting, and, as Harries states it [46], this has sometimes led to the acceptance of arbitrariness as an aesthetic principle. In recent years the constant search for novelty has taken on the character of a high-speed chase. *Avant-garde* movements following rapidly on the heels of the ones before, or existing concurrently, have led to remarks like the graffito: "*Avant-garde* is passe." The iconoclasm which, rightly or wrongly, is often attributed to artists becomes difficult to keep up if there are no icons left to break.

The advance of technology—particularly communications technology—has a lot to do with this situation. In the past, it was possible for an artist to work in comparative isolation, certainly as far as what went on in other countries or even in different areas of his own country. There were trends and he would become aware of them, but only gradually. He would have a circle of associates with whom he exchanged ideas but it would not be very large. His first objective would be to satisfy himself and his patron, if he had one. Even though he would want and expect that his art would be appreciated, it would mainly have to advertise itself. Now, in the modern age of mass communications, the media voraciously consume, then disgorge, symbols and images, constantly seeking the new and the different. The forces of commerce and technology are overwhelmingly pervasive. Art, as a reflection of its times, could not remain unaffected by them if it would.

The writer has to deal with a language threatened by emasculation, as words, lavished on trivia in a great flood of print and sound, begin to lose color and potency. The novelist must wonder if he can ever use "fantastic" or "fabulous" in his writing again. Fact outstrips the imagination daily, and the novel and the poem give way to the report. The same is true in the visual or musical arts. Popular entertainers and commercial advertisers, knowing that viewing and listening publics quickly tire of familiar faces and sounds, commit large amounts of money and talent to an unremitting search for fresh sensations. The line between modern and popular art becomes more indistinct. Of course there are impoverished artists working in run-down studios, but not in isolation from what is going on in the larger society, and not unaware of what seems to be required to make a public impression. The artist may be ignored, but on the other hand he may attain national recognition even though comparatively few people know his work except by reproduction or reputation. Not much more may be required than a chance news story or a diverting television appearance, and another personality is discovered, his commercial success assured—for the moment. Art is in the public marketplace. It is no wonder that young artists have to keep looking over their shoulders to assure themselves that what they are doing is not yesterday's fashion.

The artist who has an understanding of the history of development in his field can hardly fail to realize how difficult it is to find new ground to explore. As we stated earlier in the chapter, some writers have extrapolated this state of affairs to the conclusion that, whatever happens, art is unlikely to continue in the same forms which have predominated for several centuries. That is to say, the new visual art will be something other than the static, massive sculptures or the two-dimensional canvasses of the past; and the new music will be something other than the symphonies, chamber works or chorales. Very likely art lovers will still go to museums and concert halls to enjoy remembered treasures (assuming that the serious problem of the economics of mixed-media performances is solved), but the art of the future will be for other purposes than a balm for an elite.

Artists have never had to think more seriously about the courses open to them and the problems of choice have never been more intense. No doubt a few artists will continue to turn to nationalistic themes although these have been heavily exploited; and to exotic themes despite the fact that high-speed and inexpensive international travel has made the exotic a rarity. One wonders how long a Gauguin would remain in Tahiti had he arrived there in 1970. As for social commentary, how many paintings can compete emotionally with the instant drama of a television newscast? This is a central factor, of course. It is unrealistic to expect people, nourished as children on the continuous flow of television images, to respond quite as vivaciously to static art as they once did, at least as mere spectators. They have been bombarded with kinetic images from an early age, and to catch their attention requires something out of the ordinary.

ELECTRONICS AND MUSIC

The turn to technology is one obvious direction to consider, there to find unfamiliar materials, tools and images. The use of mechanical or chemical technologies to extend the artist's capabilities is already too familiar to spend much time discussing, but the electronic age has added another subtle dimension, offering fresh challenges, especially to the music composer.

We learn from H. H. Stuckenschmidt's book, *Twentieth Century*

Music [47], that the first electrical musical instrument, called the dynamaphone, was invented by a Dr. Thaddeus Cahill. This very large instrument, bearing some relation in principle to the later electric organs, was used, beginning about 1906, for sending concerts of standard compositions over telephone lines. This was soon discontinued because of the interference it caused with telephone service. Different forms of electronic instruments for producing sound appeared in the 1920s and 1930s and some well-known composers wrote music especially to be performed by them, e.g., Hindemith's *Concertino for Trautonium* and string orchestra, which was written in 1931. The chief characteristic of most of the instruments has been a reliance on an electronic apparatus to generate oscillations which are then made audible through amplifiers and loudspeakers. The earlier instruments produced melodic, one-note sounds but others, developed during the 1930s, were polyphonic. In general the early electronic instruments were used to replace previous instruments, while still fitting into the scheme of the Western aesthetic tradition.

Attempts to create, (as opposed to produce and perform) music with the aid of technology soon after World War I were restricted by the relative crudity of the available means. Composers experimented with such methods as scratching or otherwise mutilating phonograph records but the possibilities were limited.

In the 1950s, the science and art of electronics had reached a high level of accomplishment. The development of the tape recorder at this time gave the composer an opportunity for much greater diversification in his work with an abundance of unconventional sounds available to him. Pierre Schaeffer, an engineer with the French Radio, began to do research on natural sounds — train whistles, human breathing or dripping water, for example — manipulating them by playing them backwards or perhaps increasing the speed of the tape drive. He edited the results so that combinations of completely new sounds were formed. He used the name "musique concrète" to describe his work, considering it to be the exact opposite of abstract music. Out of this research on human and non-human sounds came a number of works. An opera concrète, "Orphée" composed in collaboration with Pierre Henry featured two female vocal parts, violin and harpsichord contrasting with the taped sounds. Edgar Varèse, who had, like George Antheil, utilized sounds from different unusual percussive devices in his compositions in the 1920s was one of many composers working with musique concrète.

In 1950, Dr. Herbert Eimert and Karlheinz Stockhausen of the Cologne Studios of German Radio initiated a parallel development in electronic composition with the difference that at first only sounds derived from an electronic source were used. They also declared that their work was theoretically related to the serial compositions of Webern rather than the chance montages of musique concrète[48]. Throughout the 1950s, musique concrète and electronic music were active areas of research but the actual physical processes were outstandingly tedious. A two or three minute composition could take months of patient work, requiring that the composer become familiar with the means of adjusting the delicate equipment, that he chop and splice the tape into hundreds of combinations, and also that he be ready to deal with any number of other mechanical problems. These were serious drawbacks but there have been other basic difficulties. Pure electronic music, once produced, is unchanging in structure. The performances do not have the variety and surprise which can come with the spontaneity of human performances. The composer may actually prefer that his ideas be presented in just the form he sought, but for the listener, who in any event tends to find the sounds less rich than those derived from standard instruments and human voices, the results are monotonous. This is one reason why composers choose to include conventional instruments or voices in their compositions.

The physical problems of composing electronic music have been simplified with the development of electronic synthesizers, the first one of which was constructed by the RCA Corporation in 1955. That machine was large and expensive, but now smaller and moderately-priced synthesizers such as the Buchla, Moog and Putney are available. Basic components include voltage-controlled oscillators to generate sine, sawtooth, triangle or square waves, or "white" sound in which all frequencies are sounded together; filters for shaping the waves; mixers and amplifiers. Voltage control confers additional subtlety on the sounds produced, but for greater complexity composers can tape the sounds as they are originally produced and combine them on multiple tracks of the same tape. There is a great variety of possibilities which they would like to know better how to handle.

COMPUTERS AND MUSIC

Of all the arts, it seems that music stands to be carried furthest by the research with computers which has been going on in many places, beginning with the first experiments about the middle of the 1950s.

Ever since the work of the mathematician, Joseph Fourier, early in the 19th Century, it has been known that all sounds are completely describable by mathematics. Sound waves, or the electrical waves into which they can be transformed, are accurately represented by number sequences showing the amplitude, (the maximum displacement of the wave from its unexcited position), at regular instants of time. In effect the groove in a phonograph record is a graph of this information. Dr. John R. Pierce indicates that about 20,000 three-digit numbers per second can theoretically describe good quality music of any kind, if the right numbers could be chosen[49].

The development of interesting sequences of numbers is one of the large challenges which composers and computer specialists have attacked. Since the computer is a most rapid and efficient number generator and manipulator, it gave early promise of being an aid in composing. Not only that, it can be used to produce musical sounds, so it can contribute to research designed to learn more about the basic nature of music. Fundamental experiments in the use of the computer to assist the composer were conducted at the University of Illinois by Lejaren A. Hiller and Leonard Isaacson about 1955. They were especially interested in the computer as a device to help them make choices among the vast numbers of symbol patterns that a composer must sift through, test and rearrange in order to arrive at a result which satisfies him. The manner of experimentation which they chose to follow was summarized in 1959 in an article in the *Scientific American*[50]. Briefly, their method may be described in information theory terms as exploring the region between complete redundancy and complete randomness. As we have already noted, music in the Western world has evolved for the most part as composers have found ways of eliminating accepted constraints; thereby reducing the tendency toward repetition of certain conventional patterns. Some redundancy is obviously necessary. A totally random sequence of notes would be achingly dull to the human ear. Nevertheless within the framework of a few established rules of composition (perhaps characteristic of a familiar style) a randomly selected succession of notes can result in the production of acceptable music. Hiller and Isaacson first programmed the computer to generate random integers and then made up a group of screening instructions which took into consideration rules of counterpoint. Setting the machine to turn out random "white" notes in four voices, they then introduced redundancy by entering the separate screening instructions one at a time. Any of the randomly generated notes which violated these rules would be rejected. Eventually the combinations of notes produced did resemble a composition in counterpoint in the 16th Century style of Palestrina, though less rhythmic. In

1957 music resulting from extension of these kinds of experiments was published as the *Illiac Suite for the String Quartet*. Additional research by Dr. Hiller led to a well-documented program called MUSICOMP by means of which a composer could set up statistical methods enabling the simulation of, say, the rules of the twelve-tone system.

The other major direction in computer music research—the production of musical sounds—originated with development work at the Bell Telephone Laboratories[51, 52]. Gerald Strang[53] describes the first public output from their project, a record called "Music from Mathematics" as being the product mainly of engineers rather than musicians. Since then, however, a respectable number of musicians have learned enough about the necessary programming requirements to feel less restricted by the demands of the computer than they undoubtedly were on first exposure to it. There is even a music programmer's manual available which has resulted from the research at the Bell Laboratories under the leadership of M. V. Mathews and J. R. Pierce.

In ordinary composition, of course, a composer is accustomed to working with musical notations which are at least roughly related to mathematical forms. To produce musical sounds by the computer he has a rather more involved procedure to go through. He can begin by "inventing" specific computer operations which he designates as "instruments." These represent groups of unit generators which can be linked together in complex circuits to produce a wide range of sound manipulation. In much the same way as standard musical notation informs a musician how to play a piece of music on his instrument, (in effect, the instrument then processes the information into sound waves), so too does the computerized system operate.

After the composer has designed the instruments, (some standard musical instrument sounds have proven extraordinarily difficult to achieve) he can specify that they make up an orchestra. This information is then compiled by means of a complex set of instructions. The composer's next task is to write a score. Aside from the fact that he must use numerical notation he can write it very much as he pleases relative to the kind of orchestra he has selected. He can specify pitch or loudness, for instance, or can allow the computer to make a random choice. If he should choose to specify loudness, he does so by setting up a numerical scale and a conversion function, e.g., punching 1 on a particular card would mean *pianissimo* and a 6 would call for *fortissimo*[54]. He finds that he can readily produce new selections of pitches or successions of tones in unprecedented combinations. The data for the score are now entered and stored in the computer together with that compiled for the orchestra, and the processing takes place according to the programmer's instructions. To get sound from the computer output it is converted by a digital-to-analog converter into a varying voltage which is then filtered and recorded on magnetic tape or transformed directly into sound by a loudspeaker.

Actually the electronic analog synthesizers have some of the same capabilities for producing a wide range of sounds that computers do, and have the additional advantage of immediacy. Although it is possible at a few research centers with advanced systems to have short passages put into the computer and played back very quickly, in most operations there is still a substantial lag between the time the score is presented and the sound heard. This prevents improvization. Another disadvantage of computer music is the cost of buying time.

In the various laboratories, it appears that attention is being focussed on theoretical or practical aspects of the analysis and synthesis of acoustical phenomena at least as much as on composition and performance of music *per se*. Even though the practical range of music being produced now is fairly limited, the computer is conceived as a universal instrument. The goal of realizing its universality to those who are involved in computer music seems to be at least as much a matter of attacking an interesting scientific problem as it is one of artistic creation. But then we must continue to wonder how well these two attitudes can now be distinguished from each other.

Progress in the computer generation of sounds points to the not unreasonable conclusion that, in time, it will be possible to generate by electronic means most familiar instrumental and vocal sounds. While the art of the composer seems sure to be greatly influenced, what the situation will mean overall to long-standing relationships between composer, performer and audience is undetermined. The computer could eliminate the need for performers of course, but there is not much indication yet that music composed with its aid will constitute a major proportion of that actually presented and heard. Record and tape sales and the standard programs offered at most well-attended concerts show that the market for serious modern music (as distinct from "popular" music), continues to be relatively small compared with that for music composed, say, forty or more years ago. The large potential lay audience as yet seems little more willing to make the considerable intellectual effort required to understand and appreciate the work of modern composers than it has to comprehend the theories of modern scientists.

COMPUTERS AND THE VISUAL ARTS

In addition to their role in the composition and generation of music, computers have shown that they can be used to help artists make rational inquiries into the characteristics of visual form. This is not the same as saying that they can create art but in association with ingenious programmers, they have assisted in the production of designs or experiences which have been judged to possess artistic value. More of the experimental work seems to have been done by engineers and scientists (often working with artists) than by artists alone because of the complexities of programming, so it may be assumed that the experiments have not yet pushed art to its creative limits.

Already by the early 1960s computers were used for drawing graphs and isometric views of solid bodies. Computer graphics was the term applied to this kind of work. The objectives were essentially practical. The images were either ink drawings made by a moving pen plotter, compositions of different letters and figures printed out by an automatic typewriter, or electron beam traces on the screen of a cathode-ray tube. Before long it was possible to have the computer assist in more complicated functions. For several years now it has been a very effective partner in engineering industrial design. By means of a light pen a design on a cathode-ray tube can be shifted to another position, diminished or enlarged, rotated or seen in perspective[55]. These and other adjustments have been shown to be most effective in envisioning and facilitating needed design changes.

The elements in computer graphics tend to be logically simple geometric patterns but even these can be arranged to produce results whose possibilities have been sufficiently interesting to enlist the attention of a small but growing group of artists. Professor Charles Csuri, an artist with a background of engineering training, has described the computer as a potential means of breaking through restricting preconceptions of what constitutes form in art. One of his interests has been in the use of mathematical functions to modify form. As an example he has produced the drawing, "The Sine Curve Man," (*see* Fig. 8-1), with the aid of J. Shaffer, a programmer[56]. The technique was to make a realistic line drawing of a man's face and then digitize the drawing line by line. The resulting X and Y coordinates were reduced to punched cards, and then a modification was introduced by applying a mathematical function. The X value remained constant and a sine curve function was placed on the Y value. Given the X and Y coordinates for each point, the computer plotted the

308 Technology, Science and the Arts

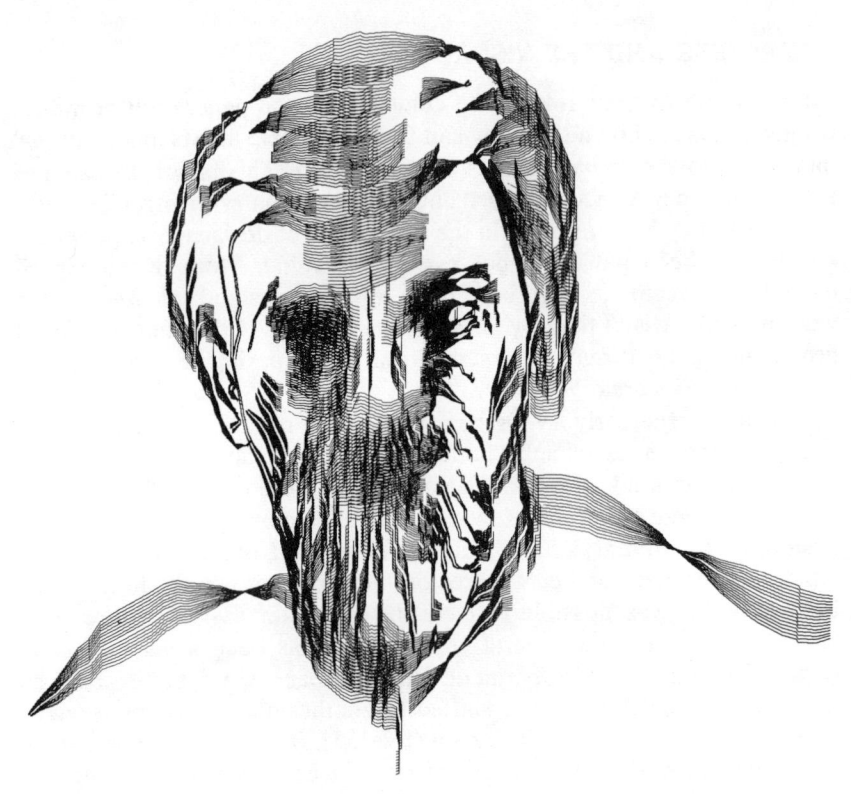

Fig. 8-1. "The Sine Curve Man." (Courtesy of Charles Csuri and AFIPS Press.)

figure from $X' = X$, $Y' = C^* \sin(X)$, where C was increased for each successive image. Csuri sees this kind of study as a means of dealing directly with visual ideas in consonance with 20th Century scientific concepts. One might apply the Lorenz transformation for example, to the drawing of a turtle to see what would happen to its shape as its movement approached the speed of light.

A. Michael Noll of the Bell Telephone Laboratories has made other studies on the use of the computer as a creative medium. He was able to produce abstract images by random or quasi-random generation of lines, which, to the lay observer at least, are virtually indistinguishable from abstract works of recognized artists[57]. One of them is an example of "op art." Op (for optical) artists rely upon mathematical forms to produce static, precise patterns on flat surfaces giving an overwhelming sense of motion through optical illusion. Certain critics have argued that op pictures are more relevant to psychological studies of visual perception than to art, but the fact remains that some of these pictures are exhibited in major modern art galleries. Noll observed that Bridget Riley's "Current" was a series of parallel lines, specifiable mathematically as sine waves with linearly increasing period. With this determination, the computer was programmed to calculate an array of points which were connected by a microfilm plotter. Other paintings less regular in design can also be reproduced by skilled copyists of course, so this experiment does not really answer any questions about the nature of the creative process that resulted in the original conception by the human artist.

Noll also carried out an interesting experiment using one of Piet Mondrian's paintings for comparison. Mondrian was one of the pioneers in modern abstract painting and is considered the originator of geometric abstractionism. He dealt with space in starkly geometric terms. His "Composition with Lines" (1917) is made up entirely of vertical and horizontal short bars within an outline that is apparently circular except for a slight flattening on the periphery at ninety degree intervals. Noll made certain observations about the pattern of placement of the bars and their relative lengths and widths sufficient to allow the programmed computer to generate a similar composition with pseudorandom numbers (Fig. 8-2). It is interesting to note that when copies of Mondrian's painting and the one produced by the computer were shown, unidentified, to one hundred people of widely differing backgrounds, fifty-nine preferred the somewhat less orderly computer-produced picture. If Mondrian had had the computer available to him in 1917 when he painted the picture possibly many of the tedious details of bringing his ideas to completion could have been eliminated, although of course we cannot ignore the effort of a different kind needed for programming. We do not know if the idea of saving time would have interested him, nor can we tell whether he would have been as satisfied with the final result.

COMPUTERS AND LITERATURE

Attempts to write literary works (mainly poems) by computer have followed along lines sketched out by students of communication theory [58]. It is possible, as we have noted earlier in Chapter 6, to construct sentences which resemble English by cutting all the words out of a long, standard English text in, say, three-word combinations, placing them in a box, and then drawing them out in random fashion. Usually "meaning" is just tantalizingly out of reach in the new text, but sometimes the results

Fig. 8-2. Computer composition with lines (1964) by A. Michael Noll.

are amusing or provocative and no less meaningful than examples of *avant-garde* poetry. The process does not require a computer but given certain patterns of input a computer can string letters and words together in a random sentence-generating program so that large numbers of word orders appear—some of them quite unusual. If one should set up a fairly extensive vocabulary containing a preponderance of words like "pallor," "sere," "haunted," "lone," and "mournful," the chances are excellent that the resulting lines will be more reminiscent of the work of Edgar Allen Poe than Ogden Nash. The additional factor that appeals to some scientists, and perhaps some poets, is that a new thought might also appear. If this should happen it often comes as a pleasant surprise, and surprise is one characteristic that art is supposed to have in far greater measure than technology. Technologists, in fact, do not like to be surprised by their machines.

One cannot help being reminded of Gulliver's visit to the Grand

Academy of Lagado and the word-frame there which was intended to give the world a complete literature of the arts and sciences. The experiments with computerized prose and poetry have no such ambitious intentions, but in any case it seems fair to say they offer little promise of far-reaching creative results. Writers and poets cannot escape the reality (if indeed they even wanted to), that unlike musicians and painters they work with a discursive form of communication. Words are their medium and words cannot be torn away from their human context. When poets began to strip their poetry of regular measure and rhythm, it may be, as Howard Nemerov has suggested[59], that this was a form of rebellion against the monotonous repetitiveness of machine processes. Much of the modern poetry has succeeded brilliantly in its new forms; but in some *avant-garde* work the experiments in randomness have become a flight from rational linearity in thought and speech patterns. These poets then become fair game for deadly irony if they see the machine as their foe, because this is one form of poetry computers would have the least difficulty creating.

OUTLOOK FOR THE ARTS

In summary, what can be said about the arts in a technological world? Many artists will continue to seek new inspiration with but little, save negative, reference to technology. If, however, they are concerned with the nature of their world, original ideas, divorced from the urban-technological milieu, will not come easily. It is clear that experiments with computers and other electronic technologies in music, and to a lesser extent in the visual and plastic arts, will bring modern science, art and technology together in a relationship closer than any they have had before. The important question remains, what kind of relationship? Symbiotic or parasitic? Reinforcing or destructive? Can art survive as a sensitive reflection of humanness if the scientist, the artist, and the engineer work together to explore its limits, or will it be overborne? One could argue that just as soon as artists cut themselves off from their traditions, art and humanity began to show separate faces. Yet science is not inherently anti-human; nor is technology until it neglects human purpose. We can perhaps gain a few clues about the possibility of a healthy fusion from the Bauhaus, but for various reasons beyond their control, (which may be immanent in society), that experiment was inconclusive.

The artist working with expensive technologies has definite economic realities to recognize. The experimentation must somehow be subsidized. No doubt the altruism of a few institutions extends to the support of certain projects for art's sake alone. Over the long term, however, if existing conditions continue, most sponsors will look for some practical outcome such as the improved understanding of synthesized speech processes or the development of new forms of computer graphics. The artist, if he is to experiment with the new methods, has little choice but to affiliate with an institution and to accept some kind of dependence on the assistance and cooperation of others who may not be artists. These conditions may affect his individual freedom to create at the same time as they expand his range. They will surely affect the forms of art in ways that are not easy to foresee. They may also have the salutary effect of re-humanizing technology. This last could be a major function of the artist in the cybernated world but it is a complex and difficult role to sustain. It requires him to pick an uncertain course between the contending forces of sophisticated commercialism, technological constraint, and individual creativeness.

Aside from providing the artist with new ways to experiment, modern technology and affluence have afforded increasingly greater numbers of people the means of enjoying art in its various forms, either as spectators or participants. Most evidence suggests that the traditional arts still command a larger public than the modern arts. The tendency for modern artists to deal in abstractions and intellectual exercises is probably inevitable in the knowledge society, but much of the art remains inaccessible to the layman even though the number of educated people who are able to comprehend and appreciate the special languages continues to grow.

Actually, many modern artists seem to want to show that art is no longer the exclusive preserve of the artist. To them his purpose should be to act as a medium, opening ordinary men's eyes to the value of the commonplace, however far removed it may be from traditional ideas of the beautiful. If the artist can thus succeed in sharing his artistic experience with the average person, there would surely be great value in it. Realizing that end was never more important than now, when the whole nature of work and leisure is being re-evaluated. In the technological world there are few more useful tasks than the one of sensitizing all men to the aesthetic experience in their surroundings. The artists who can keep their sense of direction may be the best equipped of all to point the way.

REFERENCES

1. McMullen, Roy, *Art, Affluence and Alienation*, Frederick A. Praeger, N.Y., 1968, pp. 258-60.
2. Williams, Raymond, *Culture and Society: 1780-1950*, Columbia University Press, 1958, Harper & Row Torchbooks, 1966, p. 130.
3. Levin, Harry, "Semantics of Culture," *Daedalus*, Vol. 94, No. 4, Winter, 1965, pp. 4, 5.
4. Sypher, Wiley, *Literature and Technology: The Alien Vision*, Random House, N.Y., 1968, pp. 10-12.
5. Trilling, Lionel, "On The Two Cultures," *Commentary*, June 1962, p. 461.
6. Roush, G. Jon, "What Will Become of the Past?" *Daedalus*, Vol. 98, No. 3, Summer, 1969, pp. 641-53.
7. Kline, Morris, *Mathematics in Western Culture*, Oxford University Press, N.Y., 1964, p. 274.
8. Harries, Karsten, *The Meaning of Modern Art*, Northwestern University Press, Evanston, 1968, pp. 17, 18.
9. Sypher, Wiley, *Rococo to Cubism in Art and Literature*, Vintage Books, Random House, N.Y., 1963, p. 129.
10. Hauser, Arnold, *The Social History of Art, Vol. 2*, Alfred A. Knopf, N.Y., 1951, pp. 550, 551.
11. Hauser, Arnold, ibid., p. 653.
12. Kuhn, Thomas S., *The Structure of Scientific Revolutions*, University of Chicago Press, Chicago, 1962.
13. Heydenreich, Ludwig, "Art and Science," *Magazine of Art*, April 1951, p. 140.
14. Russell, W. M. S., "Art, Science and Man," *The Listener*, January 9, 1964, p. 43.
15. Koestler, Arthur, *The Act of Creation*, (1964), Dell Publishing Co., Laurel Edition, 1967.
16. Fry, Roger, *Vision and Design*, Meridian, 1956.
17. Cassidy, Harold G., *The Sciences and the Arts: A New Alliance*, Harper & Brothers, N.Y., 1962, p. 89.
18. Cassidy, Harold G., ibid., pp. 6, 7.
19. Comfort, Alex, *Darwin and the Naked Lady*, George Braziller, N.Y., 1962, p. 9.
20. Johnson, Martin, *Art and Scientific Thought*, Columbia University Press, N.Y., 1949, p. 151.
21. Blanc, Peter, "The Artist and the Atom," *Magazine of Art*, April 1951, pp. 145-52.
22. Blanc, Peter, ibid.
23. Sypher, Wiley, op. cit., pp. 190-91.
24. Sypher, Wiley, op. cit., pp. 289-311.
25. Rich, Alan, *Music: Mirror of the Arts*, Frederick A. Praeger, N.Y., 1969, p. 258.
26. Kouwenhaven, John A., *The Arts in Modern American Civilization*, W. W. Norton and Co., N.Y., 1967, p. 103. (First published as *Made in America*, Doubleday & Co., N.Y., 1948).
27. Kouwenhaven, John A., ibid., p. 138.
28. Mumford, Lewis, *The Brown Decades*, Dover, N.Y., 1955, p. 215. (Originally published by Harcourt, Brace & Co., 1931.)
29. Kouwenhaven, J. A. op. cit., p. 125.
30. Mumford, Lewis, *Technics and Civilization*, Harcourt, Brace and World, Inc., N.Y., 1934 Harbinger Edition, 1963, pp. 345-51.

31. Sussman, Herbert L., *Victorians and the Machine*, Harvard University Press, Cambridge, 1968, p. 2.
32. Hauser, A., op. cit., p. 834.
33. Sussman, H. L., op. cit., pp. 113-17.
34. Morris, William, "The Lesser Arts," 1878, quoted in *William Morris: Selected Writings and Designs*, edited by Asa Briggs, Penguin Books, Baltimore, 1962, p. 103.
35. Sussman, H. L., op. cit., p. 129.
36. Poggioli, Renato, *The Theory of the Avant-Garde*, translated by Gerald Fitzgerald, Belknap Press, Cambridge, 1968, p. 52.
37. Naylor, Gillian, *The Bauhaus*, E. P. Dutton Co., N.Y., 1968, p. 46.
38. Naylor, G., ibid., pp. 102-07.
39. Davis, Douglas, "Mighty Machine," *Newsweek*, November 16, 1970, p. 108.
40. Tomkins, Calvin, *The Bride and the Bachelors*, The Viking Press, N.Y., Compass Edition 1968, pp. 26-27.
41. Rose, Barbara, *American Art Since 1900: A Critical History*, Frederick A. Praeger, Inc., N.Y., 1967, p. 10.
42. Rose, Barbara, ibid., p. 38.
43. Mendelowitz, Daniel M., *A History of American Art*, Holt, Rinehart & Winston, Inc., N.Y., 1960, p. 589.
44. Rose, Barbara, op. cit., p. 156.
45. McMullen, Roy, op. cit., pp. 141-42.
46. Harries, Karsten, op. cit., p. 58.
47. Stuckenschmidt, H. H., *Twentieth Century Music*, McGraw-Hill Book Co., N.Y., 1969, p. 174.
48. Stuckenschmidt, H. H., ibid., p. 180.
49. Pierce, J. R., "Computers and Music," *Cybernetic Serendipity* (edited by J. Reichardt), Frederick A. Praeger, Inc., 1969. (Reprinted from *New Scientist*, February 18, 1965.)
50. Hiller, L. A., "Computer Music," *Scientific American*, December 1959, p. 109.
51. Mathews, M. V., "The Digital Computer as a Musical Instrument," *Science*, Vol. 142, No. 3592, 1963, pp. 555-57.
52. Pierce, J. R., "Computer Synthesis of Musical Sounds," *Science, Art and Communication*, Clarkson N. Potter, Inc., N.Y., 1968, pp. 141-48. (Reprinted from Rockefeller University Review, November 1965.)
53. Strang, Gerald, "The Computer in Musical Composition," *Cybernetic Serendipity*, loc. cit., p. 26.
54. Strang, Gerald, ibid., pp. 26-27.
55. Reichardt, Jasia, "Computer Art," *Cybernetic Serendipity*, loc. cit., pp. 70-71.
56. Csuri, Charles and Shaffer, James, "Art, Computers and Mathematics," *Proc. Fall Joint Computer Conf.*, AFIPS Press, Vol. 2, 1968, p. 1295.
57. Noll, A. Michael, "The Digital Computer as a Creative Medium," *IEEE Spectrum*, Vol. IV, October 1967, pp. 89-95.
58. Pierce, J. R., *Symbols, Signals and Noise*, Harper & Row, N.Y., Torchbook Edition 1965, Chapter 13.
59. Nemerov, Howard, "Speculative Equations: Poems, Poets, Computers," *American Scholar*, Summer, 1967, Vol. 36, No. 3, p. 394.

9
Technology and Education

"The most important bill on our whole code is that for the diffusion of knowledge among the people. No other sure foundation can be derived for the preservation of freedom and happiness."

THOMAS JEFFERSON

Man, unlike the other gregarious animals, has a feeling for history. His symbols provide a continuum between the generations. Ralph Waldo Emerson once said, "A good symbol is the best argument, and a missionary to persuade thousands." However, when the time comes that many of the familiar symbols, through which communication is sustained and by means of which social activities are commonly measured and regulated, are no longer accepted; then structural upheaval and individual alienation are inevitable consequences. These conditions are notably present in 20th Century technological societies where quickly-changing perspectives have rendered many venerable beliefs untenable—at least when cast in customary forms.

It was noted in Chapter 8 that recognizable symbols were left behind when artists fled into abstraction (although Wilhelm Worringer, the German writer, observed early that some abstract works had their own kind of symbolism, which he interpreted as metaphysical expressions of anxiety). Almost simultaneously, as John McHale has described it[1], the function of the fine and folk arts as mediating channels between man and man, and man and society, was subsumed among the many new communications media. Thus the concept of art has had to be expanded to include aspects of culture which were not a part of former patterns of education in the humanities or the arts.

A basic problem for all of education, in the kindergarten or the university, is the need to find some new symbols for the technological age, to serve as well as the older ones did for so long. Many humanists, who see their province as preserving and interpreting those values and symbols in our culture which mark the distinctly human experience, have been accustomed to doing so in terms of the wisdom of the past. It seems to require a special sensitivity now to pick out the pertinent themes from among the merely antiquarian. An even more intensely demanding task is to convince students that there is anything worthwhile to be gained from a contemplation of history, especially if they believe that traditional symbols of morality, duty and heritage are at best irrelevant and at worst basically corrupt, designed to delude the majority into serving the selfish interests of the minority. The problem is not solved by suggesting that education in the humanities must somehow be made more "useful." It is in any event unrewarding to try to separate completely the useless from the useful in modern education, particularly when we must take into account rapidly altering views of work and leisure. Education, whether for ostensibly practical purposes or for leisure, is valuable. Educators need to be concerned with both aspects, although they will undoubtedly wish to distinguish, for practical and pedagogical reasons, between that training which furthers careers or the nation's economy, and the education which, in the words of Maxwell Goldberg, contributes to the individual's "enduring psychic economy." [2]

So we can put it that educators have to rethink their mission in the light of changes which contemporary technologies are producing in our way of living. An education at least through high school has for quite some time been accepted as a minimum requirement for functioning effectively in the knowledge society. Sometimes it is not enough, it is true. On the other hand, speaking in terms of training, there is reason to believe that occupations exist which can be handled very well by people without high school diplomas. Today, for a number of reasons (not the least important of which is the apparent inability of the job market to absorb all of the high school graduates) the minimum level of education to which young people are beginning to aspire is through the second year of the community college, either in terminal, technician-type programs or in lower-division preparation for further college work. This demand has produced staggering difficulties and challenges for the educational systems. The difficulties have hardly begun to be solved; indeed they are being compounded by a mounting resistance among taxpayers against shouldering the mushrooming costs of education at all levels. There is, then, an urgent requirement that education reshape its methods to find ways to take care, not only of all the young students who will be entering the schools, but also to provide for older students who are forced to change career directions or who are looking for a richer background for their leisure hours.

When it comes to the need to rethink their mission, educators are justified in pointing out that there are very few professions which are more inclined to self-evaluation than theirs. A teacher sometimes feels that almost too much of his time is spent on curriculum restudies. Nevertheless, far-reaching and lasting innovations are seldom encountered. Nor is it hard to see why. The schools are but mirrors of the larger society, especially at the grammar school and high school level and especially in the United States where universal education, locally controlled, has been so strongly championed for so many years. Accordingly they reflect the same tensions, inadequacies and strengths that characterize our other institutions and the same resistance to making adjustments for meeting change.

In the period before the 1970s, a serious obstacle to an effective continuous updating of the abilities of the public schools to meet the conditions of the new technological age was the fact that many teachers, and in particular their administrators, were brought up prior to 1945, so their outlook, unlike that of their students, was still shaped by pre-electronic paradigms. Beginning in the late 1950s, intensive efforts were made through government-supported special programs for teachers to improve their understanding of new developments in science and technology. The primary impetus for these projects was the launching of Sputnik by the Soviet Union. In retrospect, it appears that in some educational programs there was an over-reaction to this event for reasons which were not completely thought through, but at the same time many valuable courses of study in science, mathematics and engineering were developed.

Now in the 1970s, young teachers who have grown up in the electronic era are coming into positions of responsibility in the educational system. It would be a fair statement that there are among them many whose training in contemporary science and technology is first-rate; for in the reasonably affluent communities, at least, excellent science programs can be found. Yet, ironically, but perhaps not surprisingly, in view of the cyclic nature of social attitudes, science, mathematics and technology are in some disrepute among students. There are complex reasons as always, but among them one would have to list the reaction against militarily-related applied science, the presumed difficulty of the programs, and the belief that material rewards in science and engineering may be relatively modest and more at the mercy of economic conditions compared with those in certain of the professions.

How should educators react to these various conditions? What measures can they consider to handle student numbers burgeoning in terms of the percentage of the general population; to account for the relative decline of interest in classical humanistic programs as well as in science and technology; to counter growing public reluctance to meet the costs of education; and, in general, to make education more effective than it has been? In terms of practical training, if it can be clearly shown that many sub-professional jobs related to the production of hard goods and the processing of information will be eliminated, (or new ones generated) by computerized technologies, how should this be considered in the shaping of pre-college curricula? Are there now students learning typing and shorthand, keypunch or lathe operation, book-keeping or drafting with a view toward gaining employment in related work in ignorance of their eventual vulnerability to automatic technologies? Must the schools then provide, for students who cannot profit from college or who do not intend to go, some kind of preparation for employment in, say, the fields of social work or paramedical assistance; and should they emphasize, in the training programs, means of self-development in the arts of living? And if the latter, what kind of success can they expect to have with children from home environments where books have no place?

It is no wonder if conscientious teachers and educational administrators should begin to despair at the size of the task they have been asked to perform. They are beset with day-to-day crises which often forestall the most earnest efforts toward systematic planning. Some look to computerized methods and other kinds of educational technology to assist in this burden. There is by now a considerable body of information relating to experiments undertaken to explore the potential of technological devices and methods in education. An attempt will be made to evaluate some of these studies in this chapter. Since the likely success of various techniques must, in the long run, be a function of the way people learn new concepts, we shall begin by examining the sources of some of the major theories designed to lead to a scientific pedagogy.

EARLY DEVELOPMENTS IN THEORY AND METHODS OF LEARNING AND INSTRUCTION

There was a time when most educators would probably have agreed that, all things considered, the best system of instruction would be the

Rousseauian ideal of one wise, learned tutor for each child. Even now the idea has its attractions but we would recognize the unlikelihood of any single person's having a sufficient acquaintanceship with all contemporary knowledge, (or knowing where answers might be found), to satisfy the bent of each inquisitive mind. Moreover, it would also be accepted that the development of certain desirable social attitudes might best be accomplished by other means.

It is only in the highly evolved societies that the objective of expanding a student's intellectual background is considered very important. In the more primitive human societies the teaching of the young by adults occurs in informal ways. For the most part this means showing them how to perform some necessary skill. Language as a tutorial medium tends to be confined to songs and story-telling. The manipulation of abstract symbols is limited compared with what is required in modern education. A major point of disagreement between contending schools of learning theory has to do with the relative effectiveness of verbal and non-verbal methods for making clear various types of information.

Philosophers and religious leaders from the earliest times, who would not be considered teachers in the contemporary sense of having pedagogy as their primary occupation, nevertheless felt education to be a cornerstone of culture and social reform, and thus devoted much thought to the best means of passing on important moral and intellectual precepts. Most current educational theories can be acknowledged as owing a debt to these early philosophers, although only a relatively few of them tried to formulate consistent strategies to further what they considered to be the basic objectives of education.

The earliest and surely one of the greatest of these was Plato. Bruner [3] has contended that theories of education are in a real sense political theories expressing a consensus as to who should be educated, and for what roles. Plato had no doubt that education ought to be a political instrument. His doctrines were aristocratic and elitist. His Academy was devoted to the systematic pursuit of truth in philosophy and science, as a foundation for taking positions of leadership in the state. The welfare of his ideal state would be served by the rule of the wisest and the best: the philosopher-kings. It was Plato's belief that ideas, pointing to reality, were innate in a child at birth, but they remained dormant until illuminated by the dialectic processes of education.

After the Greek period, the one figure who is generally considered the most influential educational theorist before modern industrial times was Jean Jacques Rousseau, (1712–1778). He wrote passionately and eloquently, if not always rigorously, about many things, but it was the idea of naturalism as laid out in his *Emile* that meant so much to succeeding generations of educators. His own varied background included some time spent as a personal tutor, and although he had not been really satisfied with his own teaching performance, he did acquire a lively interest in educational methods.

Rousseau's doctrine was that children are naturally free from corrupt habits — the evil they may do later results from undesirable educational influences in their formative years. *Emile* begins with the words, "All is well, leaving the hands of the Author of nature; all degenerates in the hands of man." In following this statement with a description for an ideal program of education under the tutelage of a guardian — even an all-wise and compassionate guardian who cherished the freedom of the young — Rousseau constructed a paradox. Still, since a helpless child could not be left entirely on his own; nor could the complete job of education be left to the parents, (a mother, for instance, in civilized society would not have enough authority for the task) he believed some other approach must be found. The infant should first be taken into the country away from the evil city, whereupon a four stage program would begin. During the period of infancy, up to the age of five, the child should be allowed free rein for his instinctual impulses especially by being provided opportunities for motor activity and for physically experiencing a wide variety of natural objects and situations. It is the body which must be attended to in this stage. Imposition of authority in accordance with adult notions of discipline is harmful. It can only help destroy the child's innate curiosity which is what really motivates him to learn. In the stage from five to twelve, a rather similar course should be followed with expanded opportunities for self-reliant activity and a sensitive appreciation for his growing self-awareness. Rousseau felt that a negative result from an individually initiated experience is lesson enough for a child, external prohibitions being unnecessary. "Love childhood," he wrote, "encourage its games, its pleasures, its delightful instinct.... Let childhood mature in children." When lessons about morality are given at all, aside from the injunction of not doing anyone ill, it should only be at times when the child can see them to be of use to him. Not until early adolescence — the years from twelve to fifteen — is it necessary to consider the young person's ability to begin reflecting upon relationships among various things. But he still has little need of books (except perhaps for *Robinson Crusoe*) even though there is much to be learned in the natural sciences. The fourth stage, between fifteen and twenty, is marked by an awareness of his sexuality, and the importance of the social environment. Now he is ready to study the humane arts and social subjects—with instruction in religion and history and something about the nature of sex—to complete his education. At the age of twenty he can enter into society [4].

One should be cautious about reading into the work of Rousseau exact models for current educational theories and practice, but his core concepts, like those of most of the other original writers on education of centuries ago, are still very much a part of the modern discourse. His focussing attention on the needs of the child as opposed to the presumed needs of society had, because of the wide audience he commanded, an immeasurable effect on the philosophies of leaders in education who came after him. The issue of freedom versus arbitrary or even benevolent authority in the educational process is still alive, as is the related question of whether an infusion of technology for either mass or individualized instruction actually promotes or at length inhibits the freedom to learn.

EDUCATIONAL THEORIES AND METHODS IN THE 19th CENTURY

Too much has been noted about the rapid expansion of the system of capitalism and industry, the rise of nationalism and humanitarian liberalism, and the revolutionary developments in science and technology in the 19th Century for one to expect anything but profound stirrings to have taken place in the field of education. Of course the structures differed from one country to another but there were observable trends, none being more important than that control of the schools passed gradually to the state. The change did not occur anywhere without a struggle. There were continuing secular-religious controversies, and there was powerful opposition to the idea of free, publicly-supported education for all from entrenched forces who believed that it would do grievous damage to the economy to allow the working classes to climb too rapidly above their accustomed status. In England, the policy of *laissez-faire* held off government intervention until well into the century, but in France and Germany public support came earlier—first at the university and then at the secondary level. At length elementary education was provided for a majority of the children. In the United States, the need to promote common ideals among a polycultural population impelled the states and individual communities to establish programs which, in their total breadth, eventually surpassed the progress made in the other large countries.

THE MONITORIAL SYSTEM [5, 6]

An obvious need was to determine how best to accommodate the large numbers of children who would take advantage of free public education. One answer was the monitorial system. This idea, which employed older children as assistants to the teachers, had been used in India by the Jesuits, but it was most effectively developed in the secular schools (first as a private endeavor) by an English Quaker, Joseph Lancaster. Lancaster claimed that it was possible for one teacher to teach a thousand children through systematization and disciplined procedures, and proceeded to demonstrate his claim at the Royal Free School in London. In the matter of school building construction, later Lancastrian schools, which were established in many countries including the United States, were built with large, undivided rooms having ten square feet provided for each of as many as 500 pupils. Provision was made for sand tables, slates, and wall charts to save the cost of books and writing materials. As a monitor called out instructions to the ten children in his charge, each child would display his work on the slate, trace a number in the sand, stand up or sit down — all on order. Even the time-consuming ritual of hanging hats on pegs was dispensed with by having each hat suspended on a string so that it would hang down on the pupil's back after being removed from the head at a given signal.

Lancaster used ability-grouping, with assignments to sections being made on the basis of demonstrated ability in various subjects. While he rejected outright corporal punishment, the alternative methods he adopted for maintaining discipline, stressing shame instead of pain, were not always a distinct improvement. A miscreant might be suspended from the ceiling in a basket, or tied to his desk. Those unfortunate boys who had failed to wash their faces thoroughly could be subjected to the crowning indignity of having their faces scrubbed by girls before the whole school.

Whatever one may think of such a system of education in modern terms, (there are those who would advocate a return to something like it, with perhaps more sophisticated efficiency measures), it was very well accepted by many prominent educators and political figures of the early 19th Century. It is not hard to see why if we consider what they felt their needs to be. Undeniably the method was economical. Where funds and teachers were in limited supply, which was the case nearly everywhere, there may have appeared to be few alternatives to a system which seemed to produce such excellent results. Since the free schools were intended for poorer children it was presumed that there was not much point in going too far beyond the three Rs. For this kind of purpose the Lancastrian plan served very well. Many young children undoubtedly received some training that they could not otherwise have had at all. Rousseau might have said, "So much the better," but the balance between benefit and harm is hard to assess.

In one way the influence of the system was most damaging. Since the brighter, more able boys were the ones chosen as monitors, they were also the most likely to be attracted away by better paying jobs elsewhere. After a time, young women, many of whom were willing to work for a pittance, began to take the positions. So that they might be trained in the proper methods of instruction, special classes for monitors were established, some of which later became teacher-training normal schools. Generations of teachers came from these early normal schools, mostly indoctrinated with the limited concept of teaching as the parceling out of factual knowledge.

THE PSYCHOLOGIZING INFLUENCE IN 19th CENTURY EDUCATION

Although, in the eyes of the average industrial and political leader of the early 19th Century, the monitorial system could hardly be improved upon as a means of satisfying the minimal needs of education for the working class, the influence of Rousseauian humanism was too powerful to permit the monitorial concept to hold the field unchallenged. Also evolving was a counter-view of education as both process and environment in which careful, more-or-less experimentally controlled methods of observation would lead to the development of new systems of teaching, encouraging the natural growth of body and mind to proceed optimally.

There were many educators whose ideas reflected this trend to some degree. One of them, Johann Heinrich Pestalozzi, (1746–1829), a Swiss, was the first theorist who not only embraced the Rousseauian ideal, but also developed an operating system of teaching based, in part, on that ideal. Pestalozzi, as a young man, had been kept out of a potential legal career because of an association with an idealistic youth movement which had challenged corruption in the Zurich government. Turning instead to agriculture, he tried to establish an experimental farm. At this time he married, and one boy was born of the union, a sickly child who may have been an epileptic. Pestalozzi resolved to educate the boy in the manner of Rousseau's *Emile*. He had read the book when he was sixteen. Often he would take his son out into the fields to observe the animals and to note such physical phenomena as that water would always run downhill; but he was also convinced that a loving family environment was the best model for a good education. This gentle, kindly man was well-prepared to provide that much. Unfortunately, the son did not master the primary skills of reading, writing and figuring, so Pestalozzi concluded that Rousseau's theory had been deficient in not outlining practical teaching methods.

He was soon to have the opportunity to put some ideas of his own into practice when he decided to combine a school for poor children with his farming endeavor. We must recognize that at that time, in the latter half of the 18th Century, the spreading wave of industrialism was having its effect throughout Western Europe. The breaking down of organic family units and pauperizing of workers unable to cope with the vagaries of the wage economy could be found in Switzerland and Germany just as in England. There was a recognized need for educational reform. Not only were there neglected waifs to be cared for in the name of common humanity, but it was felt that through education it ought to be possible to raise their level of social and vocational skills to make them more employable as well as to turn them into respectable, God-fearing citizens.

Pestalozzi was one who had no doubt about the inherent goodness of children. The gloomy Calvinist doctrine of man born as a wretched sinner was not for him. His problem when he opened the school for the young paupers was somehow to bring the theory of natural self-development of the child into consonance with the need to train them for economic self-sufficiency. His more immediate problem, in fact, was to attend to his own economic self-sufficiency and in this he failed. He was before long forced to close the school and sell off most of his property. To help support his family, he began to write on educational theory. In 1781 he published *Leonard and Gertrude*, a sentimental novel about the regeneration of a depressed village through the application of the principles of natural education. The great success of this book accorded him a reputation which led eventually to his appointment, at the age of fifty-nine, to the directorship of the poor school at Stans. For the next quarter of a century at Stans, then Burgdorf and finally at Yverdon, he continued the experiments which brought him international recognition.

His own schools and the Pestalozzian schools started by others were based on the theory that the growth of human knowledge and awareness conforms to general laws which characterize other natural growth processes. Pestalozzi believed that the objects in the natural environment are made known to man through reflection on and clarification of impressions received through the senses. A primary function of the teacher would be to help guide the activities of the students through graduated experiences toward the state where impressions become ideas. As a result, then, the natural man would become the psychological man [7].

All lessons should begin with concrete examples, instruction being reduced to the elements of *number, form* and *language*. For instance, arithmetic, the most uncomplicated means of concept formation, begins with number. In teaching number, Pestalozzi had the learner count things in the space about him—pebbles or steps from a door to a wall—and then, after experience in counting had been acquired, proceed to counting tables where symbols which represented the number were examined. For fractions, he developed a chart of units based on visual partitioning of a square. He believed that the physical aspects of objects followed certain organizing principles according to form, and so this became a second basic part of instruction. The study of form, using the line as the simplest element, began with measuring and drawing exercises designed to improve the ability to depict the outlines of an object after it had been observed. Writing could not be attempted before drawing was mastered. But eventually, after the first impressions of nature had been absorbed through the senses, the path to knowledge had to lead through language. Here, the elementary units were sounds. Pestalozzi required the children to develop their speech patterns first by articulating simple sounds over and over again. The identification of objects by name followed. Then the language lessons went on to discuss number, form and other characteristics of familiar objects, and finally to much more complex relationships embracing concepts of geography and other advanced studies.

Critics have pointed out that in Pestalozzi's method of language instruction, logical order confronted psychological order in a paradox. It was logical for him to break words down into their component sounds, but that is not the natural order by which young children learn to speak.

One has to conclude that consistency was not among Pestalozzi's most prominent characteristics. His pronouncements were often vague and subject to many interpretations. In his own schools optimum results were not so common as he would have wished, in part because of his own deficiencies as an organizer. Nevertheless, his methods, through which he hoped to transform society through education, marked a significant step forward. His stress on the nurturing of the natural instincts of the child in an atmosphere of love and harmony was important; but equally so was his recognition that observation, sensory participation, the development of motor skills, and association were all vital to the learning process. In contemporary society where, to be sure, the grasping of abstract concepts means so much, we are coming to realize again that it can be a serious error to concentrate too exclusively on abstractions in education. Most likely there is some inherent need for learners to have a tactile involvement with physical resources at certain stages.

Among the associates and followers of Pestalozzi who made distinctive contributions to systematic thought in education were Friedrich William Froebel, (1782-1852), and Johann Friedrich Herbart, (1776-1841). Froebel is credited with the founding of the first kindergartens, while Herbart concentrated most of his attention upon education above the primary level, at which point he believed social concepts are ready to be assimilated. Herbart's work was supported by a meticulously evolved psychological theory. According to him, an individual's real self, or "soul" is the sum of the "presentations," or mental states, which result from experience through time. A presentation, such as a sound, is itself a kind of force and when it invades the soul it rises to consciousness immediately only if unopposed. More likely, however, competing or reinforcing presentations with interaction will take place. Interaction may result in a repression of feeling into the unconscious or perhaps a fusion that reaches consciousness as a more complex sensation. This abstract metaphysical system was developed by Herbart in mathematical form[8]. The primary significance of his theory for psychology was that it rejected the simpler view of the mind as a cluster of faculties which waited to be trained. Rather it placed psychology in the same realm as the natural sciences, where students of the mind could use mathematics and empirical evidence to achieve a greater knowledge of the way it functioned[9].

THEORIES AND PRACTICE OF LEARNING AND INSTRUCTION IN THE 20th CENTURY

The use of mathematical theorizing and controlled methods of experimentation, observation, and assessment of empirical evidence had brought new understanding of many physical systems in the 19th Century. It was natural that the potent promise of science and technology should now also be addressed to the human mind and to the way humans learn, if for no other reason than that the existing educational processes did not seem to be adequate for the needs of the industrial era. Research in psychological theories of learning and educational processes has generally followed two different but related directions — the first involving the observation of the measurable effects of manipulating a tightly-controlled environment upon an organism contained within that environment, and the second attempting to make the classroom itself an experimental laboratory. The latter approach which is in the tradition of Pestalozzi and his immediate successors was the one followed by two of the most distinguished educational theorists of the 20th Century, John Dewey and Maria Montessori. They were each born in the 19th Century, Dewey in 1859 and Montessori in 1870, but their reputations in education were secured in the 20th Century. Both died in 1952.

Maria Montessori

In his introduction to the American edition of Maria Montessori's book, *The Montessori Method*[10], J. McV. Hunt makes the interesting point that Montessori was offered space for her first "Children's House" by the owners of buildings in the San Lorenzo district of Rome and the director of the Roman Association of Good Building, partly in the hope that this might somehow reduce damage to their property. They felt that young children left alone by working parents to roam in the streets might be kept from vandalism by some form of custodial care. The parallel with current conditions in large cities is uncomfortably close. Today, more than half a century later, many urban schools are unable to function as much more than quasi-custodial institutions.

The strong-willed Dr. Montessori, who had left her birthplace at Chiaravalle to come to Rome with her family at the age of twelve, continued her education there at a technical school for boys in order to get the science and mathematics she wished to have. She later went on to the university and, in the face of formidable opposition from all quarters, became the first woman physician in Italy. From the very start of her medical training she was keenly interested in the problem of mentally retarded children, who were at that time being kept in the insane asylums. Without any original intention of getting into the field of education, she came to the conclusion that the special methods of instruction which she saw to be effective for improving the condition of these unfortunates (indeed bringing some of them to the point where they could pass examinations made up for ordinary schools) ought to show even better results when applied to normal children. She undertook to demonstrate this with the children in the San Lorenzo district and had an impressive success.

Montessori wrote [11] that her own work had been guided by the earlier studies of the French physicians, Jean Itard and Edouard Séguin. Itard was the writer of the remarkable treatise about "the wild boy of Aveyron." This true drama concerned a mute, mentally-retarded young boy, abandoned in a forest, who survived for years in a virtually wild state. Hunters discovered him and he was taken to Dr. Itard who specialized in the treatment of the deaf. Through patient, step-by-step advances in which he adapted his means to the spontaneous expression of his pupil, Itard eventually brought the boy to the point where he was at home in human society. The story has been sensitively portrayed in Francois Truffaut's 1970 French film, "The Wild Child."

Séguin, whom some called the "Apostle of the Idiot," had started by studying the experience of Itard but had carried the training of mentally deficient children much further on a systematic basis. He subscribed to the St. Simonian belief that the purpose of education was to adapt each new generation to a place in an ideal social order. While also agreeing with the Rousseauian idea that the training of the organic and motor functions is of vital importance in early education and should be undertaken before any attempt to impart knowledge, he was nevertheless convinced that there could be no complete separation. Training muscles and senses also developed the intellect and the will [12].

Montessori understood that the didactic materials that Séguin used for early training were also intended to act upon the spirit, and thus consistently emphasized the need for the teachers to be mentally and spiritually well-prepared for their work. A highly trained and responsive teacher is essential to the success of the Montessori method. Nothing however could be more misleading than to conclude that the teacher is in controlling command. Instead it is the spontaneous interest of the child in the challenges presented by the various didactic materials which controls the atmosphere of the classroom. After once demonstrating a proper use of materials, the teacher in effect withdraws to observe and to help the individual child only when he needs her.

A large part of the genius of the Montessori method derived from her recognition that the motivation for learning results from what Hunt [13] characterizes as a discrepancy between an input of information to the child and certain standards of expectation established through his earlier experience in the environment. If the discrepancy is too great the tendency is to withdraw from the situation; if too little the motivation is minimal. After a long and painstaking period of trial-and-error experimenting, Montessori had been able to develop materials and an approach which went far toward providing experiences to match the learner's need to maintain an optimum interaction with the environment. The materials are organized into a carefully graded developmental series to allow for growth. It is especially important to note that it is the child himself who selects the materials which match his interest and abilities at the moment. He chooses the rate at which he wishes to progress, but advancement can occur only after mastery of primary concepts essential to understanding of higher-level concepts.

Montessori believed strongly that it requires active work for the child to reach his full potential as an adult. By this she meant real work, not aimless play. For that purpose he must have real equipment. He must have chairs light and small enough to arrange as he wishes. The tables should not be too heavy for two children working together to move. Kitchen utensils are realistic as are materials such as cleansers, sponges, brushes and dusters for him to keep his area clean and neat. Dr. Montessori emphasized that children should start their training in her system at the age of three, or no later than four, or otherwise they would be past the critical sensitive period. Beginning with children at the ages which we normally consider pre-school, her method starts with lessons in grace and courtesy leading to social development; and with exercises training each of the senses to discriminate differences and recognize similarities through isolating one dimension in each piece of equipment. For an exercise in size, everything is the same shape, texture and color. Plain and sandpaper-covered boards demonstrate various grades of difference between rough and smooth. In a later exercise the child may have letters and numbers with a sandpaper surface to trace with his fingertips. The visual color sense is stimulated by using tablets of different colors. In Montessori's words, "Looking becomes reading, and touching becomes writing."

It did happen this way, and children working at their own pace, were soon reading, writing and doing arithmetic, sometimes two or three years before children of the same age going to the regular schools were able to reach the same capabilities. This led to one objection which critics raised about the system. They felt, and some still feel, that learning these skills at too early an age may be wrong for the child's proper psychological and social development. It is true that the role of play is not emphasized as it was in the Froebelian kindergarten, or in many modern primary schools. But neither does one get the feeling of grimness or that the child's natural impulses are held down by lock-step methodologies in a properly run Montessori classroom. Rather, the atmosphere which may be initially disordered tends to become one of serenity and purposefulness. Pupils play some group games ordered to a definite end, and often collaborate with each other on their own initiative.

Perhaps more telling are the criticisms that in certain instances, revival of the Montessori system has taken on some of the aspects of cultism, with a too slavish employment of her didactic materials precisely as it is believed she directed them to be used. Some of this kind of tendency to misunderstand and to reshape the original message seems to be inevitable among the followers of any innovator, but in any event there is no compelling reason why new materials culturally related to the experiences of contemporary children cannot also be adapted to the system. Montessori considered the didactic materials to be part of a whole "prepared environment" which included the activity of the teacher, the practical exercises and the classroom furniture—all sharing a relationship to the experience of the child. McDermott[14] contends that the newer technologies which prove themselves within the total fabric of learning, after a period of careful observation and study, can also become a part of the prepared environment designed to stress expansive experience options.

Montessori is important, then, because her painstaking construction of a program of teaching remains one of the best examples of a working educational system combining methods and devices, and permitting self-learning without anarchy. She has something helpful to say to contemporary advocates of individualized instruction.

There has been a resurgence of interest in the Montessori system in the United States in the past few years; although in the second decade of this century, when her ideas had already strongly affected European elementary education, early enthusiasm had quickly waned in America. Even though she had some influential spokesmen supporting her concepts in this country, they were not by and large among the leaders in educational philosophy or psychology. Rather, the prevailing opinion came to be made up of theories of Darwinian predetermination, stimulus-and-response psychology, and Freudian psychology. A science of learning, strongly influenced by Edward L. Thorndike's studies of learning in animals, had begun to emerge; but perhaps more significant was that the United States had its own pioneer in John Dewey. From his experimental school and his writings came the ideas which were to transform education in this country.

John Dewey

John Dewey was a philosopher—one of the leading exponents of the peculiarly American philosophy of pragmatism. Rather than being a doctrine, pragmatism is a method of deciding metaphysical disputes by trying to look ahead to see what would be the practical consequences of adopting one mode of thought over another. Dewey believed that education in the 20th Century would be valuable to the practical extent that it would train young people for fully rounded living in a society whose hallmark was rapid social change caused by unprecedented advances in science and technology. Educational methods should thus be based on an analysis of the aims and needs of that society. This was in the spirit of progressivism which assumed that the time had come to break out of the restrictive social and intellectual practices of the past. It meant elevating ideas derived from the experimental method above ethical absolutes. He felt that each school ought to be an embryonic community characterized by involvement of the students in activities reflecting the problems of the larger society. He identified those problems as the same ones, for the most part, which we are still trying to analyze and solve.

Dewey has been called a progressivist as well as a pragmatist—and also an instrumentalist to reflect his interest in methods—but it would do a disservice to his philosophy to try to pin any label on it. His thoughts on education are contained in a number of works beginning with *My Pedagogic Creed* (1897) and *The School and Society* (1899), and including *Democracy and Education* (1916), *How We Think* (1933), and *Experience and Education* (1938). These and other writings cannot be conveniently capsulized, but there was a fundamental assumption of an organic relationship between the psychological makeup of the individual child and the social environment. He accepted the newer psychological concepts which viewed the mind as a maturing process, rather than a fixed set of separate faculties waiting to be trained by formal intellectual exercises. With changing degrees of responsiveness at different periods of growth, the individual organism, functioning as mind, continuously adjusts to the changing environment. The teacher, by being aware of the psychological needs and impulses of the child and the milieu in which he must live, can adapt the educational experience to the child so that he will progress naturally to a useful role in society. The basic educational process was not to be considered as preliminary training toward a specific vocation, but a means of sustaining "the social continuity of life." Every child, according to Dewey, has inherent impulses: to communicate socially, to make things, to assuage his curiosity by investigation, and to express himself in some creative or artistic way. The school thus needs to provide activities containing within them seeds of motivation and interest. The carrying out of these activities then becomes a challenge and a pleasure for the child instead of a dull chore.

In the primary years, from four to eight, play was to be emphasized expressly to expand the child's social consciousness. Activities would begin with experiences reflecting the familiar life at home and then proceed to encompass the larger community. Reading and writing would not be taught until near the end of this period, when presumably there would be a store of experiences providing the motivation to learn. By the time of the period of spontaneous attention, from eight to twelve, the child would have reached a state of development distinguished by an ability to separate ends from means. With more confidence in his own powers he would be ready to acquire different skills and apply them according to logical rules. Dewey called the third period, from the age of twelve and beyond, the period of reflective attention. By that stage, the child would have achieved a degree of mastery of methods of reflective thought, investigation and activity which would allow specialization in distinct areas with intellectual and technical ends in view.

Dewey became one of the most controversial figures in American education. Many of the charges which have been leveled against him reflect an inadequate understanding of his ideas. To some extent this is because his writing was often obscure. It is possible that his concepts would never have had such a widespread effect on 20th Century educational philosophy had there not been others to interpret (and sometimes to misinterpret) his message. He did not, for example, champion unbridled chaos in the classroom as it is sometimes claimed. Although he was opposed to an authoritarian atmosphere with the teacher dispensing facts as irrefutable dicta, he also wrote: "It is equally fatal to an aim to permit capricious or discontinuous action in the name of spontaneous self-expression"[15].

Dewey's progressivism was more hard-headed than romantic; it was committed to drawing conclusions from experimental facts. He thought it would be possible to adjust to the increasingly complex environment wrought by technology only through controlled and systematic thought. Even though he wrote: "At times it seems as though we were caught in a contradiction; the more we multiply means the less certain and general is the use we are able to make of them,"[16] he still hoped that further development of science would offer ways to tame the complexity.

The primary aim which Dewey sought—that of affirming the worth of a democratic society—was a serious one, and if it meant anything at all it could be demonstrated in an educational system where democratic principles prevailed. Even at this late date, some two decades after his death, his influence is still being blamed in certain quarters for much that is wrong with the schools in the United States. By the 1950s, dissatisfaction with the results of public education was widespread, reaching a peak when it was learned that Russia, a country which was presumed to have a much more primitive technology than the United States and a backward educational system, had become the first nation to send a man into space. Somehow, it appeared, the children had not been learning to read or write properly nor was the general academic preparation in the secondary schools at the level required for a society in which growing numbers of young people would be wanting a college education. It was natural that a great deal of the opprobrium settled on the methods of progressive education. Probably we are still too close to the time when that movement was in the ascendancy to assess its strengths and weaknesses justly, but still, if we look at the situation as it existed when Dewey and his popularizers were trying to clarify their ideas, the methods they favored were not unreasonable. The schools were faced with a huge increase in the numbers of students from a heterogeneous population. Many of these were second- or first-generation Americans who did not aspire to vocations requiring a level of education beyond the beginning of high school, if even that much, and they were in school because that was what the law was now demanding. If the objective of the educational system was to capture their attention and especially to mold them into the American way of life, progressivism seemed to offer some answers. And if we now tend to question their basing the methods and projects largely on a white, middle-class value system, that is because our sensibilities are those of a different generation.

EDUCATIONAL TECHNOLOGY IN THE 20th CENTURY

Each of the educators whom we have discussed in the foregoing paragraphs looked at the educational process in a certain fresh way which has had a lasting effect on theories and methods of instruction. Doubtless each of them felt that his concept of instruction was somehow more logical—more "scientific"—than the concepts of his predecessors. Yet it seems clear that a true science of instruction, universally and timelessly applicable, has still not emerged. This does not mean that no progress has been made, through experimental and theoretical science, in uncovering keys to the processes by which humans (or the lower animals), learn. Since the last two decades of the 19th Century, psychologists working in the field of learning theory have brought much interesting data from the laboratory. But, overall, one is forced to conclude that the results have been too fragmentary, and derived from experiments too narrowly directed toward limited objectives, to be transferred effectively to an average classroom. Many of the results have been seen to be confirmations of observations which had been made countless times before by individual teachers, but never cast in a generalized form. Learning theorists themselves disagree about how well-prepared they are to apply the results of their experimental work to a broad revision of the educational system. Many do not even accept that a solution of classroom problems should be any significant part of the goals of their research. Even if there were wider agreement on basic principles of learning and instruction it is highly problematical that most teachers would be ready to employ them as an infallible prescription for all the difficulties arising in their own situations; or that most parents would be willing to go along with new directions in education for their own children that appeared to be too experimental or strange to their own experience. In any event, the psychological laboratory, and the standard classroom (even if considered as a kind of laboratory) are fundamentally different in more ways than one [17]. The bridging of those differences in a formidable task.

Many, if not most, of the technologies which are now used regularly in the classroom, (for the purposes of this discussion we shall include organized teaching techniques and simple materials as a part of educational technology), have been introduced gradually over many years largely through the influence of such innovators as we have already mentioned. Manufacturing technologies also evolved in a similar fashion in earlier times but more recently the process has often been accelerated, either by systematically borrowing new technologies from one area of application for use in another, or by developing them directly from theoretical principles. Each of these two approaches has also been followed in attempts to establish an effective contemporary technology of education. Admittedly the approaches are sometimes related since it is possible to employ a device borrowed from one field in some special manner in a classroom, according to postulates of experimental science, such that a new technology of instruction appears. However, this rarely happens. It is more common to find that films, television or computers are used essentially as variations of existing methods of instruction [18].

We shall want to describe in somewhat more detail various types of technologies, especially those products of 20th Century engineering or behavioral science which have been proposed for or actually used in contemporary schools. It should be emphasized that it is the use to which the device is put as a medium of education which qualifies it for our interest. Our attention will encompass the educational system as a whole — school management as well as classroom instruction — for in the modern school the learning atmosphere in the classroom cannot help but be affected by the way in which the school system is being run. We shall be forced to generalize and simplify. Obviously the enormous variety that exists in the decentralized American educational enterprise — the private, public and parochial schools from the pre-primary years to the high school, private and public universities, training institutions for special purposes run for private profit or to meet the needs of industry or the military, adult education, and a host of peripheral businesses which serve education — renders a definitive picture out of the question. The alternative is to select a few examples of specific technologies and to explore their present utility or promise in the light of what we have already discovered about principles of learning and classroom practice.

Audio-Visual Methods

A study of the emergence of the motion picture film as an instructional medium gives an excellent basis for judging the effects of a new technology upon educational practices. It is not new anymore, of course, but even the earliest attempts to use films in the classroom, going back to the first and second decades of this century, are sufficiently near in time that a clear and quite extensive record of progress and setbacks exists for anyone to examine. Saettler[19] has published an interesting documentation of the development of instructional films as part of the whole audio-visual movement; and Brown, Lewis and Harcleroad[20] have also given an authoritative account of programs of audio-visual instruction. While motion pictures as a form of public entertainment quickly captured the imagination of a mass audience, their adaptation to the classroom came much more slowly. For one thing, students, however diverse, were considered as a specialized audience — specialized from the standpoint of being young and impressionable and needing to be edified. Due regard had to be given to improving their moral attitudes while dispensing worthwhile factual information.

There were some real enthusiasts ready to proclaim the potential of motion pictures. Thomas A. Edison, for example, stated in 1913 that they would transform the schools in a decade. There was also considerable interest in the commercial prospect of producing and distributing films and manufacturing special equipment for educational use. Most professional makers and distributors of films for the mass public in Hollywood and New York were not quick to share this enthusiasm, reasoning that they would prefer to have young people pay their way into the theaters rather than to have them see similar pictures without cost in the schools. Of course, the national film industry in spite of itself was a powerful educational force. It did much to shape the habits of film-going generations just as television does today even more ubiquitously with home viewers. But the early educational films shown in the classroom were almost uniformly dreary. They were largely culled from portions of out-of-date commercial motion pictures, government training films and industrial films testifying to the excellence of a certain product or corporation. True, there were a few companies, including Edison's own and the Ford Motor Company, which made a serious attempt to produce relatively inexpensive films especially for the schools. Some of these efforts were of good quality, but the whole educational film movement suffered from certain fundamental weaknesses which plagued it for decades. The most talented film makers were working for the large commercial studios. Those who were involved in making the inadequately financed educational motion pictures too often lacked not only some of the technical expertise necessary for consistently superior results; but also they were not themselves educators as a rule and had little understanding of what might or might not be effective in the education of children.

In the classroom there were often mechanical difficulties with the early equipment even if the teacher were familiar with its correct operation. But there were also the additional logistical problems of previewing, ordering and scheduling of films; of seeing to it that the necessary components for a showing would arrive in the classroom at a time when its subject matter could add an effective dimension to the topic being studied. These are chores with which many teachers still refuse to be bothered. The introduction of sound films, arriving just about the beginning of the Depression years did little at first to enhance the use of motion pictures as an educational medium. It cost more to produce and show them just when little money was available for educational innovations of any kind; and they were a novelty at a time when it seemed safer to cling to accustomed ways. They did surely offer one advantage which could be exploited for the education of young children, i.e., they did not require that the audience be literate, as silent education films to a degree did.

Gradually from isolated individual or group efforts to use films and other visual instruction aids in the schools, educators began to bring together a body of techniques and organizations comprising the visual (later with the utilization of sound films and radio, to be called audio-visual) instruction movement. Although there were attempts to relate methods of presentation to certain theoretical rationales of instruction as they were understood at the time, not until World War II was it really correct to say that instructional films had started to achieve their full potential. It is perhaps useful to make a distinction between films, (or for that matter radio or television programs) produced for educational and those produced for instructional purposes. Instructional films are designed to introduce specific subject matter to individuals, groups or even a mass audience following a formal course of study designed to enhance a particular expertise. Educational films are intended to impart information to a seriously-inclined but more general audience, although they can also be suitable for specialized groups.

During World War II the need to train military and industrial personnel literally by the millions for performing many different tasks led to an unprecedented effort to expand instructional technologies to serve these purposes. Larger sums of money than any educational systems had ever had to work with before were made available; but probably more important for the overall quality of the results was that professional film producers, artists and educators were brought together to work cooperatively. They proceeded to do this after an initial period of mutual adjustments. It is generally agreed that valuable contributions to instructional technology came out of this period. Besides perfecting techniques of using the camera as if it were the eyes of the operator of the piece of equipment whose use was being demonstrated, and adapting dramatic techniques from the field of entertainment, the instructional technology specialists also developed supplementary means of enhancing the training programs. They devised exceptional graphic displays and three-dimensional models, wrote handbooks for instructors and made realistic mockups with auditory enhancement to permit simulation of actual operating conditions for trainees. Cartoon series were developed for illiterates, and there were also notable examples of self-instructional devices.

Undoubtedly much of the overall success of the armed services instructional programs (and to a lesser extent the training programs for industrial workers) can be laid to the high degree of motivation of the trainees in wartime, and to the fact that the skills being taught were nearly all straightforward and readily definable in functional terms.

However one cannot overlook the significance of a new sense of the possibilities for instruction based on organized use of a variety of technologies either singly or in support of each other.

Instructional Radio and Television

We shall briefly examine the use of radio and television for instructional purposes; although considering them both under the same heading does not necessarily signify that their respective effects as instructional media have been similar. Some students of educational technology have suggested that the promise of instructional television has been no more fully realized than that which was originally held out for radio or even for films, but in fact there is reason to believe even though a great deal of research has been conducted that neither the overall potential nor the effects of instructional television have yet been completely assessed.

Undoubtedly instructional radio had its period of maximum influence in the years before World War II; but as early as 1925 the Coolidge administration, by making the decision that national radio would not be governmentally controlled, had insured that commercial broadcasters would essentially dictate radio programming in the United States. Although in the following decade or so there were numerous educational programs, some with federal and some with private or local government support, there was never a strong effort to maintain instructional radio on a national scale. We should not discount the "schools of the air" which the national broadcasting systems sustained as regular short programs available to the schools and to home audiences, but the primary attempts to innovate were largely initiated by stations developed by universities and other local educational systems. Overall, however, educators were more indifferent than not about radio in the schools. It must be granted that too many teachers who tried to give their lectures on radio made inadequate attempts to adjust pedantic classroom styles to the demands of the medium.

Language courses and music, by their nature, seemed best suited for instructional radio. Plays and dramatic readings as a means of education were often imaginatively produced, and social studies and science as well as literature were also offered in the form of courses, but altogether the instructional programs were generally spotty. Most children who were pupils in schools with broadcasting facilities during the 1930s probably remember radio, if at all, as a means by which occasional newsworthy events of broad interest were piped into the classroom. Instructional radio all but faded out of the national educational picture by World War II, and has not been revived.

In the period following the war there was ample reason to feel that television was about to become a very important instructional medium. Few could doubt that many new possibilities would be opened up by the opportunity to stimulate a student visually as well as aurally at the very moment an event of educational value was going on.

From a practical standpoint, another development of equivalent significance took place. By 1948 the federal government was about to consider the assigning of channels for public telecasting. This time, however, a group of dedicated educators and concerned citizens were ready to work to insure that the mistake that had been made in giving commercial broadcasters virtually complete control of the A.M. frequencies for radio broadcasting should not be repeated. At first there was little indication that the Federal Communications Commission was prepared to grant any exclusive channels for educational uses. However, the aforementioned group, led by Frieda Hennock, one of the commissioners, succeeded in marshalling an impressive array of expert testimony to the effect that the television stations then operating were offering programming that had little or no educational value and was often tasteless besides. As a result, the commission in 1952 reserved 80 VHF and 162 UHF television channels for educational purposes[21].

This was a crucial battle to win and in the following ten or fifteen years there were other victories especially from the point of view of financing. During the Kennedy administration, construction of educational television stations was funded through the Communications Act, passed in 1962. In addition, for a number of years large supporting grants were made by private foundations. Support from the Fund for Adult Education of the Ford Foundation permitted setting up the National Educational Television and Radio Center (NET) which has been the source of educational programs of exceptional merit. Besides being widely shown on educational television stations throughout the country, NET programs have also been made available on film for use in schools.

Instructional television programs designed to constitute courses of study related to specific subject material, as distinct from those programs produced for more general educational purposes, have been offered on commercial channels as well as on educational stations. In the universities and public school systems the initial attempts to provide a format comparable to that used on commercial television were not successful. As time went on the emphasis shifted to cable-connected closed-circuit systems ranging from hookups connecting several rooms in a single building to systems spanning an entire state. Published reports comparing the effectiveness of instruction by television with ordinary classroom teaching have not been conclusive enough to lead to general recommendations strongly favoring one method over the other. This, in spite of the fact that the studies have often been conducted with the need to find a remedy for overcrowded classrooms and shortages of teachers as a primary concern; and that the tendency has been to select teachers with outstanding reputations for instructional television experiments.

One of the advantages ascribed to instructional television, of course, is that it is supposed to permit the best teachers to reach the maximum number of students. Presumably, too, if the same teacher can reach more students at the same time, or at different times by means of videotape, the cost of instruction per student will be decreased. However, it is not quite that simple. For one thing, the expense of installing an adequate system for either open or closed-circuit telecasts is still so high that many financially harried school administrators are unable to justify taking that initial step. True, the tendency is for costs of new technologies to come down as they become more widely used and perfected; but one still gets what he pays for, and for programs of quality it is necessary to have first-rate equipment operated by talented personnel. Instructional television is a team effort demanding the closest kind of cooperation between the teacher and the technical staff.

Balanoff[22] has pointed out that separate groups of technical specialists may favor either an audio-visual or a professional broadcast approach to instructional television and the resulting effort will reflect the difference. The audio-visual approach considers television as one of several means of communicating information, and presumes it should be used in conjunction with tape recorders, slide projectors, etc. The television equipment functions as an eye (and an ear) disseminating information, which is generated in the classroom, to a large outside group. Television can enhance teaching effectiveness by enlarging a display or displaying microstructures, for example; but the audio-visual specialist, unlike the professional broadcaster, does not insist upon extensive pre-production preparation, nor on a carefully polished performance from the instructor. He may even allow the instructor to handle the camera by remote control. The broadcaster is used to commercial standards. To achieve them he must retain greater control over all phases of the presentation, during which the instructor must react at the right time to the proper cues. Some teachers have no objection to adjusting their teaching to these kinds of requirements; but others resent what they consider to be unwarranted limitations placed on their natural classroom styles and do not feel comfortable teaching in that manner. There is presumably a place in education for both the audio-visual and the professional broadcasting approaches, although the latter is undoubtedly the more costly to mount.

Many instructional television programs are offered strictly in the lecture format. This is the easiest type to produce, and with a knowledgeable and entertaining speaker may serve its purpose up to a point. One fault with the straight lecture presentation is that it does not take full advantage of the medium. Commercial broadcasters are well aware that a single speaker discussing a scholarly subject for a considerable length of time without interruption will have fewer listeners at the end of his talk than when he started. There is no reason to think that the situation would be any different in a large classroom full of students watching a televised lecture except that the tuning out might be only mental. Thus it seems important, whenever possible, that there be audio-visual enhancement of the lecture by a variety of means, and an opportunity for question-and-answer communication between the lecturer and the students. The lecture should also be supplemented by more extended discussions in smaller groups with an adjunct teacher. It is only fair to mention here, though, that some instructional television research has shown no clear cut differences in the results of examinations taken by college students whether or not special visual aids were used or two-way communication was established in the television presentation. This may say something about the validity of examinations or the ability of college students to do critical thinking independently. If we consider television programs at the pre-school level we cannot fail to note the success and apparent effectiveness of the remarkable "Sesame Street" which has been imaginatively visual but which has not employed a two-way communication loop. It has used motivational techniques which advertisers have claimed to be rewarding.

There are a few considerations — sometimes subjective and sometimes supported by factual data — that one has to think about in determining whether or not a given course should be offered on instructional television. Obviously, subjects which have a substantial visual content lend themselves, on the whole, to this form of presentation more so than those which deal almost exclusively with abstract verbal concepts; although creative people have shown themselves capable of doing interesting things with the most unpromising material. Then, too, it goes without saying that the effectiveness of a teacher in this medium may have little relationship to his scholarly reputation. What must also be said, however, is that not every teacher who has shown consistent ability to motivate children or adults in an ordinary classroom can accomplish the same results on television. The reasons why teachers succeed or fail in a given situation are often unaccountable. Some teachers have qualities of understanding and personal warmth enabling them to establish a one-to-one communication without which many students cannot be reached. This may not always come through in the context of television and it may be much more important in pre-primary education than at the higher levels.

Observations of these kinds are founded on studies that are still so incomplete and sometimes contradictory that it is not surprising that decisions to establish extensive instructional television systems are based on criteria which are judged to be more practical. Only those fundamental subjects which are standard offerings for large numbers of students can justify television presentation on a regular schedule; the specialized courses in which only a few students are involved cannot. When programs are developed of sufficient value to be distributed to a wide area, then it is necessary to consider alternatives among relatively expensive transmission technologies. Videotape has some of the advantages and disadvantages of film and the newer video magnetic tape recorders are relatively inexpensive; but if Marshall McLuhan is correct, television with its inferior detail carries a lesser degree of information than the motion picture. It has been a common assumption that school buildings should be designed to utilize instructional television to best advantage and this feature is incorporated into many new constructions. However, some educational specialists have chosen to think of the medium as a possible means of eliminating the need for schooling to be exclusively conducted in clusters of buildings reserved for educational purposes. Some universities, for example, have been offering regular courses for graduate credit to engineering employees of industrial firms in the area who meet at regular hours in television classrooms on the company's premises. This system permits the employees to update their professional capabilities without the need to consume excessive time away from work traveling to and from the campus. The services are rather expensive however, and the tendency is for only the larger companies with many eligible employees to subscribe. The economic factor is bound to be critical in determining how far educational and instructional television systems develop.

Teaching Machines and Programmed† Instruction

There are few examples of unique mechanical or electrical systems specifically developed for instructional purposes. One of them would be the simple teaching machine represented in its early phase as a piece of equipment first demonstrated in 1925 by Sydney L. Pressey, a psychologist at the Ohio State University. His initial device was not designed as a teaching machine at all but as a means of scoring objective tests. However, he was quick to observe its possibilities as a teaching system. Pressey's machine displayed in a window four possible answers to multiple choice questions. The student could press one of four keys corresponding to the answer he selected, and if it was correct the next question would appear. If, however, he guessed wrong the initial question remained displayed until he made the right choice. An automatic record was kept of all attempts. It is also interesting to note that an attachment to the machine caused it to present a candy lozenge to the learner if he achieved a pre-set score indicated on a reward dial[23]. Although Pressey made a good many improvements on his devices and methods of auto-instruction in the succeeding years, and although he felt confident that his work presaged what he termed an "industrial revolution" in education, there was little effort on the part of others to carry his work further until the decade of the 1950s. Pressey's teaching machine and contemporary elaborations of the idea would now be considered one manifestation of programmed instruction. Programmed instruction methods, which also have antecedents in earlier educational innovations (for example in the self-learning aspects of the Montessori system), now range from the simplest linear programmed textbooks to adaptive teaching machines and complex computer-oriented systems, and they also incorporate many new audio-visual techniques.

The reasons for the ferment in educational circles in the 1950s have already been mentioned. At this time educators were more than willing to take second looks at innovative techniques of instruction. It seemed as though the technological revolution might at last invade the classroom to such a degree that its devices and methods could begin to reshape the instructional processes.

Programmed learning was a key development which caught the imagination of many in the teaching profession. This advance was an example (in fact one of the few) of an instructional technology based on precepts of psychological learning theory. Professor B. F. Skinner of Harvard

† Professionals in the field in the U.S. usually favor the spelling, *programed*.

University is generally acknowledged as the person who supplied the major stimulus to the programmed instruction movement. His paper "The Science of Learning and the Art of Teaching,"[24] published in 1954, focussed attention on the possibilities of a practical learning device based on "operant conditioning," an empirical approach to learning theory which had won him an international reputation. Operant conditioning differs from the concept of classical conditioning, described in 1906 by the famous Russian psychologist, Ivan P. Pavlov, in the following way. In Pavlov's well-known experiment, dogs learned to salivate to the sound of a tuning fork (the conditioned stimulus) after it had been consistently presented simultaneously with meat paste (the unconditioned stimulus). Conditioned avoidance behavior was similarly produced by coupling a stimulus with an electric shock. This type of conditioning depends on the contiguity in time or space of a reflex behavior with an extraneous stimulus. Operant conditioning is determined not only by the stimulus but by the rewarding of a certain response initiated by the organism being trained. A naturally occurring response such as a pigeon's turning its head in one direction can be immediately rewarded with food after successively greater turns and thus the pigeon can be trained to turn in a complete circle.

As the result of an extended program of laboratory investigations, Skinner has described some rules governing the formation of operantly conditioned responses [25, 26]. He proposed that emitted responses (for which the preceding stimuli may not always be recognizable) constitute a major part of the behavior of the lower animals and humans; and that learning is simply a change in the frequency, called the operant rate, with which these responses occur. The operant rate can be changed in a controlled manner by reinforcing (rewarding) these responses whenever they take place. The probability that responses will be positively conditioned depends upon the time interval between the response and the reward, and upon the reinforcement schedule. Those responses closest in time to the reward are the most reinforced. Different patterns of response are produced by different schedules of reinforcement. If the response is continuously rewarded it is learned most quickly. However, when reinforcement is stopped it is also extinguished most rapidly. Extinction is the term used to describe the loss of a response when it is no longer reinforced. For extinction to occur there must be the opportunity for the response to be given and to go unrewarded. If intermittent reinforcement is given on a *fixed ratio* schedule—after a certain number of responses—a stable high rate of response is produced. In humans this schedule is applied to workers who are paid by the piece or quantity of output. If, on the other hand, rewards are meted out at a *fixed interval*, the response frequency shows a characteristic variation. Immediately after a reward the activity is low, but it increases as the time for the next reward approaches. This anticipatory spurt can also be seen in some human activities—in drivers waiting for a traffic light to turn green or in housewives putting the house in order before their husbands come home.

Social learning in humans usually consists of a sequence of operants that can be altogether quite complex. Skinner believes that the procedure, which he calls *shaping*, suitable for teaching a dog to retrieve a newspaper, for example, might also apply to teaching a child to write his name or an adult to perform even more demanding tasks. When the task can be described as relatively simple components linked together, the teacher armed with a knowledge of the final desired sequence can begin by individually reinforcing each step in the sequence, in order, beginning with the first. Next the second step has to appear before reinforcement is offered; and then this progression continues throughout the entire sequence with reinforcement achieved shortly after the last step. When the introduction of strong reinforcers throughout the sequence of responses is time-consuming and inconvenient, as it may often be, secondary reinforcers can serve a similar purpose. The common way of training animals is to have them seek food under controlled conditions, and the snapping of fingers may be a signal to insure rapid reinforcement of the desired response in the initial stages of training. Later this secondary reinforcer on its own can do the work of the primary enforcer. In teaching humans the teacher's smile, attentive nod or brief remark offered at the proper moment as the steps in the learning process progress may act as secondary reinforcement. The skilled teacher would hope to discover the kind of secondary reinforcement that works best for each learner, and to administer it appropriately throughout the learning chain.

Some writers including Skinner himself[27] have suggested that relationships to earlier theories of learning can be found in the principles of programmed instruction which he derived from his research. Socrates, for example, led his students along through small steps and aided their progress by verbal prompting; but Skinner feels that the Socratic method differed in that the prompts always had to be present, i.e., the pupil was not carried to the point where he was on his own. Johann Comenius, a reforming educator of the 17th Century, also recommended that material to be taught be broken down into small steps each of which should be assimilated before the learner proceeds to the next. Unquestionably, the Montessori didactic materials strongly foreshadowed modern concepts of programmed instruction; and in the work of her contemporary, Edward L. Thorndike (1874–1949) who made important contributions to theories of learning, one finds a striking anticipation of later developments. He wrote in 1912:[28] "If, by a miracle of mechanical ingenuity, a book could be so arranged that only to him who had done what was directed on page one would page two become visible, and so on, much that now requires personal instruction could be managed by print. Books to be given out in loose sheets, a page or so at a time, and books arranged so that a student suffers only if he misuses them, should be worked in many subjects." Learning as a function of rewards is a prevailing theme in Thorndike's writing and Skinner's work is probably more closely related to his than to any other predecessor's.

In the belief that he had outlined a method for establishing consistent behavior patterns, Skinner devised a teaching machine to implement it. His first machine, built in 1953, incorporated the same basic ideas that are still being used in today's linear teaching machines. The term, linear programming (or programing) whether used in the context of teaching machines or programmed textbooks refers to an approach having the following characteristics. The educational material is broken up into small related steps each of which, (in Skinner's machine) is presented to the learner as a question displayed in a small window. In a second, adjacent window the learner is required to make a written response to the question by filling in a blank representing a word missing from a sentence, or by some other such means. After he has indicated the answer, the release of a lever carries the answer to a third window where it can immediately be compared with the correct answer. Whether or not the response was correct the next frame is presented, so that there is always the same sequence in the set. However, it is possible for the student to have the machine again show a question which he failed to answer properly. The programs are logically devised to present minimum difficulty to a student trying to answer the separate questions. Aiming toward a high percentage of correct responses is a natural consequence of Skinner's concept of reinforcement, since only correct answers can be reinforced positively. If a high ratio of correct to incorrect answers is not achieved then this is considered to be an indication that the program needs revision.

A linear program of the kind described truly represents a technology of instruction, although it does not necessarily rely on the use of a mechanical contrivance. Any medium of instruction—television, audiotape, film or book—can be adapted to the method, and the programs can be widely used. They are usually arranged with several frames on the same page to save space, but not in sequential order. Having consecutive frames follow each other on consecutive pages offers the advantage of better control over any tendency for the student to look ahead. Machines can probably control "cheating" better, and they have the additional advantage that they can greatly facilitate record keeping. It has also been suggested that they may insure stricter attention than programmed textbooks, but this assertion is not easy to prove.

Some very expensive types of hardware and some that are relatively inexpensive have been developed for the commercial market as the range of topics taught through programmed instruction has increased. Some of the systems are based on concepts which depart from Skinner's fundamental ideas to a greater or less degree. N. A. Crowder[29] for example, conceived the method of *branching*, or *intrinsic* programming, which recognizes that students will make errors; and furthermore that the error response, far from being the sign of an inefficient program, can be an important part of the learning experience as a piece of information having remedial value. With Crowder's method, the choice which the student makes of one among several likely-looking answers to a multiple choice question will automatically lead to a specific body of material, different from what would appear if he had chosen otherwise. With a correct answer the program moves automatically on, but if the answer is wrong, another frame will be presented to him, or he will be directed to another part of the program explaining, often in considerable detail, why he was wrong. He then returns to the original frame for another try. The branching program, according to Crowder[30], is more responsive to the activity of the individual learner than the linear program. He believes that the writing of linear programs, designed to cause the student to make the fewest number of errors, wastes the time of better students and dissipates much of the vitality of the subject matter by reducing it to fragments. The branched-programmed machines have been quite successful, although the adaptation of the technique to book form, as "scrambled books," has not always been so well received. In the "scrambled books," pages are not read consecutively. Rather, the multiple choice questions direct the student to different page numbers. He turns to the page which he thinks will have the correct answer, and thus progresses through the book in irregular stages.

Should an educator wish to compare programmed instruction with the more traditional forms he might logically begin by asking what can be accomplished through mechanization or systematic programming that cannot otherwise be managed so well. The most distinctive feature claimed for programmed instruction is that it focusses on individualized learning, such that the student can progress through a course of study at his own pace. This can appear as an attractive option for an educational system forced to deal with motivating and instructing large classes full of individuals having diverse objectives and abilities, especially when only a limited number of expert teachers is available. There is not much doubt that if a well-devised system of programmed instruction is inaugurated and carried through, the distribution of work in the teacher's load can be significantly altered. Ideally, a fair portion of the teacher's time which is ordinarily spent in various aspects of classroom routine, sometimes merely imparting information, can be devoted to better preparation for more demanding instructional problems.

Individualized instruction through programming offers other apparent advantages to the learner in addition to the self-pacing feature. He can begin his instruction at a stage consistent with his background of achievement, and he can also utilize the instructional media at such times as he finds convenient. Granted that some students rarely find it convenient to attend to formalized learning unless conditioned to it by an externally imposed schedule, it can be assumed that one who is properly instructed ought not to have this difficulty. Also, under the best circumstances, the learner can have his choice among many instructional media, each treating the same subject matter. Theoretically he would choose the medium which is best suited for mastery of the given material, although the evidence to show that this usually occurs is not convincing[31]. For difficult new material, learners appear to prefer turning to methods of presentation to which they are habituated.

What has not yet become apparent, in spite of a large body of research results bearing on comparisons between programmed and traditional instruction, is any indication that one method (or philosophy) or the other shows consistently superior results as a general guide to good instruction. Perhaps this should come as no surprise, given the great variety existing in subject matter and styles of teaching, and the differences in individual abilities and perceptions. Nevertheless, advocates of faster progress toward technologizing the instructional process, recognizing the intractability of many teachers when it comes to revising their accustomed classroom techniques, are sometimes disappointed that the results of research on teaching methods infrequently provide unchallengeable evidence to support their contentions more strongly.

Computers in Education

Although programmed learning devices, from the crudest to the very intricate and expensive, have been developed for the commercial market, and some have served admirably as alternative means of facilitating the learning of certain defined skills under given conditions, the systems have limitations when it comes to adapting to differences between individual learners. Some educational specialists believe that the simpler ideas for individualizing instruction must give way to adaptive systems utilizing the computer. The logical capabilities of the computer make it a device that can recall previous responses of the learner and relate them to an inquiry under current consideration. As we have learned previously, computer experts do disagree about the subtlety of the relationships which can be constructed into a workable system.

The concept of individualized instruction through the use of computers (or by means of simpler programmed learning methods) is not without its questioners. Anthony A. Oettinger, in his witty, astringent book, *Run, Computer, Run* [32], has stated that representing individualized instruction as a panacea for educational problems ignores certain realities. He notes, first, that the concept has not really been defined in a generally acceptable manner. If the student can advance at his own pace, guided by a technique which he prefers, this may indeed be an advantage for him; but since, to assess attained performance levels, objectives must be clearly defined by the programmer, it would be a fiction to suggest also that the individual pupil really has a decisive say about where the program should ultimately lead him. This situation is not necessarily unacceptable, except in the eyes of proponents of unconditional freedom for the student in education. Human societies, in order to hold together, must try to achieve some uniformity when goals are set. To do this without stifling individual aspirations is the basic problem they have to face and no ideal solution is in sight. Oettinger makes the additional assertion that the practical difficulties of scale involved in allowing each student in a large system to move at his own pace (even when his objectives are similar to the majority's) are very great—although they become less serious at the higher levels of instruction as students attain maturity. The question of whether or not a concentrated application of computer technology will make these logistics problems more manageable is still unresolved.

Educational systems at the present time are often very large and complex enterprises, so it is entirely natural, even necessary, that they employ digital computers or other data-processing machines for routine clerical functions to the fullest extent just as any business will do. In Chicago, for example, computers have been handling data on more than 600,000 students in over 600 schools since 1960[33]. The information includes complete cumulative records on each student, the financial condition of each school, the supply of textbooks on hand, updated class rosters as students transfer from one class or school to another, and many other kinds of statistics which would take thousands of hours of teacher-clerical time to keep current. In some smaller educational institutions where the budget for computer operations is limited, teachers who hope to use the computer extensively for instructional purposes have often been dismayed by the limited time made available to take care of their needs after administrative record-keeping chores have been accommodated.

Still there are and have been broad ranging studies intended to improve the educational situation for the individual learner through computer-managed instruction (CMI), or computer-assisted instruction, (CAI). In computer-managed instruction, the computer helps the teacher or student to make macro-decisions about the educational process. As Brudner[34] describes it, the computer is used as a "man Friday," providing the information necessary to plan whole instructional sequences either in a conventional or automated mode. The information relating to each student is made available in summarized form, and it is possible to conceive programming which suggests to the teacher courses of action based on this information[35]. The description of a student's background and learning aptitudes can be stored in the computer, and there may also be information listing guides to available instructional materials in major subject areas appropriate to individual needs. All of the teaching units for each learning module, however, have the same ultimate objectives in terms of the students' attainments. Brudner[36] has described the actual functioning of a computer-managed system in the classroom beginning with the student's taking a group of tests which, when analyzed, form the basis for selecting appropriate objectives. The ultimate decision about a method of approach here is left up to the teacher but if the choice recommended by the computer is to be rejected or modified, a carefully reasoned explanation should be provided. The student then proceeds to the individual learning module (which may be computer-aided or not) and his progress is eventually evaluated again by the computer's reporting to the teacher a record of answers missed or objectives either completed or unfulfilled. In the case of the latter result, another remedial course of action can be recommended.

In computer-assisted instruction, the student interacts directly with a computer system containing stored material required for an instructional program which may be considerably more diverse than a step-by-step presentation of small increments of information. The interaction is in the form of a dialogue between the student and the computer, ordinarily by means of an electric typewriter, although other or supplementary modes of communication are also employed. The cathode-ray tube with a light pen is widely used, or for spelling lessons, for example, the words can be presented through an audio system. Programs which have been developed range from complicated simulations in college-level subject material to so-called "drill-and-practice" routines. Dean[37] has pointed out that drill is one thing, and practice, another. From the standpoint of learning theory they should not be confused. Thus, a "drill" can be defined as the formation of correct habits under routine conditions, while "practice" means utilizing the correct habits resulting from drill to achieve various results under diverse conditions until the complex responses become "comfortable." The instruction strategy for the drill is not the same as that for practice.

Ample evidence exists to show that computer systems can very efficiently present words for a spelling drill from a given list and then evaluate the answers statistically. Based on these results and on certain assumptions about the meaning of the kinds of response given, a presumably optimum strategy for re-offering the words can be selected by the computer. The information storage capacities, data analysis capabilities and feedback characteristics of computers used for drill or practice confer much greater flexibility on these approaches to instruction than that made possible with, say, a simple Skinner linear programmed machine. The hard question which remains, however, is, will whatever advantage to be gained justify the widespread adoption of this still relatively expensive form of technology? Suppes and Morningstar[38] have published results from an experimental arithmetic study in the lower grades which showed, in one case, that twenty-five extra minutes per day of regular classroom drill in a control school was more beneficial to students in terms of improved performance than five to eight minutes per day of extra individualized computer-controlled drill for an experimental group. Nevertheless, the fact that the improvement for the experimental group was positive and resulted in saving twenty-five minutes of the teacher's time, which otherwise would be devoted to arithmetic drill, was cited as an indication of the value of the computer-assisted program as a supplement to the teacher's regular instruction.

The general effectiveness of the computer for drill or practice is now fairly well accepted, but the most skillfully wrought routines still really use the computer in a relatively elementary way. There is nothing wrong with this. The computer excels at performing time-consuming, repetitive, symbol-manipulating tasks. Still, in view of the additional expense required to install such a system on a large scale in the public schools, (relative cost figures reached by numerous investigators vary according to their different assumptions), it seems hardly likely that many educational administrators will soon be willing to accept computerized drill or practice as an important departure from, or even supplement to, current educational methods. Very likely, most would prefer to see more dramatic evidence of the computer's ability to go beyond the interchange of facts and interact directly with many individual students as a versatile, knowledgeable tutor in diverse fields.

In this direction, computer specialists speak of storing information as simulated models of complex systems, such as the human circulatory system, and then having the computer produce relevant open-ended responses to a wide range of student inquiries [39]. Ingenious programs in an "inquiry" mode have been designed and reported in the literature, wherein the student is confronted with a problem whose solution he may approach in any of various sequential ways, none of which can be uniquely foreseen. He may, for instance, progress through a program in analytical chemistry, learning a correct scheme for identifying an unknown substance. He can proceed by specifying certain tests or calculations to be made, and then when the computer indicates what the effects would be, he chooses a subsequent procedure based on that response. His selection of a reagent to be added to a specific solution, for example, could result in the computer's displaying a picture of a precipitate of a given color in the bottom of the solution. The appearance of the precipitate will give the student a clue to the next test he ought to specify. If his choice will lead in an erroneous direction the display will give him some specific indication that this is the case, and that he should select another course of action. No such description of a program of this kind, however, can even suggest the enormous difficulty of anticipating the possible variations of student responses in order to produce learning materials capable of doing a reasonable job of adaptive instruction.

What can be said about the potential of computer-based or other technical strategies in education? In the first place, it is useful to recognize that the computer programs, which have been reported in the literature and extensively discussed in many places, have actually involved small bodies of researchers working with a relatively few students out of the total student population. Although attempts have been made to incorporate testing and research programs as an integral part of school routines rather than in an artificial, isolated setting, experiments of the kind which have been carried on invite suggestions that the so-called "Hawthorne" effect may account for at least some of whatever comparatively favorable results have been reported for certain computer-based programs. The reference is to a celebrated study conducted by Elton Mayo at the Hawthorne Plant of the Western Electric Company. Mayo, sometimes referred to as America's first industrial sociologist, was attempting to establish a relationship between the efficiency of women workers and the amount of illumination in the plant. After a long testing program lasting several years, he was led to the conclusion that, almost irrespective of the amounts of illumination or any of a number of other variables related to working conditions which he isolated and manipulated, the output of the workers increased as they experienced a sense of group involvement in the experiment. Apparently the feeling that someone is taking a special kind of interest in the performance is a strongly motivating factor. While this is a sufficiently well-known phenomenon that any careful researcher would try to take account of in his overall evaluation of results, the question of whether extensive cybernated instructional programs would become routine to the students and less motivating over a long period of time remains unanswered. Some students have been reported as finding programmed instruction (computer-assisted or otherwise) very quickly boring, and other students respond enthusiastically to those same programs. But it is the same with human teachers. There are those who are considered dull and long-winded by the majority (though rarely all) of the students who have had them; while other teachers covering the same subjects are generally judged to be inspiring. To educational researchers, the frustrating thing is that the inspirational ingredient cannot be reduced to any well-defined formula to serve as a model for designing programs.

As a case in point, Oettinger refers to an elementary course in biology which has been developed by Professor Samuel Postlethwait at Purdue University [40]. In 1962, to meet a large increase in enrollment, Postlethwait set up a program combining group and individual activities. He was aided by one other instructor, a secretary, and several student assistants, and had the backing of a university well-endowed with instructional resources. In this particular instance the computer has been used only in connection with session-scheduling on the course-scheduling system of the university, but other technologies have been effectively employed. The laboratory contains thirty-two carrels, each of them having a pre-recorded tape covering the material for the week and offering guidance to the individual student. The carrels are made available throughout the day and evening on week days, half of them by reservation and half on a first-come, first-served basis. Every week-end the members of the staff introduce new material for the coming week, so each student will have had to reach the previous week's objective before then. The students are referred to various reference sources including books, journals and film loops, some in the carrel and some obtained elsewhere. They may be asked to examine or manipulate specimens or equipment in the carrel or in the central laboratory, and then to record observations in workbooks. There are also general assembly sessions in which students meet together for special presentations, and integrated quiz sessions in which eight students at a time meet with a staff member, each student presenting orally his explanation of the week's work. Then they take a written quiz. The tapes, which set the stage for the week's activities, are recorded by Postlethwait as if he were conversing with a single student. He erases the tapes at the end of the week, and thus keeps the material fresh and open to current developments in biology.

This description suggests what is, in fact, the case; that educational technology is really a way of thinking to which properly utilized "hardware" may contribute. It shows something of what a dedicated, master teacher with talented associates and ample resources can do to combine the best features of individual and group learning. Many large educational institutions from the elementary to the college level are blessed with at least one such person and often many comparably gifted people. When they are supported by a well-organized audio-visual department, and have adequate financial backing from the administration or special grants from government or private agencies, with reasonable credit given for teaching preparation, impressive results can be obtained. The results are not always conveniently assessable in terms of quantitative criteria, but anyone who has had the opportunity to observe the progress of a course planned along similar lines and to sense the absorption and enthusiasm of typical students can have few doubts that some valuable educational experiences were taking place. A conscious structuring of the programs according to specific theories of learning would be the exception, yet one who was familiar with the works of the earlier teaching masters would not fail to see unmistakeable imprints of their most durable concepts; and the matching of educational devices to instructional strategies would at its best be accomplished according to basic principles of a systems approach.

The system we have briefly described is a relatively small one serving mature students who ought to be able to exercise a reasonable amount of individual initiative. Although a fair number of public elementary and high schools in the United States have been designed to incorporate some of the most innovative features of contemporary educational technology, there is no promising likelihood that methods used in individual courses and programs can soon be scaled up to work equally well for entire curricula in much larger systems. We must note that it takes many exceptional people, convinced of its worth, to make educational technology function effectively in a system—it does not happen automatically. Even if there were enough exceptional people to meet the demand, the harrying problems of maintaining equipment, of sustaining complicated networks of communication, of coping with eruptions which are a part of the turmoil of society, of reconciling community differences about the proper way to conduct an educational enterprise, of matching resources, abilities and needs, and especially of garnering long-term, high-level support that an aspiring educational program must have—none of these would disappear. With regard to the need to obtain support for education, for example, it is ironical that one of the most effective arguments for purchasing automated educational equipment was less persuasive in large areas of the United States in 1970 and 1971 than it had been before. With fewer students entering the schools than some earlier estimates had predicted, with taxpayers in revolt, with unemployment relatively high, and given the growing tendency of college graduates to look to careers in education, there was suddenly a surplus of teachers in many disciplines. Nobody knows how many promising teachers will have been lost to the profession because the various educational systems have been unable to muster the financial resources to put these young people to work. Thus we see again the dynamic, changing nature of the relationship between technology and employment.

Most significantly of all, it is almost dismaying to realize there is little certainty about what the isolated, if encouraging, successes of newer educational techniques mean when considering overall objectives for an educational system. Perhaps there can be no consensus about the best ways to prepare youth for the adult role since that role is pictured so variously among individuals and the separate generations. Young people all over the world are increasingly making their own decisions after the primary years about the kinds of educational experiences they wish to have. Certainly, they will continue to learn much from each other, and they will choose from an unparalleled variety of possibilities within and apart from the classroom milieu. As for the teachers, it is not unlikely that their activities will be changing from "tellers" to learning resource coordinators, counselors and tutors, but this, of course, is not really a new concept at all.

Some critics of public education are proposing that, where public systems are deemed to have failed, they be replaced by private, profit-making corporations guaranteeing to bring students up to a specified level of achievement. Some such programs are already functioning, but there is no real indication that the approach has any more potential than have the present systems for clearly foreseeing what young people will need to know for the preservation of freedom, happiness and wisdom in the technological society.

REFERENCES

1. McHale, John, "Education for Real," Transnational Forum, World Academy of Art and Science Newsletter, June 1966, p. 3.
2. Goldberg, Maxwell, "Automation, Education and the Humanity of Man," in *Automation, Education and Human Values*, edited by William W. Brickman and Stanley Lehrer, School and Society Books, N.Y. 1966, p. 14.
3. Bruner, Jerome S., "Culture, Politics and Pedagogy," *Saturday Review*, Vol. LI, No. 20, May 18, 1968, p. 69.
4. Hendel, Charles William, *Jean-Jacques Rousseau: Moralist*, Vol. II, Oxford University Press, London 1934, Chapter XVI.
5. Saettler, Paul, *A History of Instructional Technology*, McGraw-Hill Book Co., N.Y., 1968, pp. 28-31.
6. Gillett, Margaret, *A History of Education: Thought and Practice*, McGraw-Hill Book Co., Toronto 1966, pp. 206, 207.
7. Gutek, Gerald Lee, *Pestalozzi and Education*, Random House, N.Y., 1968, p. 85.
8. Boyd, William, *The History of Western Education*, 8th ed. Revised by Edmund J. King, Barnes & Noble, Inc., N.Y., 1966, pp. 340-42.
9. Frost, S. E., Jr., *Historical and Philosophical Foundations of Western Education*, Charles E. Merrill Books, Inc., Columbus, 1966, p. 384.
10. Montessori, Maria, *The Montessori Method*, Introduction by J. McV. Hunt, translated from 1912 Italian edition by Anne E. George, Schocken Books, Inc., 1964, p. xi.
11. Montessori, Maria, ibid., pp. 31-42.
12. Boyd, William, op. cit., pp. 364-66.
13. Montessori, Maria, op. cit., pp. llvii-llviii.
14. Montessori, Maria, *Spontaneous Activity in Education*, Introduction by John J. McDermott, translated from the Italian by Florence Simmonds, (1917) Schocken Books, Inc., N.Y., 1965, p. xviii.
15. Dewey, John, *Democracy and Education*, The Macmillan Co., N.Y., 1916, p. 118.
16. Dewey, John, "Psychology and Social Science," in *The Influence of Darwin on Philosophy*, N.Y., 1910, p. 71.
17. Bugelski, B. R., *The Psychology of Learning Applied to Teaching*, The Bobbs-Merril Co., Indianapolis, 1964, pp. 21-24.
18. Travers, Robert M. W., "Directions for the Development of an Educational Technology," from *Technology and the Curriculum*, edited by Witt, Paul, W. F., Teachers College Press, Columbia University, 1968, pp. 86-87.

19. Saettler, Paul, op. cit., pp. 95-249.
20. Brown, James W., Lewis, Richard B., and Harcleroad, Fred F., *A-V Instruction: Materials and Methods*, 2nd ed., McGraw-Hill Book Co., N.Y., 1964.
21. Saettler, Paul, op. cit., p. 230.
22. Balanoff, Neal, "New Dimensions in Instructional Media," in *The New Media and Education: Their Impact on Society*, edited by Peter H. Rossi and Bruce J. Biddle, Anchor Books, Doubleday & Co., Garden City, N.Y., 1967, pp. 66-68. (Aldone Publishing Co., 1966.)
23. Saettler, Paul, op. cit., pp. 251-52.
24. Skinner, B. F., "The Science of Learning and the Art of Teaching," *Harvard Educational Review*, 24: 86-97. Spring 1954.
25. Skinner, B. F., *Cumulative Record*, New York: Appleton-Century Crofts, Inc., 1961.
26. Bugelski, B. R., op. cit., pp. 209-11.
27. Skinner, B. F., "Reflections on a Decade of Teaching Machines," in *Teaching Machines and Programed Learning II: Data and Directions*, edited by Robert Glaser, Department of Audio-Visual Instruction, National Education Association of the United States, Washington, D.C., 1964, pp. 15, 16.
28. Thorndike, Edward L., *Education*, The Macmillan Co., N.Y., 1912, pp. 164-66.
29. Crowder, N. A., "Automatic Tutoring by Intrinsic Programming," in Lumsdaine, A. H. and Glaser, R. (eds.), *Teaching Machines and Programmed Learning*, National Education Association, Washington, D.C., 1960, pp. 286-98.
30. Crowder, N. A., "On the Differences Between Linear and Intrinsic Programing," in *Programs, Teachers and Machines*, edited by Alfred de Grazia and David A. Sohn, Bantam Books Edition, N.Y., 1964, pp. 77-85.
31. Mitzel, Harold E., "The Impending Instruction Revolution," *Engineering Education*, March 1970, p. 750.
32. Oettinger, Anthony A. (with Sema Marks), *Run, Computer, Run*, Harvard University Press, Cambridge, 1969, pp. 117-55.
33. Oettinger, Anthony A., "Computers: New Era for Education?" *Education U.S.A.*, National School Public Relation Association, Washington, D.C., 1968, p. 19.
34. Brudner, Harvey J., "Computer-Managed Instruction," *Science*, Vol. 162, November 29, 1968, p. 971.
35. Cooley, William W. and Glaser, Robert, "The Computer and Individualized Instruction," *Science*, Vol. 166, October 31, 1969, p. 575.
36. Brudner, Harvey J., op. cit., p. 974.
37. Dean, Peter M., "Learner Versus Teacher Controlled Arithmetic Practice," Paper presented at DAVI Conference, Portland, Oregon, April 28, 1969, pp. 4, 5.
38. Suppes, Patrick and Morningstar, Mona, "Computer-Assisted Instruction," *Science*, Vol. 166, October 17, 1969, p. 346.
39. Alpert, D. and Bitzer, D. L., "Advances in Computer-Based Education, *Science*, Vol. 167, March 20, 1967, p. 1583.
40. Oettinger, Anthony A., op. cit., pp. 150-55.

Recommended Bibliography

Sidney G. Tickton (ed.), *To Improve Learning: An Evaluation of Instructional Technology*, Vols. I and II, R. R. Bowker Co., N.Y., 1971.

10

Intimations of the Future

"Who heeds not the future will find sorrow close at hand."
CONFUCIUS, *Analects*. Bk. xv

"For my part, I think that a knowledge of the future would be a disadvantage."
CICERO, *De Divinatione*, Bk. ii

Much of life is anticipation. Each living thing is programmed for reproduction. In the behavior of all animals there is an orientation toward the future demonstrated by the storing of food, by the urge to migrate as the seasons start to change, or by a variety of other responses. Only in man, however, has the sense of the impermanence of the present, heightened by remembrance as well as anticipation, led to a conscious attempt to conceive of future possibilities in some fully reasoned, systematic way.

This habit of thought seems notably characteristic of individuals or groups in positions of responsibility in modern Western nations. In business, politics, military affairs, education or any other large enterprise the decision maker tends to believe he has too much at stake to permit impending events to be shaped by others' whims. Besides, each new acquisition brings its own burden of responsibility. Without planning and the "stitch in time," the concatenation of devices and events soon threatens to get out of hand.

According to one point of view, the present for modern man is already so unstable as to bring him to a point of acute neurosis. Alvin Toffler[1] writes of a mass disorientation he calls "future shock," caused by the onrush of technologies, thrusting forward new perspectives before the average individual has had time to become accustomed to the old. A general educational commitment to orderly studies of the future instead of the past is suggested as a partial remedy for the phenomenon.

There is no doubt that interest in forecasting probable future developments has been resurgent in the past decade. During World War II and for some time thereafter the problems of the moment were so obsessing that few social thinkers found time to occupy themselves seriously with long-range speculation. Daniel Bell[2] has reasoned that, once relationships between major powers had seemed to arrive at an acceptable, if precarious, point of balance, there was time to give more attention to possibilities inherent in contemporary technologies. Now the era of space exploration has given the public an awareness of new concepts of space and time; expansion and modernization of transportation and communication networks have strengthened national and international interdependencies, (although not without sharpening important differences) suggesting other forms of political institutions in the offing; and the new "intellectual technologies" have focussed attention on the formulation of interesting models of future systems. Then too the end of the millennium is approaching. The realization spreads that the great majority of people now living will still be alive in the year 2000 A.D. barring some cataclysm, and there will be many, many more sharing the planet and its resources.

In Western intellectual circles, at least, it has become almost a psychological necessity to feel that there be a reliable, organized means of forecasting to keep dangerous or disruptive combinations of events from catching us all by surprise. The idea is that there is no such thing as a single, predestined future but rather a range of possible futures. Some are more probable than others, but each is more or less probable depending upon the kinds of actions taken in the present. With a foreknowledge of those many possibilities, planners in government or private organizations can then use forecasting as a guide to decision making. The decisions taken, obviously, would be those which would tend to favor the more desirable probable outcomes.

EARLY VIEWS OF THE FUTURE

Ancient soothsayers also predicted favorable or unfavorable events which would occur either inevitably or as a result of taking certain courses of action. It is natural that we are reminded more of their successful prophecies than their failures. Indeed, a two-sided prediction such as that which the Delphic Oracle made in telling Croesus a mighty empire would be destroyed if he crossed the Halys River, could hardly go awry. How much inspired intuition, shrewd analysis of certain portentous conditions or sheer charlatanry shared in the art of the prophets of old, we cannot say. The name of Nostradamus, who lived in 16th Century France when the study of astrology was at a peak, has become almost synonymous with mystical prognostication of the more apocalyptic sort. His vision of the future was so vivid and so filled with intimidating detail that his contemporaries could hardly help being impressed. The rhymed prophecies were also enigmatic, which makes it possible for us today to read a great deal into them, perhaps more than was intended.

Thomas More's *Utopia* also appeared in the 16th Century. There is some question to what extent it was meant to be a political statement, but in any event it seems to have been more that than a prophecy. It had relatively few counterparts for another two centuries or more, since the visions of the future, such as they were, were still predominantly in a religious vein. In the 17th Century, the belief in the Christian millennium — the establishment of the Heavenly Kingdom on earth — influenced writing about the future; but by the 18th Century the major prophets were also scientists and scholars who were convinced that coming changes would be determined by natural laws and could be divined through a knowledge of those laws. They were filled with boundless optimism that the new science would produce unfailing progress toward an ideal society. They were more concerned with projecting the course of social and political systems than with forecasting the future of technology. Some of these men and their notions have already been mentioned in Chapter 5. Condorcet, who is well known for his work in the mathematics of probability, is particularly worth reconsidering as a future forecaster. He was interested in the extrapolation of future events from data already at hand; and he also addressed the problem of the probability of arriving at the truth by a vote of the majority. The relation to the modern Delphi technique of future forecasting will shortly become apparent. According to de Jouvenel[3], Maupertuis was the first in this period to assert that anticipations of the future were closely related to knowledge of the past. Thus since it was possible to improve our knowledge of the past, so was it also possible to perfect forecasts of the future. Voltaire strongly attacked him for suggesting there was any such symmetry between future and past.

From this time on, religious prophecy became less prevalent, and its place to some extent was taken by the beginnings of science fiction. An early example was *L'An 2440* (The Year 2440), written by Louis Sébastien Mercier in 1771. Its references to moving pictures and inextinguishable lamps were rather pedestrian—certainly requiring no more imagination than that shown by much earlier writers. Friar Roger in the 13th Century had written about great mechanically-propelled ships guided by one man, flying machines and submarines. But while these early projections were probes of the imagination, with little or no real logical basis, in the 19th Century the rate of technological advancement was so obvious that no serious writer who sought to peer into the future could avoid considering the possibilities inherent in scientific and technological change.

The earlier science fictions tended to be conservative, consistently underestimating the pace of change. By the latter part of the century, though, writers of future fiction, of whom Jules Verne was perhaps the outstanding representative, had a relatively sophisticated knowledge of the contemporary technology. Verne's predictions, which enthralled generations of readers, were wide-ranging and acute. Arthur C. Clarke, himself a gifted extrapolator of technology, believes that only Hugo Gernsback, a comparatively unpublicized American editor and inventor of the 20th Century, has outdone Verne in the dimensions and accuracy of his predictions [4].

New archeological and geological discoveries in the 19th Century sharply modified man's concept of time, expanding the future to an infinity in which, as H. Bruce Franklin points out [5], the biological events allowed by Darwin's theory, and social events according to models of Marx, Spencer or others could transpire. Future fictions thus began to incorporate these elements along with the technological possibilities, although before H. G. Wells no writer combined them all with much authority. Edward Bellamy's *Looking Backward* (1888), for instance, was an immensely popular picture of utopian socialism, but his descriptions of new technologies were rather pallid.

The difficulty of fashioning a realistic fusion of technological, sociopolitical and biological possibilities has continued to hamper most writers of the 20th Century who have tried their hand at future fictions. A profound disillusionment with the early promise of modern technology has been characteristic of the works of many of the best known authors. In Aldous Huxley's *Brave New World* (1932), George Orwell's *1984* (1949), and Kurt Vonnegut's *Player Piano* (1952) the themes are antiutopian, stressing that technology, if given free rein, must rob man of his individuality. The influence of these writers has been great, although in each case the actual descriptions of potential technologies or their effects are less detailed or convincing than those that can be found in works of men such as Frederik Pohl, Ray Bradbury or Arthur C. Clarke who are more closely identified with science fiction.

At first thought it might appear that future fictions (written mostly as pure entertainment, but also often with a more serious purpose in mind) are something quite different from scholarly attempts to develop reliable methods of forecasting. Actually, neither future-fiction writers nor, say, economic or technological forecasters are dealing with future facts since verifiable facts exist only with regard to the past or present. Rather, they are offering opinions about what might happen. The probability of their being right is presumed to be related to how carefully they consider the facts on hand. The forecasters are committed to doing so very judiciously. Oftentimes their opinions serve as the basis for the committing of large sums to a given project. The fiction writers are less constrained. While they will generally avoid violating what is accepted as a scientific principle, by allowing their imaginations free play a few of them have proven more prescient about the scientific and technological future than distinguished but cautious scientists who were their contemporaries. In any event there is a substantial area in which scholarly conjecture and science fiction may overlap.

METHODS OF TECHNOLOGICAL FORECASTING

Serious studies of the future are not limited to forecasting. David C. Miller[6] has distinguished between definitions of forecasting and other means of looking toward the future by suggesting that a forecaster is willing to assert a particular event has a certain probability of happening within some assigned period of time; a prophet proposes that an event will occur without specifying a time; and one who indulges in speculation more conservatively suggests that a given occurrence might take place, but with no time period or odds mentioned. One might also work with a conditional model for which future consequences may be projected, *if* preceded by given occurrences.

In the modern era, many find it natural to consider that forecasting of quantitative effects from accurate compilations of current data ought to be the most reliable way to plan for the future. The great cost of significant errors in assessing future probabilities has prompted businesses and governments to invest considerable resources on the identifying and perfecting of forecasting techniques. An assumption which has gained favor is that technological forecasting, i.e., the forecasting of technological change (as new invention, innovation or diffusion of a technology) is a promising means of insuring against uncertainties [7, 8]. This assumption seems to be based in part on the belief that there are causal relationships linking the growth of technologies (that the invention or diffusion of one kind of technology will obviously require others to service or complement it); and also that the essential features of technologies clearly lend themselves to being described quantitatively. There is an additional implication which would not be accepted in every quarter; that uncertainty is minimized because technologists have a relatively large degree of control over the direction that their technologies will take. If the implication is accepted, it is often with more concern than satisfaction, since it might suggest that technological decisions are made in the atmosphere of a closed, authoritarian system. In any case, somewhat the same points could be made about forecasting to facilitate economic or industrial management decisions. Leonard E. Schwartz [9] contends that the chief factor separating technological forecasting from, say, economic forecasting is, simply, technology; and he then asks the question why so much of the literature on technological forecasting concentrates on the descriptions of the forecasting techniques and has relatively little to say about specific technologies. One reason must be that attempts to project probable new combinations of technology can quickly become excessively complex, particularly if the potential social implications of technological change are taken into account. Some writers assert that technological forecasting ought to be kept separate from social and political factors but there is no convincing evidence that the resulting forecasts would be realistic. In the following sections, after a few of the most common forecasting techniques have been described, an attempt will be made to explore some projected futures in the light of certain interrelationships between technological, social and political possibilities.

Trend Projections

In the absence of an adequate general theory for the analysis of systems interacting with each other over a period of time, the most common forecasting technique has been to examine the available technical data relating to a given variable, to try to detect a pattern of continuous progress. A curve reflecting the data can then be extrapolated to some future time. Different assumptions can be made about the observable trend, but the most obvious one is that the rate of change will continue into the future.

If, for example, we examine the semi-logarithmic plot of data on the graph in Fig. 10-1 we would feel fairly safe in stating at what average rate the production tonnage of copper would continue to grow for some time to come—for another few years at least and probably much longer. Obviously there would be yearly fluctuations from the increasing rate of growth shown by the plotted relationship; but there would appear to be no

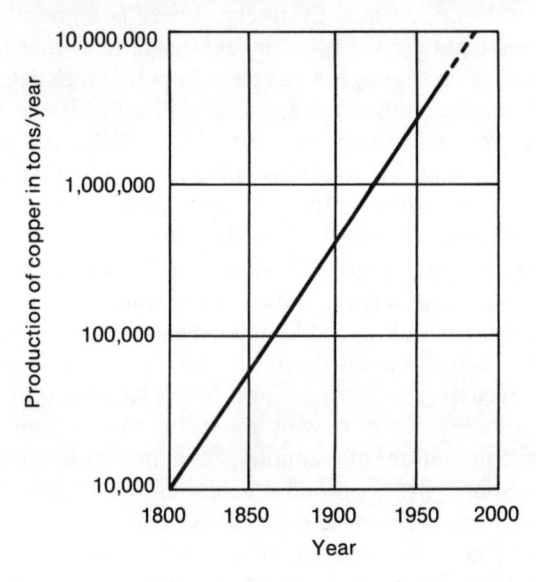

Fig. 10-1. Growth of world copper production since 1800.

surprising trend in store for as far ahead as most technological, military or corporate planning strategists would find it practical to be concerned. Nevertheless, acceptance of growth as a simple exponential process can lead to seriously faulty conclusions. Even assuming that the data are adequate (which would seem not to be a matter of doubt in the above instance) the advent of sudden unforeseen demands as a result of important technological developments or the drastic curtailment of production operations because of long-lasting political upheaval in countries rich in a given resource could materially change the pattern of a trend. A full-range forecast, of course, can try to take such possibilities into consideration. Moreover, extrapolation of growth curves to an indefinite future may very well conceal the fact that a constant increasing rate of change is not necessarily normal. Often what appears to be just that is only the straight line portion of the so-called S- or sigmoid curve.

In Fig. 10-2 we see such a curve. If the data were available only for the period of time between T_1 and T_2 they would not reveal the true form of the growth pattern; although other considerations might cause the forecaster to anticipate a change. The S-curve is characteristic of the growth patterns of biological populations where each element has the capacity for multiple replication. After an initial period of barely perceptible

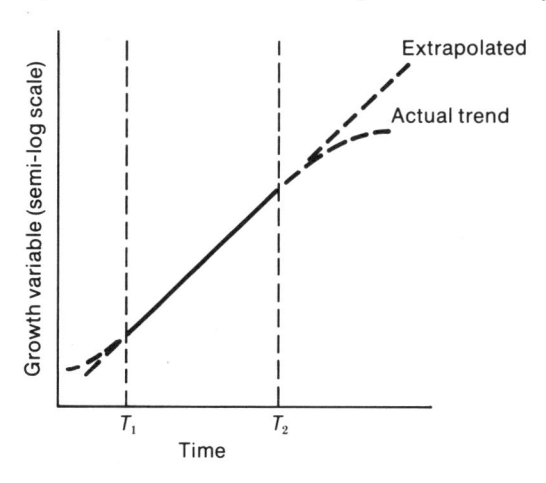

Fig. 10-2. S-curve.

population increase, a point is reached at which the growth rate takes off in an exponential fashion. But at some point the available nutrient begins to be exhausted, and the competition for the existing supply, along with other phenomena associated with the suffocating proximity of all the other organisms in the huge population, begins to cause a slowing down of the growth rate. It has often been noted that growth patterns in general, such as the rate of increase in efficiency or application of a certain technology, exhibit a similar behavior. However this does not necessarily mean that technological progress along given lines must essentially grind to a halt when the plateau region of the curve is reached. Experience also shows that at such periods research to develop replacement technologies appears to be stimulated, and the resulting innovations often produce new growth trends. As Wills points out [10], managements can anticipate what functional capabilities they might expect from their technologies at some given period in the reasonably near future. If the probability of a currently employed technology's reaching that level is low then efforts should be strengthened to work toward a new technology.

366 Intimations of the Future

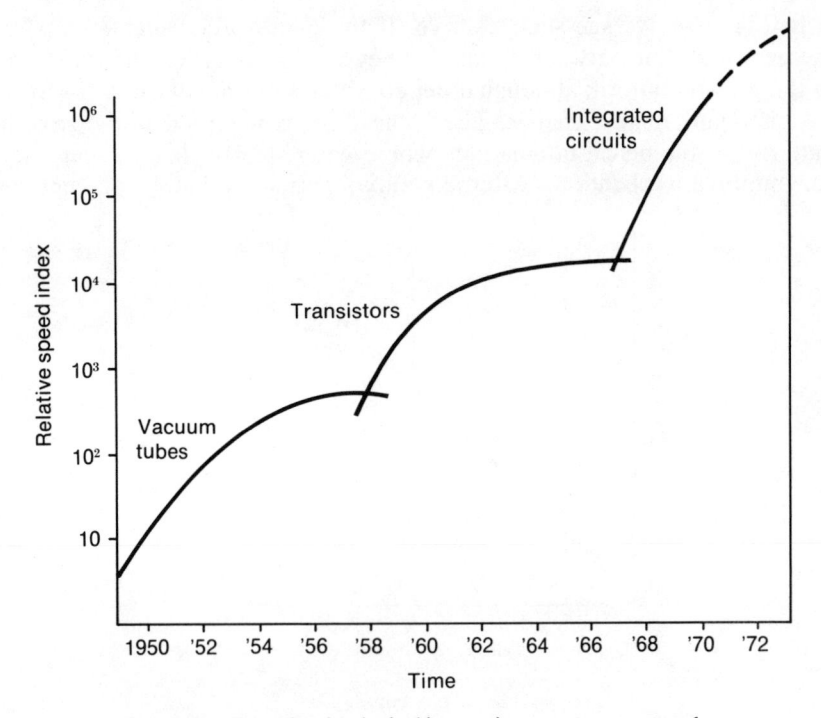

Fig. 10-3. Effect of technological innovation on computer speed.

Figure 10-3 offers a simplified picture of the way in which technological innovation has affected the rate of increase in the computation speed of modern computers. A curve drawn roughly tangent to the separate curves for the individual technologies would represent the overall rate of progress. This is called an "envelope" curve. The significance of such curves for the analysis of trends has been discussed by Ayres[11]. Many writers have commented on the futility of trying to predict the advent of any specific innovation. It would have been hard to find experts ready to predict the introduction of a transistor-like device and its rapid integration into computer designs, five years before one was actually invented. An envelope-type curve, however, might point out the approximate period in which some new technology could be expected to heighten the growth rate of computer speeds without actually identifying the specific characteristics of the technology. But obviously, too, an envelope curve could be thought of as a magnified or more detailed form of a simple growth curve which, when extrapolated naively could just as easily lead to erroneous results.

Trend correlation can be considered a variation of the methods of projection described. This technique attempts to compare a trend in a given technological field with the identifiable and similar advance of another factor, related to it by virtue of certain causal connections. For example, the development of bombing planes for the military has in the past been sufficiently in advance of the development of civilian transport planes to allow an acceptable forecast of the speeds of the latter[12].

Consensus Forecasting

Polling has become a highly organized and extensively practiced method for predicting the tendency of a group of individuals to take a certain course of action between or among given alternatives. It is based on the expressed or inferred intentions of the majority as determined from responses to questions prepared by the pollsters.

Ordinary opinion polls have shown little value for long-range predictions. They have, however, been generally successful in predicting the outcome of imminent elections except when the race was too close to call, and except for certain notable failures. Still the method does not really qualify as a quantitative forecasting technique precisely along the lines of the definition previously offered.

A significant attempt has been made by Dr. Olaf Helmer and associates to improve the rigor of polling techniques for forecasting. His procedure, called the "Delphi" method[13] is generally thought to be a workable way to obtain an explicit forecast from expert opinions rather than from a quantitative data base alone. The first study was founded on the assumption, implied at least, that data are increasingly unreliable tools as the time of the forecast is pushed further into the future. For one thing, total reliance on figures overlooks the important contribution that personal expectations of individuals may make toward influencing actual later developments.

In its original form, the Delphi technique used consensus estimates of technological change from a panel of intellectually qualified people from all over the world. The first panel, consisted of twenty engineers, seventeen physical scientists, fourteen logicians and mathematicians, twelve economists, nine social scientists, five writers, four operations analysts and one military officer. Essentially most of the more than one hundred Delphi experiments conducted since the first one have been directed toward establishing a chronology of scientific and technological events and forecasting how soon they might be expected to occur. Briefly, the panelists, who remain anonymous to each other, are interrogated by a program of sequential, individual questionnaires. A list of hypothetical events—focussed about such major areas of concern as population growth, future weapons systems, automation and space exploration—is drawn up by asking specialists what important developments they foresee. The questionnaire which goes to each of the panelists solicits their estimate of the probability of the various events occurring by a given date. A typical question involving computers might appear: by what year would you anticipate the cybernation of office work and services leading to displacement of twenty-five percent of the work force?

All of the answers to the questions are collated and expressed as distributions in terms of medians and quartiles. A second round of questioning then begins in which the distributions are fed back to the respondents, along with comments from individuals in support of their responses to the first questions, and perhaps with additional questions to be answered. These responses to the second round are treated as before, and the feedback process may continue through several subsequent rounds. It is assumed that the process will bring the respondents to a point of refining their original thinking, that it will help clarify ambiguities, and that it will bring about a convergence of expert opinion, all without the drawbacks which are often a part of committee dynamics, e.g., the tendency for an individual to hold fast to his initial opinion in the face of arguments by another for whom he feels antipathy, or on the other hand, the possibility that one person will move others to his side by his skill in argumentation. However, the assumption that eliminating face-to-face interactions will diminish the tendency for individuals to be persuaded by specious arguments is rather tenuous, since it seems as likely that the questionnaire technique may not provide the opportunity for the kind of penetrating analysis needed to expose common misconceptions about a given propostion[14]. An abbreviated example of a typical Delphi forecast is shown in Fig. 10-4. An actual forecast would contain many more items than this.

The Delphi method is considered to be an example of exploratory forecasting, which is a statement of what ought to be possible at a given future time without offering a detailed indication of the process of development of the event from present or past conditions. The original Delphi experiments concentrated on developing ranges of probabilities for future events, while more recent experiments have tended to inquire about such factors as the desirability of the forecast events, or their effects, should they take place.

1. Computerized diagnosis for medical profession.
2. Large scale use of household robots.
3. Machine which will score 140 on standard I.Q. test administered in English.
4. Automated highway system.
5. Direct interaction between human brain and computers.

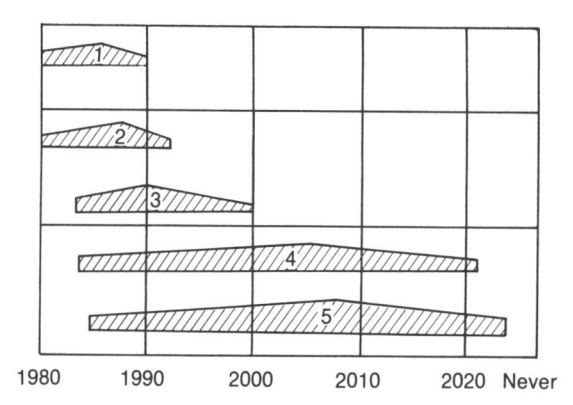

Fig. 10-4. Consensus of panel on use of computers and automation (medians and quartiles).

In discussing the significance of Delphi as a forecasting tool, W. T. Weaver[15] has suggested that a singular weakness appears in its lacking an explanatory quality. Since an accepted major purpose of forecasting is to influence the future course of events by indicating promising present programs of action, the forecast should be able to convince people of the desirability of acting. A consensus, even of highly trained people, does not of itself confer plausibility on a forecast, particularly if there are no clear statements of the various bases for making the choices. Panels of distinguished experts may, by their nature, tend to look for familiar patterns of technological and scientific progress, and overlook occasional signs of radical departures to which representatives of a younger generation might be more alert.

On the other hand, it should be noted that the Delphi approach was meant to be an experiment and much of the work along these lines has continued to be an experimental exploration of the validity of certain techniques for identifying courses of action. Procedures are still evolving. Since Delphi can also be used to shed light on the past and present, perhaps its major value may be as a teaching tool.

Combination Approaches

Methods of forecasting can be combined—indeed they often are—whether with formal recognition of the fact or not. One can hardly doubt that Delphi panelists take recognizable current growth trends into account in their responses.

The *cross-impact matrix* approach may be considered a variation of certain features of Delphi-like forecasts based on the assumption that a given event either enhances or diminishes the chances of other events taking place (or has no effect); and that the strength of the relationships can be assessed. A matrix is constructed tracing the effects of each event in terms of each of the other events considered. Then, assuming the advent of some random event, a computerized analysis of the matrix indicates how the probability of each cross-impacted event is modified. Granting the desirability of using the computer to handle large numbers of interrelated events, the whole process is of such forbidding complexity that progress along these lines has not been very promising.

One of the difficulties with forecasting the long-range future is that a wide range of possible events may seem to be equally plausible. This can be especially disconcerting if the occurrence of one would seem to rule out the possibility of another. If a forecaster wishes to contribute to the formulation of practical policies looking toward the future, he is almost forced to go into detail in his projections. This means he must focus on specific alternatives, probably more or less arbitrarily. One way to do this, is by means of *scenarios*. This technique may start by identifying a trend and making up a projection (usually modified by additional theoretical considerations) which seems consistent with the trend. Within this framework one can then specify certain "worlds" with which other possibilities may be compared. It is also desirable to determine logical steps that may lead to the projected worlds.

Herman Kahn and Anthony J. Wiener have been notably identified with the formulating of future scenarios. In their book, *The Year 2000; A Framework for Speculation*[16] they have explored, as a reference base, a "standard world" reflecting their "surprise-free" projections of an alternative future which most forecasters would consider least improbable; and "canonical variations" of that world, which are theoretical possibilities affording useful comparisons. While a good deal of the speculation relates to the forecasting of new military technologies, a field with which Kahn has been associated for many years, their projections are very much involved in economic, political and cultural issues as befit studies connected with government policy research. The various elements comprising their work are so detailed that only the outlines can be traced here. Briefly, they begin by identifying certain long term trends which students of Western society have commonly observed and discoursed upon for many years. They also cluster significant events by thirty-three year intervals to see what combinations give rise to new clusters, and attempt to develop statistical baselines projecting key societal variables.

We have already discussed some of the trends in preceding chapters. Growth of scientific and technological change, increasing urbanization, bureaucratization, spreading affluence and the decreasing importance of occupations in the primary industries are all hallmarks of the industrial or early post-industrial societies. Kahn and Wiener have assumed that these trends are representative of what Pitirim Sorokin[17] has called an increasingly sensate culture, marked by empirical, rational, secular, materialistic, humanistic and erotic tendencies and shading in its later phases toward vulgar exhibitionism, hedonism, sadism and tumultuous protest. According to Sorokin, the sensate culture is one stage in a three-stage quasi-circular pattern through which all societies seem to progress, the other two stages being respectively identified by characteristics of mysticism, allegory, anonymity and traditionalism; or grandeur, heroism, public morality and patriotism. It is a manifestation of the metaphor of the life cycle of an organism—of genesis, growth and decay—to which social philosophers from the time of Plato, Aristotle and Heraclitus have referred.

By concluding that the materialistic and rational aspects of the sensate tendency will continue throughout this century and spread out over ever wider areas of the earth, Kahn and Wiener could assume a standard world reflecting their own expectations. In this context they could justify anticipating specific developments in technology and science, as well as in politics and social relations.

One hundred scientific and technical innovations considered "very likely" for the last third of the century were listed (this was in 1967), together with twenty-five others which seemed less likely but which would be important if they happened. In the first category many of the innovations proposed, such as: (a) extreme high strength and/or high temperature structural materials, (b) ground effect machines, VTOL (vertical take-off and landing machines), giant or supersonic jets, (c) widespread use of cameras for mapping and geological investigations, (d) more reliable and longer-range weather forecast, and (e) extensive and intensive centralization of current and past personal and business information in high-speed data processors, all seem relatively ordinary from the point of view of technical possibility. Perhaps a later projection sensitive to more recent changes in the public's tolerance of environmental disturbances might have lowered the probability of the supersonic jets becoming a major factor in air transportation.

Other potential innovations counted in the very likely category such as: (a) extensive use of robots and machines "slaved" to human beings,

(b) use of nuclear explosives for excavation, mining and generation of power, (c) individual flying platforms, (d) inexpensive world-wide transportation of humans and cargo, and (e) widespread use of computers for intellectual and professional assistance, (including translation, teaching or literature search), tend to be either more demanding technically or to require for their development and implementation somewhat different political or economic circumstances than those existing either in 1967 or at the present time. That is to say, one could not accept certain forecasts as simple extrapolations from the given base point, but would have to take it on faith that some fundamental public attitudes or concepts of social organization would change, or that new key technical insights would emerge. To give one illustration from the list, the individual flying platforms are a technological reality, but there is no transportation system now in sight that conceives of a practical way in which large numbers of these machines could be accommodated in an urban milieu. Considering another projection by the authors in this same category, as an example — i.e., that there will be high-quality medical care for underdeveloped areas — one would presumably have to accept the likelihood that good medical care will first be available for all in the United States, the most affluent of nations. This is not now the case and for it to become a reality, major restructuring of the medical system must occur, almost certainly in the face of potent political opposition. But of course it could happen, and one would not wish to quarrel with its being listed, particularly if it helped serve as a spur to making the changes required now to arrive at that alternative future situation. It must also be noted, however, that the list of probabilities is not limited to those which would be termed desirable by a majority of the people. The authors agree that many of their projections, including the last one cited, are controversial. Other examples of controversial forecasts in the very likely category, such as: (a) genetic control over the "basic constitution" of an individual or (b) capability to choose the sex of unborn children are of the kind which prompt some responsible scientists to question very seriously the legitimacy of the research promising to lead to those events.

There are individuals today who actively oppose many, if not most, attempts to innovate and change the physical and social environments, presuming sometimes that such efforts may even violate immutable "laws." In some cases this position would not be unreasonable. To take an illustration that could perhaps be thought trivial: Kahn and Wiener list the development of unspecified methods for permitting "non-harmful over-indulging" among the very likely but potentially controversial innovations. In one frame of reference this could be considered a clear contradiction in terms — out of the question in any overall scheme of things because it violates inherent principles of order. Individuals or groups can never extract anything from a system of which they are a part without either impoverishing another part of that system, or borrowing from the larger environment. The piper must be paid. Sometimes the payment for material advantage will be in terms of social disorganization somewhere else, but in any event, to overindulge is to boost the cost which must somehow be met. In a world of finite resources who can say for how long the extra cost can be borne without harm being done?

This position taken to an extreme would carry little weight in most contemporary technical circles, if for no other reason than that it smacks of a moral determinism steeped in the Puritan ethos. It could also be seriously inappropriate if used as an argument for stopping all technological innovation, but there is still an element of truth to it that cannot be ignored. Technical people need to approach their work conscious of the narrowness of a euphoric vision which sees only the benefits of material progress. This means constantly reminding themselves and each other that every major innovation has had, in time, some unexpected consequences. Had forecasting reached a state of advancement that would permit foreseeing the unforeseeable, the task of deciding how far to proceed in a given direction would be somewhat less formidable. Since it has not, the best that engineers, scientists and forecasters can do is to assume that, as a result of each innovation introduced for a positive purpose, the condition of some person or thing will be diminished somewhere, sometime. The negative effects may very well be considered less damaging or significant in the overall plans of the decision makers than the potential good which can result. For example, no one would suppose that the development of penicillin would have been stopped had it been known there would be incapacitations and deaths from allergic reactions, although the extent to which it was administered might well have been sharply modified. Research and development plans should thus be deemed unready for implementation as practical systems if no potentially harmful effects have yet been studied, identified and weighed since the presumption that there will be such effects is a very strong one.

This is a sober, perhaps dogmatic, way to look at the future. If encouraged, it could lead to a frugal approach which might seem contrary to the spirit and practice of modern technology. It is surely not easy to picture engineers and scientists receiving long term subsidization from either government or industry if their studies should consistently uncover practical or theoretical reasons why various lines of development would be unpromising or unwise. Yet it could be argued that something of this attitude has been implied, if not always encouraged, in sound engineering practice, since that is supposed to rely on a careful analysis of major variables and their consequences. Some industries, it should be noted, now have groups of engineers spending a substantial portion of their time doing work that might be defined as in the public interest. Perhaps at some future period the cultural aspect of engineering as represented by a search for understanding irrespective of the promise of gain will become as important a part of the overall function of the profession as the practical is now.

These are a few considerations about the future of engineering innovation which are clearly positioned in the social matrix. It is contended here that forecasts of the future which list exciting and innovative ideas for possible new combinations of technologies serve little purpose unless they also give some thought to conceivable processes leading to a form of social organization in which those technologies could function usefully, to the ultimate benefit of man.

THE CYBERNETIC SOCIETY

The reader will not fail to have observed that the concept of technological forecasting in the short term has much in common with technological assessment and the systems approach. There are, to be sure, differences in emphasis. The systems approach is more closely directed toward identifying, analyzing and synthesizing an attack on a given problem, while technological forecasting throws out a broader net. Unlike the simple forecast, the systems approach involves a plan; but (depending on the persons conducting the systems study) it may have less to say about what "ought to be" than technological assessment does. In any case, they are all efforts to look toward the future in a technological world by rational means. Much of the discussion in the previous chapters has had to do with man's attempts to seek direction in this age of perplexing contradictions through methods of analytical reasoning, now aided by new devices for processing and disseminating information. Since clear answers to present socio-technical problems have not readily been found except perhaps in theoretical form, the focus of the efforts must inevitably shift toward the future. What may emerge from an orderly study of current problems and future possibilities is difficult to say but, hopefully, it may at least lead to a more mature understanding of certain basic questions about man's true potential.

As an example of such questions, each of the four earlier chapters on the industrial order, government, the arts and education, respectively, has alluded to the relation between the increase of affluence, power or opportunity made possible through systematization of organizations and processes on the one hand, and the concomitant reduction of important aspects of freedom for the individual on the other. It has been implied that such a fundamental dichotomy cannot adequately be addressed without exploring its ramifications in the many sectors of human activity. The dichotomy may, indeed, be inherent in the human condition; yet there are those who are reasonably confident it is a negligible problem, or, if more than that, at least it is one that can be resolved in the knowledge society.

One aspect of the last argument proposes that since the demands of the societies whose enterprises are rapidly becoming more mechanized and cybernated appear to be for professional workers with increasingly higher levels of education, those demands must be met by expanding and strengthening the various educational institutions. Then, this argument goes, as the general level of education increases so should productivity and affluence increase comparably, along with the flexibility to meet the complex demands of the technological society. This kind of flexibility is presumably one of the marks of growth of individualism and freedom.

In fact, things do not work out quite that logically. True, in spite of serious difficulties encountered in trying to organize the educational systems to optimum effect, it has been possible to raise both the level and amount of graduate training in the universities in the past decade. But, as we know, for the present at least and probably for the next decade in some fields, the economy does not seem to be able to make room for all of these exceptionally trained people, especially at the Ph.D. level. Quite apart from the dilemma this poses for the universities, the effects on the morale of the highly trained in the knowledge society is bound to be immeasurable. A disgruntled intellectual elite is a bone in the throat of any country. One wonders what sort of systematic planning, in hindsight, would have led to a happier outcome.

There are signs of a growing disenchantment with the benefits of higher education. If the primary purpose of that education continues to be training for people whose sophisticated skills are not always wanted, the disenchantment could spread. Attempts to compose a believable forecast of future events might have to deal with a radical cultural shift which rejects scientific rationalism as a guide to a life-style. This could lead to an alternative future whose characteristics would be most difficult to foresee. "For it is written I will destroy the wisdom of the wise, and will bring to nothing the understanding of the prudent," said St. Paul (I Cor. 1:22).

As we have stated before, the conventional wisdom of the 20th Century technological society accepts, as eminently sensible, the proposition that organized knowledge can allow man to devise plans to counteract disharmony and limitations to human striving; and eventually to institutionalize and control social change. Nevertheless, an important minority among the young appears to have lost faith in a coherent picture of the world; or at least one coherent only by virtue of the orderly imperatives of science and technology. That it is still a minority there can be little doubt. For the majority there seems to be no immediate predisposition to renounce the technology whose artefacts are taken so much for granted. Erik Erikson suggests that technology may serve youth as the link between a new culture and new forms of society[19]. Still, one cannot ignore the extraordinary tendency of so many of the newer generation to take *all* experience as equally valid, rejecting the primacy of objective knowledge. However unrewarding it may be to generalize about the young, (since exceptions to any assertions are so easy to find), their attempts to experiment indiscriminately out to the boundaries of human possibilities and their relative freedom to do so are clearly documented. Perhaps these hedonistic experiments, initially conceived as tentative probes, have been forced to extremes by the distorting pressures of the public media, but there is little reason to question that they now represent a distinct cultural phenomenon.

Theodore Roszak[20] writes of this "counter-culture" as an unprecedented generational disaffiliation from the traditional assumptions of the technological society, and sees in it the beginnings of a new sense of reality which is in fact a return toward the mystical and the occult. No doubt the recognizing of visionary consciousness as one acceptable form of human awareness relies a good deal on the spreading influence of the hallucinatory drug experience, which in the form of obsessive, uncontrolled experimentation, has hugely destructive possibilities for minds too dull or immature to sense its potential trap. Pharmacological technology has made the production of drugs such a lucrative enterprise and the drugs themselves so assimilable that we cannot rule out, as one alternative picture of a future society, a general, drug-induced docility similar to that described by Aldous Huxley in *Brave New World*, or later, more optimistically, in *Island*. But the youth culture in its more provocative manifestations, (leaving out the violent fringe), does not really seem to require any artificial stimulation to cause it to embrace fantasy and hedonistic spontaneity. To be young at one of history's crucial turning points is stimulation enough.

Are we at one of those turning points? Every generation in every era must, after all, believe that it is the hinge upon which civilization swings. Could this disaffected minority, skeptical of authority cloaked in pompous scientism, be merely an expression of the anti-institutionalism typical of all younger generations; or is it the advance guard of a general revolt against the systematic constraints of a society committed to rationalism? Will professions in science and technology be bypassed by significant numbers of the intellectual elite who are disillusioned by lack of opportunities, or, more critically, by the intolerable thought that methodical science can perhaps never fully mirror reality?

It would, in fact, be misleading to suggest that the philosophical positions of most of the populace, even in the technological societies, owe more to scientific objectivism than to homespun moralizing or superstition. Their investment in the forms of existing institutions is emotional and so those institutions can remain strong even if logically obsolete. Yet society does change, as often as not through the actions of people who prefer to have things remain as they are, and who take seemingly sensible steps to see that they do. When a businessman in the United States decides to bring his operations up to date by purchasing a computer system, his basic concern is for stability. He wishes to keep his business abreast of the competition, and to maintain his own position in the social hierarchy. At the same time, thousands of other businessmen are making similar decisions. The outcome is that the stability each seeks is threatened by a remorseless flood of accelerating change.

The pace of technological change contributes impartially to the malaise of the older as well as the younger generations. If the numbers of those who have lost faith in the stability of a system ordered by rational priorities should grow, it is conceivable that the isolated experiments of alienated youth could become a pattern for the future. This could see Western man turning away from the systematic structure of the technological society, and back toward tribalism, accepting some form of transcendental religiosity as a pillar to cling to. Beryl L. Crowe has suggested that the return to small community living, away from the influence of the electronic media, may be a functional response related to perpetuation of the species[21]. Such a response would not be inconsistent with theories of the cyclic nature of social development first proposed by the Greek philosophers and restated in various ways by writers so diverse as Oswald Spengler, Arnold Toynbee and Pitirim Sorokin in the 20th Century. However, a monistic philosophy of cyclic change for a pluralistic world is not much more satisfactory overall than the theories of linear progress typical of the 18th and 19th Centuries. The spread of a pre-scientific (or post-scientific) tribalism, accompanying a reduction in the rate of growth of technology and science is not beyond reason; but it requires quite a leap of the imagination to see a new form of primitivism predominating on a world scale except in the aftermath of a nuclear holocaust.

Let us ask the same kind of question we have posed before. Can we assess the probability of this type of social trend's coming to pass, just as we hope to forecast technological probabilities? Raymond Aron, in his book *Progress and Disillusion* [22] has surmised that the current interest in forecasting may reflect what he calls the unresolved antithesis between history and technology. For the most part, exclusively technical activities are such that the objectives and approaches can be defined formally and thereby determined, but these objectives, or ends, are restricted. They are different in some intrinsic way from the truly historical ends which relate to the many subsystems of human existence. There may be inherent limits in the extent to which rigorous examination can be applied to the latter systems, but since no one knows what the limits are it is surely a legitimate intellectual pursuit to seek to identify them. In this, we find a *raison d'être* for the philosophical orientation in cybernetics and systems studies; this and the fact that they offer valuable metaphors to aid in conceptualizing the vast diversity of social change. Behind the veil of ignorance there are always tantalizing glimpses of certain basic ideas which would help to bring order to confusion if only they could be seen more clearly.

Cybernetic metaphors applied to social systems in a condition of change are helpful adjuncts to other methods for studying the future. Perhaps they are more prophetic than predictive in proposing a broadgauged, long-range view of man's predicament rather than offering a close-up, year-by-year forecast of probable events. One can prudently assume, though, that the survival of mankind may well depend on priorities arranged according to the long term trends rather than the short. The concept with the most obvious implications for the future is that, on all levels, there is a tendency toward ultimate disorder. Since its finality is so thoroughly remote, however, we can be more impressed by the negative feedback metaphor as an aid to envisioning the potential course of future events. As we have seen earlier the core idea relates to the control of dynamic processes. By the act of diverging from a given control level, a process in change generates information, thus motivating forces which work to slow down, and, in time, reverse or defeat the original trend. This can be viewed as a reliable, self-regulating control mechanism in systems for which a control point has been established. When social systems are examined in these terms, the difficulty lies in establishing an unequivocal goal for them, unless one is prepared to accept that it is merely the survival of the human species. There are many feedback loops in large social systems and overall response is sluggish, but in time the self-limiting aspect will tend to become manifest. For example, the processes of technology must be edging toward self-limitation by using up resources or creating pollution. Also, the technological society, based on science and rationalism, could be spawning a generation of mystical idealists. Careful analysis ought to be able to detect countertrends in any process in full flower.

It would be easy to mistake the implications of this metaphor as arguments for a *laissez-faire* political posture. Indeed, the *laissez-faire* assumption, in contending that natural economic forces ought to be allowed to work themselves out without the distorting effect of government interference, overlooked the considerable imbalance in the remaining contending forces. If, theoretically, those forces would have equalized over the long term; the economic conditions were often thoroughly out of control over the short term. Consequences of many socio-technical movements can well be so intolerable that men must intervene—often, for critical periods of time, to good effect. But what happens too commonly is that plans to establish control do not sufficiently acknowledge all the counterforces already acting on the system, and the intended solution causes an overshooting of the mark. Professor Jay W. Forrester of the Massachusetts Institute of Technology, whose work was cited in Chapter 6, has concluded that many socio-technical solutions to basic problems are predictably self-defeating because humans tend to look only to the recent past for the genesis of a dilemma. One example we have only recently come to recognize is the attempt to cope with the transportation problem simply by building more freeways. It took too long to learn that more freeways permit more traffic, and thus magnify all the ills that accompany too many automobiles. Social problem solving on the basis of single cause-and-effect relationships must be self-limiting and futile. Forrester's studies utilize the computer to analyze and plot huge amounts of data depicting both short and long term trends of a maze of interconnected projections such as natural resource availability; growth of capital investment, population and pollution; and an attempt is even made to plot data showing how the quality of life will decline or improve. Some of the conclusions are documented restatements of old wisdoms. It becomes clear, for instance, that the reduction of a stress or combination of stresses at one place allows the building up of a new stress at another, and so problem solving becomes a trade-off between short term and long term benefits or threats.

Developing a picture of the future on the basis of a broad metaphor offers limited promise as a guide to specific practical answers to the most critical human difficulties in the next few decades. Rather it fosters a recognition that there are no absolute solutions to the fundamental social problems. If there were apparent solutions, changes in socio-technical equilibria would still invalidate them with time. This need not be a despairing philosophy. It ought instead to lay a groundwork for a more sober and realistic approach to the future.

With regard to the progress of technology, for example, this point of view would argue against the more enthusiastic projections of unlimited technological growth leading to a utopia of affluence and leisure. Technology in the Western nations will become more and more self-limiting. That does not mean that one can look ahead to a world-wide renunciation of either mechanical and electronic devices or rationally planned systems. Technology is in the very fabric of Western civilization and it is also the hope of the underdeveloped countries. No other path to widespread affluence has yet been discovered. As long as the majority sees its benefits outweighing the disadvantages, no aroused minority, however articulate and dedicated, will succeed in exorcizing it. And yet, while man rejoices to find that his activity options seem to be expanded by technological innovation, the price is always high. At its best the symbiosis between technology and man is an uneasy one. There are no technological Polynesias. If, as some foresee, there could be a gleaming, smoothly-running, computer-monitored future, man would first have to change himself to become something very different from what he is. Only by forcing him into a collective mold which smothers some of his uniqueness can technology function at its peak efficiency; and if he resists according to his nature, the social torsion becomes intense.

Can then, or should, man adapt himself to technological imperatives; or is it possible to redirect and limit technology to humane purpose? Not many would consciously elect the former alternative, and if man is to strive for the latter he will not find a simple way. He must learn to apply his devices and his techniques moderately, even parsimoniously, and he must think of technology as a cultural as much as an economic activity; but more fundamentally he must come to a more mature understanding of himself in relation to his universe. To seek this understanding he will have to continue to travel back and forth between the poles of scientific rationalism and other-worldly romanticism—between abstraction and sensualism, in a rich, if bounded milieu; and as always his vision will fall somewhere short. He will find no sure answers at either pole, but hopefully he will never give up, for in the striving is the only lasting satisfaction.

REFERENCES

1. Toffler, Alvin, *Future Shock*, Random House, N.Y., 1970.
2. Bell, Daniel, "The Study of the Future," *The Public Interest*, Vol. 1, No. 1, Fall 1965, pp. 119–30.
3. De Jouvenel, Bertrand, *The Art of Conjecture*, (translated by N. Lary), Basic Books, Inc., N.Y., 1967, pp. 12–15.
4. Clarke, Arthur C., *Profiles of the Future*, Bantam Books Science and Mathematics Edition, 1967, (Harper & Row, Inc., N.Y., 1963), p. xii.
5. Franklin, Howard Bruce, *Future Perfect: American Science Fiction of the 19th Century*, Oxford University Press, N.Y., 1966, p. 402
6. Miller, David C., "Comprehensive Long-Range Forecasting Systems for Management," *Mankind 2000*, Robert Jungk and Johan Galtung (eds.), Allen & Unwin, London, 1969, pp. 291–98.
7. Quinn, James Brian, "Technological Forecasting," *Harvard Business Review*, March–April 1967, pp. 89–106.
8. Bright, James R., "Can We Forecast Technology?" *Industrial Research*, March 1968, pp. 52–56.
9. Schwartz, Leonard E., "Another Perspective," in *Technological Forecasting*, edited by R. V. Arnfield, Edinburgh University Press, Edinburgh, 1969, pp. 74, 75.
10. Wills, Gordon, Ashton, D., and Taylor, B., (eds.) *Technological Forecasting and Corporate Strategy*, American Elsevier Publishing Co., Inc., N.Y., 1969, pp. 4, 5.
11. Ayres, Robert U., *Technological Forecasting and Future Planning*, McGraw-Hill, Inc., N.Y., 1969, pp. 20, 21, 102–109.
12. Lenz, Ralph C., "Technological Forecasting," (2nd ed., Report ASD-TDR-62-414), Aeronautical Systems Division, U.S.A.F.
13. Gordon, T. J. and Helmer, O., *Report on a Long Range Forecasting Study*, P-2982, The Rand Corporation, September 1964.
14. Ayres, Robert U., op. cit., pp. 148–50.
15. Weaver, W. Timothy, "Delphi As a Forecasting Tool," *Notes on the Future of Education*, Education Policy Research Center, Syracuse University, Vol. 1, Issue 3, Spring–Summer 1970. pp. 7–10.
16. Kahn, Herman and Wiener, Anthony J., *The Year 2000: A Framework for Speculation*, Macmillan, N.Y., 1967.
17. Sorokin, Pitirim, *Social and Cultural Dynamics*, Vol. 4, N.Y., 1962, pp. 737ff.
18. Kahn, Herman and Wiener, Anthony J., "Technological Innovation and the Future of Strategic Warfare," *Astronautics and Aeronautics*, December 1967, p. 35.
19. Erikson, Erik H., "Memorandum on Youth," *Toward the Year 2000—Work in Progress, Daedalus*, Journal of the American Academy of Arts and Sciences, Summer 1967, p. 863.

20. Roszak, Theodore, *The Making of a Counter Culture*, Anchor Books, Doubleday & Co., Garden City, 1969, pp. 42–44, 159.
21. Crowe, Beryl L., "The Tragedy of the Commons Revisited," *Science*, Vol. 166, November 28, 1969, p. 1106.
22. Aron, Raymond, *Progress and Disillusion*, Mentor—The New American Library, N.Y., 1968, p. 274.

Index

Abacus, 40–41
Abstractionism
  geometric, 309
  see music
  see visual arts
Ackoff, Russell L., 184
Act of Creation, The, 272
Action potential, 232
Acts of enclosure,
  see enclosure
Adams, Henry, 29
Adams, John, 36
Aesthetics, 266–271
Aid to Families With Dependent Children Program, (AFDC), 101
Aiken, Howard, 53
Algorithm, 238
Alt, Franz L., 70
America, 18, 33
  see United States
American Chemical Society, 97
American Gothic, 298
American System, 33, 37–38
American Telephone and Telegraph Company, 164
Ampère, André, 28, 202
Analytical Engine, 2, 49–52, 237
And-or circuits, 58
Antheil, George, 293, 302
Anti-ballistic missile, (ABM), 154

Appalachia, 78
Apter, Michael J., 205
Aristotle, 56, 267, 270, 371
Arkwright, Richard, 14–16, 34
Army Ordnance Corps, 53
Arnold, Matthew, 264–265
Arnold, Thomas, 265
Aron, Raymond, 378
Arp, Jean, 297
Art, 6–7
  definitions of, 266–271
  engineering and, 284–285
  relation to science, 271–283
  relation to technology, 283–312
Art, Affluence and Alienation, 262
Artificial Intelligence, 207, 230, 236–237, 246, 256
Ash Can School, 297
Ashby, W. Ross, 253
Assembly lines, 39, 73–74, 100
Atom, fissioning of, 2, 29
Atomic bomb, 62, 152
Atomic war, 106
Audio-visual instruction, 335–342
  motion pictures, 335–337
  radio, 338–339
  television, 339–342
Austria, 71
Automatic Language Processing Advisory Committee, 249

383

384 Index

Automatic Sequence Control Calculator, 53
"Automatic" writing, 296
Automation, 39, 66, 70, 74, 96
Automaton, (automata), 51, 233, 251–253
Automobile industry
   see Industry
Avant-garde, 297, 300, 310–311
Avram, Henriette D., 251
Ayres, C. E., 102
Ayres, Robert U., 366

Babbage, Charles, 2, 46–53, 88, 237
Bacon, Francis, 148
Bakunin, Mikhail, 150
Balanoff, Neal, 340
Ball, W. W. R., 224
Ballet Mechanique, 293
Bardeen, John, 54
Barlow, Fred, 224–225
Bartholdi, Frederic Auguste, 288
Bauhaus, The, 293–295, 311
Baumgarten, Alexander, 266
Bay Area Rapid Transit, (BART), 131–132
Becquerel, Henri, 59
Bell, Daniel, 67, 359
Bell Telephone Laboratories, 163, 167, 305, 308
Bellamy, Edward, 361
Bellman, Richard, 190, 229
Belper mill, 34
Benet, Stephen Vincent, 222
Bentham, Jeremy, 196
Bentham, Sir Samuel, 32, 33, 196
Benz, Carl, 28
Berg, Alban, 282
Bernstein, Jeremy, 224
Bible, The, 78
Bidder, George Parker, 225–226
Biderman, Albert D., 136, 138
Binary digits, 210
Binary numbers, 55–56
   see number systems
Birmingham, England, 21
Black, Joseph, 20
Black box, 160
Blackett, P. M. S., 152, 164
Blake, William, 262, 289
Blanc, Peter, 276

Blast furnace, 23–24
Blue Horseman, The, (Der blaue Reiter), 279
Blue-collar workers, 66, 70, 76–77
Boguslaw, Robert, 194–195
Bolivar, Simon, 26
Boole, George, 57
Boolean algebra, 57
Boulding, Kenneth, 161–162
Boulton, Matthew, 21
Boulton and Watt, 21–22, 25, 27
Bowers, Dan M., 248
Bradbury, Ray, 362
Brain, the, 230
   information processing in, 236
Bramah, Joseph, 32
Branching, 53–54
Braque, Georges, 279
Brattain, Walter H., 54
Brave New World, 361, 376
Breton, André, 296
Breuer, Marcel, 294
Briggs, Henry, 43
British Small Arms Commission, 37
Broadbent, D. E., 215
Brougham, Lord, 85
Brown, James W., 335
Brown, Moses, 34
Brown, Sylvanus, 34
Brudner, Harvey J., 351
Brunel, Marc Isambard, 32–33
Bruner, Jerome, 319
Bruno, Giordano, 256
Building codes, 122
Bull, Edward, 25
Bundy, McGeorge, 113
Bureau of Labor Statistics, 136
Bureau of Public Roads, 118
Bureaucracy, 94, 105, 107, 147
Burks, Arthur W., 54
Bush, Vannevar, 152
Butler, Samuel, 222
Byron, Lord, 50, 221

Cahill, Thaddeus, 302
Calculator, mechanical, 43–52, 55
Calculus, 194
   "felicific", 196

California, 115–116, 118
Čapek, Karel, 253
Capitalism, 66, 81, 85, 91, 150, 290
Carlyle, Thomas, 10, 16, 289
Carson, Rachel, 140
Cartwright, Edmund, 17
Cassatt, Mary, 287
Cassidy, Harold G., 273
Census Bureau, 137
Century of Inventions, 17
Cézanne, Paul, 277, 279
Charter, S.P.R., 140, 199
Chase, Stuart, 151
Chemical Abstracts, 246
Chemical processing, 74–76
Cholera, 88
Churchill, Winston, 153
Churchman, C. West, 199
Cicero, 358
Clarke, Arthur C., 222, 229, 361–362
Clegg, Samuel, 28
Clement, Joseph, 47–48
Clermont
  see steamship
Closed-loop control, 126
  see systems, control
Coal in iron-making, 24–25
Coalbrookdale, 24
Cockcroft, John, 152
Coketown, 88
Colburn, Zerah, 224–225
Comenius, Johann, 345
Comfort, Alex, 274
Commission on National Goals, 138
Committee on Science and Astronautics, 142
Communication
  between artist and scientist, 273–275
  and control, 205–214
  man-machine, 5, 244–245
  in technological age, 300–301
  technologies of, 3, 108, 111, 113
Communications, Act of 1962, 339
Communism, 91
Community Fund, 97
Composition with Lines, 309–310
Compton, Karl T., 152
Computer-assisted instruction, 351–353

"drill-and practice", 351
inquiry, 352
Computer graphics, 307, 312
Computer-managed instruction, 350
Computing-Tabulating-Recording Company, 53
Computers, 2, 3, 5
  analog, 43
  art and, 263, 286, 307–309, 311
  as enemy, 223
  automation and, 368–369
  chemical processes using, 74–76
  costs of, 70
  demographic studies with, 118
  digital, 43, 53
  educational use of, 349–353
  employment and, 65–70
  game-playing with, 193, 236–243
  history and development of, 40–59
  information systems, 169–170, 178–179
  installation of, 67
  integrated systems using, 65
  in language translation, 245–249
  literature and, 309–311
  logic and, 54, 56–58
  and the mind, 220–221
  music and, 303–306
  numerical control with, 73
  politics and, 108–111
  privacy and, 145–147, 179
  and self-regulating systems, 201
  simulation by, 109, 110, 180–183, 234–235, 243–244
  and sociological data, 136
  speeds of, 366
  theory of, 215
  and transportation systems, 125–129, 132–133
Comte, Auguste, 150, 216–217
Concept of Equilibrium in American Thought, The, 216
Concerto for Trautonium, 302
Condenser, steam, 20–21, 26
Conditioning
  classical, 344
  operant, 345–346
Condorcet, Marquis de, 360
Confucius, 358

Continuous flow processes, 65
Control, 5
    see feedback
    see fly-ball governor
    see systems
    in steel-making, 72
Coolidge Administration, 338
Copernicus, 256
Cornish boiler, 26
Cort, Henry, 24
Cotton gin, 35–36
Council of Economic Advisors, 135–136
Counter-culture, 376
Cowper, William, 193
Coxe, Tenche, 34
Craftsmen, 30
Craik, Kenneth J. W., 203
Creativity, 6, 272
    see *The Act of Creation*
Croce, Benedetto, 270–271
Croesus, 360
Crompton, Samuel, 16–17
Crowe, Beryl L., 377
Csuri, Charles, 307
Cubism, cubists, 277–279, 295, 298
Culture,
    standards of, 265–266
Culture and Anarchy, 265
Curie, Marie, 25, 59
Cybernation
    definition of, 40
    and industry, 66–67, 74–75
    processes of, 99
    and unemployment, 70, 96
Cybernetic Anthropomorphous Machine, 255
Cybernetic society, 374–381
Cybernetics, 223
Cybernetics, 5, 201–203, 205, 212, 378
    relevance of, 214–218
    and government, 217–218

Dadaism, dadaists, 295–299
Daddario, Emilio Q., 139
Daimler, Gottlieb, 28
D'Alembert, Jean le Rond, 149
Dali, Salvador, 297
Dalton, John, 276

Dante, 226
Darby, Abraham, 24
Darwin, Charles, 256, 361
"Darwin Among the Machines", 222
Data bank(s), 109, 145–147
Data processing, 4, 53, 169–173, 178–179
David, Jr., E. E., 191
Dean, Peter M., 351
Debussy, Claude, 262, 276
de Caus, Solomon, 17
de Chirico, Georgio, 297
Decision making, 166, 179, 185, 190
Decision tree, 240–242
Declaration of Independence, 10
Degas, Edgar, 276
De Grazia, Sebastian, 99
Degrees of freedom
    in transportation, 130–131
de Jouvenel, Bertrand, 360
Delphi method, 367–369
Delphic Oracle, 360
Democratic party, 110
Democritus, 256
De Morgan, Augustus, 57
Demuth, Charles, 298
Department of Defense, 94, 154
Depression
    see Great Depression
Descartes, René, 148, 221, 256–257, 268
Deserted Village, 81
de Toqueville, Alexis, 38
Deutsch, Karl W., 218
de Vaucanson, Jacques, 32
Dewey John, 139, 327, 330–333
Dickens, Charles, 88, 289
Dickinson, Emily, 286
Difference Engine, 47–49
Disneyland, 131
Divine Comedy, The, 227
Division of labor, 30, 216, 290
DNA, 252
Document indexing, automatic, 250
Doran, F. S. A., 258
Doxiadis, C. A., 119
Drucker, Peter, 65
Duchamp, Marcel, 295–299
Dynamaphone, 302
Dynamic programming, 188, 190

Eakins, Thomas, 287
Eckert, J. Presper, 2, 53
École Polytechnique, 149
Economic indicators, 135
Economic theories, 83–85, 90–95
    see capitalism
    see Keynes (General Theory)
    see Marx
    see mercantilism
Economy of Manufactures and Machinery, 48
Edison, Thomas A., 336
Education,
    for democratic society, 7
    early developments in, 318–327
    in 19th Century, 321–327
    problems in, 316–318, 375
    psychological theories of, 323–326, 331, 343–345
    technology of, 8, 315–356
Educational technology, 333–355
    see audio-visual instruction
    see computers
    see programmed instruction
    see teaching machines
Eimert, Herbert, 302
Einstein, Albert, 152, 271, 275, 278
Eisenhower, Dwight D., 106
    administration, 123, 141
Elections, political, 108–112
Electromagnetic induction, 28
    dynamo, 29
Electronic synthesizers, 303, 306
Electronics and music, 301–303
Elite of scientists and enginers, 107
Ellis, David O., 159
Ellul, Jacques, 107, 147
Emerson, Ralph Waldo, 287, 315
Emile, 320, 323
Employment
    projections, 119
    totals (in U.S.), 64, 69
    see unemployment
Employment Act (Full Employment Act), 62–63, 135
Enclosure, acts of, 80
Energy
    certificates, 151
    consumption by industries, 63
    electrical, 29
    new sources of, 27–29
    transformation of, 2, 29
Engels, Friedrich, 150
Engine, engines
    internal combustion, 28, 39
    Lenoir, 28
    see steam engine
Engineering (*also* engineer)
    definition of, 162
    studies in, 6
England, 18, 23, 26, 33–34
    elections of 1970, 109
    population growth in, 84
    social discord in, 79–83
    Victorian, 85, 88
    War with France, 36, 81–82
ENIAC, 2, 53–55
Enlightenment
    in France, 148
Entrepreneur(s), 10, 16, 20, 26, 150
Entropy, 212–214, 216
Environment
    problems of, 88, 95, 99, 108, 119, 141
Equilibrium
    definitions of, 216
    economic, 92, 217
    social, 28
Erewhon 222
Erikson, Erik, 376
Erlang, A. K., 180
Ernst, Max, 297
Evans, Oliver, 26–27, 37, 39
Experiments in Art and Technology, Inc., 285
Expressionism, 277, 279

Factory system, 3
Faraday, Michael, 28–29
Featherbed, 97
Federal Communications Commission, 339
Federal land grants, 116
Feedback, negative, 215, 378–379
    control systems 21, 161, 203–204, 217–218, 255
Fein, Louis, 101, 259
Ferkiss, Victor, 112, 154

388 Index

Fernandez, Juan, 216
Fisk, Miles B., 248
Flying platforms, 372
Fly-shuttle, 11–13, 15
Ford Foundation, 339
Ford, Henry, 39, 115
Ford Motor Company, 336
Forrester, Jay W., 218, 379
Foster, David, 75–76
Fourier, Charles, 197
Fourier, Joseph, 303
Frankenstein, 221, 254
Franklin, H. Bruce, 361
Franklin Institute, 140
Free enterprise, 92
Freud, Sigmund, 256, 296
Friar Roger, 361
Friedman, Milton, 101
Froebel, Friedrich William, 326
Fry, Roger, 272
Fulton, Robert, 27
Future studies, 8–9, 358–374
  early views, 359–361
  "future shock," 358
Futurists, futurism, 9
  in art, 293, 298

Galileo, 158, 256
Game theory, 188, 191–193
Gauguin, Paul, 276, 279
Gauss, Carl Friedrich, 226
General Electric Company, 67
General Problem Solver, 243–244
General Purpose Systems Simulator (GPSS), 181
Generational transfer, 78
George, Frank H., 206
Georgetown University, 246
Gernsback, Hugo, 361
Gerry-mandering, 110
Gide, André, 279
Gilbert, William, 28
Gilded Age, The, 297
Goddard, Robert, 158
Goldberg, Maxwell, 316
Goldsmith, Oliver, 80
Goldstine, Herman H., 53–54
Golem, The, 221

Government and technology, 105–155
Governor, fly-ball, 21–22, 202
Grand Academy of Lagado, 310–311
Great Depression, 62, 67, 90–93, 298
Green, Mrs. Nathaniel, 34–35
Gropius, Walter, 294–295
Gross National Product, 77
Guaranteed income, 81, 100–102
Guilbaud, G. T., 211
Gulliver, Lemuel, 310

Hall, Arthur D., 163, 168
Hamilton, Alexander, 34, 107
"Hand-eye" experiments, 255
Harding, Warren G., 105
Harcleroad, Fred F., 335
Hargreaves, James, 13–14
Harries, Karsten, 268, 300
Hartley, R. V. L., 209
Harvard University, 53
Hauser, Arnold, 268–269
Hawthorne effect, 353
Heat death, 214
Hebb, D. O., 233
Hegel, G. W. F., 270
Helmer, Olaf, 367
Hennock, Frieda, 339
Henry, Pierre, 302
Heraclitus, 371
Herbart, Johann Friedrich, 326
Herodotus, 11
Herschel, John, 46
Herzen, Alexander, 150
Heuristics, 228, 247
Heydenreich, Ludwig, 271
Highway location analysis, 129
Hiller, Lejaren A., 304–305
Hindemith, Paul, 302
Hollerith, Herman, 52–53
Hoover, Herbert, 152
Humanoids, 255
Human Use of Human Beings, The, 5, 214
Hunt, J. McV., 327–328
Huxley, Aldous, 140, 361, 376
Huxley, Thomas, 256, 265

IBM Corporation, 53, 125, 240–241, 246
Idiot savants, 223

Iliad, The, 266
Illiac Suite for the String Quartet, 305
Impressionism, 276–277
Industrial Age, 27, 29
Industrial design, 290
Industrial relations, 87
Industrial Revolution, 1, 4, 10–11, 31, 61, 66, 71, 77–78, 86, 88, 99, 264
Industry,
  automobile, 39, 62, 70–74, 115
  early American patterns of, 89–90
  "knowledge", 63
  steel, 62, 70–72
Information,
  bits, 210–211
  capacity, 209
  and choice, 209–212
  coding, 206–207
  as element of technology, 3
  and entropy, 212–214
  gathering of, 136–138, 145–147
  handling, 169–179
  industries, 65
  as organization, 205
  retrieval of, 249–251
  theory of, 206, 208, 212, 215, 304
Input-output studies, 195–196
Instruction, individualized
  see programmed instruction
Intelligence Quotient (I.Q.), 229
Interchangeable parts,
  see manufacturing
International Exhibition of Modern Art, (1913), 297
Iron-making, 10, 23–25
Isaacson, Leonard, 304
Island, 376
Itard, Jean, 328

Jacobs, W. W., 223
Jacquard, Joseph Marie, 49–50, 52
Jacquard loom, 49
James, Henry, 287
Jefferson, Thomas, 36, 107, 315
Johnson, Lyndon B., 94, 113, 136
Johnson, Samuel, 158, 226

Kahn, Herman, 370–372

Kandinsky, Wassily, 279, 294
Kant, Immanuel, 268–269
Kay, John, 11–13
Keats, John, 270
Kelso, Louis, 101
Kennedy, John F., 93–94, 109–110, 113, 158
Keynes, John Maynard, 91–95, 96
Kitty Hawk, 59
Klee, Paul, 294
Kluver, Billy, 285
Knight, Charles, 85
Koestler, Arthur, 272
Korean War, 153
Kranzberg, Melvin, 106
Kubrick, Stanley, 222
Kuhn, Thomas S., 271

Laboring class
  in the U.S., 61
Lady or the Tiger, The, 191
Laissez-faire, 265, 268, 321, 379
L'An 2440, 361
Lancaster, Joseph, 322
Lancastrian schools, 322–323
Land use, 117, 119–120
Language, concepts of, 257
  machines, 244–245
  and mathematics, 268
  see symbolic language
La Rochefoucauld, 8
Latent heat, 20
Lathe, 31–32
Leavis, F. R., 264
Leibniz, Gottfried, 43, 45–46
Leisure
  and work, 98–100
Leonard and Gertrude, 324
Leonardo da Vinci, 31, 275
Leontief, Wassily, 195
Les Fauves, 298
Letwin, Shirley, 106
Lewis, Richard B., 335
Leyden jar, 28
Library of Congress, 251
"Light-Space Modulator," 295
Lindemann, F. A., 153
Lindsay, Robert K., 250

Index

Lindsey, Robert B., 205
Linear programming, 125, 188–190, 195
Linguistics, 245–246, 249
Local Government Information Control System (LOGIC), 169–179, 184
Logarithms, 42–43, 46
Logic
  see computers
  diagrams, 181–183
  machines, 229
Looking Backward, 361
Lovelace, Lady, 50–52
Lovelace, Lord, 51
Luddites, 82
Ludwig, Fred J., 159
Luftwaffe, 152
Luria, A. R., 226–227

Macauley, Thomas Babington, 226
McCormick reaper, 38
McCulloch, Warren S., 202, 233
McDermott, John J., 330
McHale, John, 315
Machiavelli, Niccolo, 193
Machine tools, 31–33
Machine translation, 245–249
Machines,
  automatic processing, 72–73
  boring, 21
  data-processing, 4, 53
  game-playing, 236–243
  "intelligent", 6, 220, 223, 229, 257, 259
  learning, 237
  see machine tools
  pattern-recognizing, 235
  self-reproducing, 251–253
  see teaching machines
  see textile machines
  threshing, 10
  tracer-controlled, 73
Machlup, Fritz, 63
Machol, Robert E., 198
McKay, Donald, 203
McLuhan, Marshall, 115, 342
McMullen, Roy, 262, 299
Maholy-Nagy, Laszlo, 294–295
Mallarmé, Stephane, 277
Malthus, Thomas, 84

Man Made World, The, 191
Management science, 165
Manet, Edouard, 276–277
Manheim, Marvin, L., 129
Mantoux, Paul, 80
Manufacturing
  method of interchangeable parts, 10, 35–36, 39
  see mass production
  pin-making, 30–31
  automobile, 70–74
Marinetti, Filippo, 293
MARK I, 53–54
Marquis of Worcester, 17
Marsh, George Perkins, 140
Marx, Karl, 85, 88, 150, 361
Mass production, 29–33, 288
  see American system
Materials handling, 179–183
Mathematical models, 75, 125, 166, 188, 194–195, 215, 252
Mathematical programming, 188
  see linear programming
Mathews, M. V., 305
Mauchly, John, 2, 53
Maudslay, Henry, 31–33, 47, 73
Maupertuis, Pierre Louis, 360
Maxwell, Clerk, 273
Mayo, Elton, 353
Mazlish, Bruce, 256
Mechanization, 66, 76, 78
Melville, Herman, 286
Menabrea, L. F., 50–51
Mercantilism, 83
Mercier, Louis Sébastien, 361
Metallurgical technology (metallurgy), 23–25
Meynaud, Jean, 154
Michael, Donald N., 40
Michelson-Morley experiment, 276
Microelectrode, ultrafine, 232
Military-industrial complex, 106
  MAGIC, 106
Miller, David C., 362
Miller, Herman, 77
Miller, Phineas, 34–35
Mind-body dualism, 258
Mind, human, 255–259

Miner's friend, 18
Minsky, Marvin, 236
Miró, Joan, 297
Mnemonics, 226, 228
Mondrian, Piet, 309
Monet, Claude, 276–277
Monitorial system, 322–323
"Monkey's Paw, The," 223
Monorail, 130–131
Montessori, Maria, 327–330
Montessori Method, The, 327
Moore, Edward F., 252
More, Thomas, 360
Morganstern, Oskar, 191
Morningstar, Mona, 351
Morris, William, 289–292, 294
Morse, S. F. B., 209
Mosher, Ralph S., 255
Mozart, Wolfgang A., 272
Mumford, Lewis, 140, 287–288
Murdock, William, 22, 28
Music
  abstractionism in, 280, 282, 302
  atonality in, 282
  counterpoint in, 281
  modality in, 281
  tonality in, 280–281
Music from Mathematics, 305
MUSICOMP, 305
Musique concrète, 302

Nader, Ralph, 144
Napier, John, 42–43
Napoleonic Wars, 37, 196
Nash, Ogden, 310
National Academy of Engineering, 142
National Academy of Sciences, 134
National Commission on Technology, Automation, and Economic Progress, 136
National Educational Television and Radio Center, 339
Nazis, 294
Negative Income Tax, 101
Nemerov, Howard, 311
Nerves of Government, The, 218
Nervous system, 215
  axon, 231
  dendrites, 231

imitations of, 236
  neurons, 230–233, 235
Neural networks, 230, 233
New Atlantis, 148
New Deal, 93
New Lanark, 87
New York City
  welfare in, 77
  transportation in, 132
Newcomen, Thomas, 18–19
Newell, Allen, 243
Newton, Isaac, 29, 148, 158, 194, 271, 273–274, 278
"Nightmare Number Three", 222
Nineteen Eighty-Four, 361
Nixon Administration, 101
Nixon, Richard M., 94
Noll, A. Michael, 308–310
Nostradamus, 360
Nuclear power, 2, 59
Nuclear Science Abstracts, 246
Nude Descending a Staircase, 295, 297
Number systems
  decimal, 41, 55
  binary, 55–56
Numerals
  Arabic, 41
  Roman, 41
Numerical control, 73

Oersted, Hans Christian, 28
Oettinger, Anthony A., 349, 353
Onions, Peter, 24
Op art, 308–309
Operations Research, 48, 162–167
Opinion polls, 108–109, 367
Optimization, 188, 190–191, 196, 198
Orange County, 118
Organization, 3
Orphée, 302
Orwell, George, 361
Owen, Robert, 87
Owen, Wilfred, 114, 123

Page, Thornton, 164
Panopticon, 196
Papert, Seymour, 236
Papin, Denys, 17–18

Pareto, Vilfredo, 217
Parliament, 80, 87
Parzen, Emanuel, 186
Pascal, Blaise, 43-46, 226
Paul, Lewis, 13-14, 16
Pavlov, Ivan P., 344
Penn-Central railroad, 117
Penny post, 51
Penrose, L. S., 252
Penydaren Iron Works, 26
Perceptions, 233-236
Personal transit, 132-133
Pestalozzi, Johann Heinrich, 323-326
Pesticides, 140
Petty, Sir William, 194
Philadelphia Centennial Exhibition, 288
"Philosophes", 148
Phonetic writing, 207-208
Picabia, Francis, 295
Picasso, Pablo, 279
Pickard, James, 22
Pierce, John R., 304-305
Pike, E. Royston, 86
Pirandello, Luigi, 279
Pitts, Walter H., 202, 233
Planck, Max, 275
Planning
   for transportation, 118-135
Plato, 107, 194, 266-267, 319, 371
Player Piano, 361
Pledge, H. T., 41
Poe, Edgar Allen, 287, 300, 310
Poggioli, Renato, 293
Pohl, Frederick, 362
Poisson distribution, 187
Polanyi, Karl, 83
Polhem, Christopher, 33
Political Arithmetik, 194
Political campaigns, 108-112
Poor laws, 79-83, 100
Pope, Alexander, 267
Population growth
   in England, 84
   in U.S., 118
Post-industrial age, 2, 371
Postlethwait, Samuel, 353-354
Poverty, 4, 77-83, 98, 137
Powers, James, 52-53

Pragmatism, 331
Precisionists, 298
Pressey, Sydney L., 343
Primitivism, 292, 299
Prince, The, 193
Principia Mathematica, 194
Privacy
   see computers
Private enterprise, 116
Probability theory, 51, 186-187
Prodigies, mental, 223-228
Production, automatic, 3, 37, 39, 62
Productivity, 39, 62-63, 67
Programmed instruction, 343-348, 353
   intrinsic (branched), 347
   linear, 346-347, 351
Programming
   see linear programming
Progress and Disillusion, 378
Puddling process
   see iron-making
Pumps, water, 17-18
Puritan ethic, 99, 373

Quantum theory, 275
Queen Elizabeth I, 23, 28, 79
Queen's College, 57
Quesnay, Francois, 195-196
Queuing theory, 125, 180
Quidditism, 299

Radio, 108, 111-112
Radium, 29
Railways, 27, 38, 116-117
Ramsden, Jesse, 32
Randomness, 186-187, 190, 304
Rauschenberg, Robert, 285
Real-time, 198
Rede lecture, 264
Redundancy, 208, 304
Reimars, Paul R., 251
Reinforcement in learning, 345-346
Relativity theory, 275, 278
Renaissance, 267, 275
Renoir, Pierre Auguste, 276
Republic, The, 194
Revolution
   American, 37, 149

computer, 66
French, 1, 10, 37, 81, 148, 269
see Industrial Revolution
second Industrial, 2
Keynesian, 91
technological, 114
Triple, 67
Ricardo, David, 84–85
Riley, Bridget, 309
Rimbaud, Jean Arthur, 277
Rivett, Patrick, 184
RNA, 252
Roberts, Paul O., 129
Robots, 253–255, 371
Roebuck, John, 21
Romanticism, romantics, 268–270
 and classicalism, 272
 in Germany, 268–269, 279
Roosevelt, Franklin D., 62, 93
Roosevelt, Theodore, 297
Rose, Barbara, 298
Rosenblatt, Frank, 233
Roszak, Theodore, 376
Rote learning, 235, 238, 242
Roush, G. Jon, 265
Rousseau, Jean Jacques, 268, 319–321, 323
Royal Society, 17, 46, 268
Run, Computer, Run, 349
R.U.R. (Rossum's Universal Robots), 253–254
Ruskin, John, 264, 287, 289–290
Russell, W. M. S., 272
Russett, Cynthia E., 216

Saettler, Paul, 335
St. Augustine, 267
St. Paul, 374
Saint-Simon, Louis de Rouvroy, 148–150
St. Thomas Aquinas, 267
Salomon's House, 148
Salton, G., 250
Samuel, Arthur L., 240–242
San Francisco, 116, 131–132
San Jose, (California), 125–126
Santa Clara County, (Calif.), 169
Sapir, Edward, 257
Savas, E. S., 217–218
Savery, Thomas, 18

Say, Jean Baptiste, 85
Say's Law, 85, 91
Schaeffer, Pierre, 302
Schlager, K. J., 163
Schoenberg, Arnold, 279–280, 282
Schopenhauer, Arthur, 269
Schuetz, George, 48
Schwartz, Leonard E., 363
Science, definition of, 3
Science fiction, 360–362
Scientific Management, 151
Scientific method, 265
Scott, Howard, 151–152
Sea Island cotton, 34
Séguin, Edouard, 328
Servomechanism, 203
"Sesame Street", 341
Seurat, Georges Pierre, 276
Shaffer, J., 307
Shaft furnace, 23
Shannon, Claude, 57, 206
Sheeler, Charles, 298
Shelley, Mary Wollstonecraft, 221
Shelley, Percy B., 88
Shockley, William B., 54
Siemens, Werner, 29
Signac, Paul, 276
Silberman, Charles E., 67
Silent Spring, 140
Simon, Herbert A., 243, 252
Simulation
 gaming, 193–194
 of human thought, 243–244
 in instruction, 352
 materials handling system, 180–183
 political, 109–111
 traffic, 125
Sine-Curve Man, The, 307–308
Six Characters in Search of an Author, 279
Skinner, B. F., 343–347, 351
Slater, Samuel, 33–34
Slide rule, 43
Smith, Adam, 10, 30, 83, 196
Smuts, Jan Christian, 226
Snow, Sir Charles, 153, 264, 266, 284
Social indicators, 135–139
Social science, 6
Socialism, 150

Society for the Diffusion of Useful Knowledge (Steam Intellect Society), 85
Sociology, 150, 353
Socrates, 345
Soothsayers, 359
Sorokin, Pitirim, 371, 377
South America, 26
Speenhamland, 81, 102
Spencer, Herbert, 136, 216–217, 361
Spengler, Oswald, 377
Spinning (textiles), 11
  spinning machines, 12–17
  see textile machines
Sprague, Richard E., 147
Sputnik, 317
Statistical analysis, 186
Steam carriage, 26–27
Steam engine, 1, 10, 17–22, 24–25, 27–29
  see condenser
  corliss, 288
  high-pressure, 25–27, 37, 140
  Newcomen engine, 18–19
  parallel-motion device, 21
Steamship
  boiler explosions, 140
  Clermont, 27
Steel industry
  see Industry
Steel-making processes
  basic open-hearth, 71–72
  basic oxygen, 71–72
Steiger's "Millionaire", 52
Stella, Joseph, 298
Stephenson, George, 27
Stieglitz, Alfred, 297
Stockhausen, Karlheinz, 302
Stockton, Frank R., 191
Stockton and Darlington Railway, 27
Stored program, 54
Structural unemployment
  see unemployment
Structure of the American Economy, The, 1919–1929, 195
Structure of Scientific Revolutions, The, 271
Stuckenschmidt, H. H., 301
Sub-optimization, 198
Suhrbier, John H., 129
Sumerians, 207

Sun-and-planet gear, 22
Sundstrom, J. F., 180–183
Supersonic transport, (SST), 141–142
Suppes, Patrick, 351
Surrealism, surrealists, 296–298
Sussman, Herbert L., 289, 292
Sweden, 23, 33
Syllogism, 56
Symbolic language, 56
Symbolists, 277
Sypher, Wiley, 268, 276, 279
Systems analysis, 170–171, 185, 198
Systems approach, 5, 124, 159, 164, 167–168, 194, 197–199, 375
  see systems engineering
  see systems analysis
Systems engineering, 162–167, 198
Systems, system
  adaptive, 160–161
  classes of, 159–162
  closed, 160, 198
  control, 39, 40, 74, 202–205, 208, 216–217
  definition of, 159
  design of, 168, 194–197
  information handling, 169–179
  materials handling, 179–183
  models, 75, 125, 166, 180
  Montessori, 330, 343, 346
  Newtonian, 276
  open, 160, 162
  problems, 167, 184–194
  self-organizing, 235
  self-regulating, 199, 201
  self-reproducing, 251–253
  see systems approach techniques, 185–194
  telephone, 65, 163–164, 180
  traffic control, 126–128
  transportation, 74, 204
Szilard, Leo, 152

Table of differences, 47
Tableau Économique, 195
Taylor, Frederick W., 151
Teaching machine(s), 343–348
Technocracy, 151–152
Technological elite, 107, 147–154

Technological forecasting, 139, 362–374
  by consensus, 367–369
  cross-impact matrices, 369–370
  see Delphi method
  envelope curve, 366
  by scenario, 370–373
  Sigmoid curve, 364–365
  trend correlation, 367
  trend projections, 363–367
Technological Man, 112
Technological Society, 77, 100, 112, 356, 375
Technology
  and art, 262–312
  definition of, 3
  and education, 315–356
  energy-intensive, 66
  information-intensive, 66
  progress of, 108
  and social change, 2–9, 66, 106, 135, 263
Technology assessment, 108, 139–145, 374
Telecommunications, 208–209
Television, 108, 111–112
Teller, Edward, 154
Textile machines
  mule, 10, 16–17
  spinning jenny, 10, 13–16
  water frame, 10, 15
Textiles, 11
Thales, 256
Theobald, Robert, 101
Theories of learning, 318–334, 345, 354
  freedom vs. authority, 321
  operant conditions, 344–345
Thermodynamics, 212
  2nd Law of, 212, 259, 264
Thompson, Francis, 274
Thomson, J. J., 276
Thorndike, Edward L., 330, 346
Tic-tac-toe, 237–239
Toffler, Alvin, 358
Toulouse-Lautrec, Henri de, 279
Townsend, William, 216
Toynbee, Arnold, 377
Trade(s) Unions, 88, 90
Traffic control
  by computer, 126–129
Traffic engineering, 124–125

Transistor, germanium, 54
Transportation, 74, 106
  financing of, 123
  technology and, 124–133
  terminals, 121–122
  travel patterns, 120–121
  urban, 114–135
Tredgold, Thomas, 162
Trevithick, Richard, 25–26
Truffaut, Francois, 328
Truman, Harry S., 62
Truth tables, 57
Turing, Alan M., 229, 244
Twain, Mark, 287
Twentieth Century Music, 302
Two Cultures, The, 263–266, 283–284
Two-state code, 209–210
Tyndall, John, 216

Underdeveloped nations, 115
Unemployment, 3, 4
  in education, 355
  of engineers and scientists, 97
  in England, 38, 78–79, 92
  structural vs. aggregate demand, 95–96
  in U.S., 67–70, 76–77, 91–98
  in Western nations, 85
United States, 2, 34–35, 37, 39, 115, 116
  Air Force, 158
  art in the, 285–288
  Bureau of the Census, 52
  Bureau of Standards, 141
  Constitution, 146
  early industrialization in, 89–90
  Food and Drug Administration, 141
  Navy, 140
  transportation problem in, 118
University of Glasgow, 19–20
University of Illinois, 304
University of Pennsylvania
  Moore School of Engineering, 53
Ure, Andrew, 87
Utilitarianism, 196
Utopia, 360
Utopia(s), 148

Van Gogh, Vincent, 279
Van Musschenbroek, Pieter, 28

Index

Varèse, Edgar, 302
Veblen, Thorstein, 150–151
Verhaeghe, Oscar, 225
Verlaine, Paul, 277
Verne, Jules, 361
Vesalius, 275
Victorian art, 289–292
Victorian doctrine, 264
Videotape recorders, 342
Vietnam, war in, 94
Visual arts, abstraction in, 279–280, 293, 299, 309
Vogt, William, 102
Volta, Allesandro, 28
Voltaire, 148, 360
Von Guericke, Otto, 28
Von Neumann, John, 54–55, 191, 208, 233, 251–253
Vonnegut, Kurt, 361

Wagner, Richard, 281–282
War gaming, 193
Ward, Lester Frank, 216–217
Water power, 15, 25
Watt, James, 10, 19–22, 25, 28, 202, 272
Wealth of Nations, 10, 83
Weaver, W. T., 369
Weaving, 11
Webern, Anton, 282
Welfare, 77, 100–102
Wells, H. G., 361
Wesley, John, 28
Westin, Alan, 146

Westminster Abbey, 26
What is Cybernetics?, 211
Whistler, James McNeil, 287
White-collar workers, 68–70
Whitman, Walt, 287
Whitney, Eli, 35–37, 73
Whorf, Benjamin L., 257
Widrow, B., 235
Wiener, Anthony J., 370–372
Wiener, Norbert, 5, 31, 202, 205, 206, 214–215, 223, 237
Wiesner, Jerome, 154
Wild boy of Aveyron, The, 328
Wilkinson, John, 24–25
Williams, Raymond, 263
Wills, Gordon, 365
Wolcott, Oliver, 36
Wood, Grant, 298
Woolwich Arsenal, 32
Workhouse, 80, 82
Working class, 62, 85
 blue-collar, 62
Working poor, 66
World War I, 92, 151
World War II, 61–63, 71, 93, 106, 144, 153, 158, 163, 201, 337, 359
 Battle of Britain, 165, 167
Worringer, Wilhelm, 315
Wyatt, John, 13

York, Herbert, 154

Zola, Emile, 276
Zoning, 122

TITLES IN THE PERGAMON UNIFIED ENGINEERING SERIES

Vol. 1.	W. H. DAVENPORT/D. ROSENTHAL – *Engineering: Its Role and Function in Human Society*
Vol. 2.	M. TRIBUS – *Rational Descriptions, Decisions and Designs*
Vol. 3.	W. H. DAVENPORT – *The One Culture*
Vol. 4.	W. A. WOOD – *The Study of Metal Structures and Their Mechanical Properties*
Vol. 5.	M. SMYTH – *Linear Engineering Systems: Tools and Techniques*
Vol. 6.	L. M. MAXWELL/M. B. REED – *The Theory of Graphs: A Basis for Network Theory*
Vol. 7.	W. R. SPILLERS – *Automated Structural Analysis: An Introduction*
Vol. 8.	J. J. AZAR – *Matrix Structural Analysis*
Vol. 9.	S. SEELY – *An Introduction to Engineering Systems*
Vol. 10.	D. T. THOMAS – *Engineering Electromagnetics*
Vol. 12.	S. J. BRITVEC – *The Stability of Elastic Systems*
Vol. 13.	M. NOTON – *Modern Control Engineering*
Vol. 14.	B. MORRILL – *An Introduction to Equilibrium Thermodynamics*
Vol. 15.	R. PARKMAN – *The Cybernetic Society*

THE ONE CULTURE
Pergamon Unified Engineering Series, Volume 3
By **William H. Davenport**, *Harvey Mudd College, Claremont, California*

It has now been a little over a decade since C.P. Snow wrote his famous work, *The Two Cultures and the Scientific Revolution,* in which this eminent British novelist and scientist describes the dangerous split existing between our literary and scientific communities. Using Lord Snow's controversial premise, both as a departure point and a springboard for his own philosophy, William H. Davenport in **The One Culture** re-examines the state of culture — particularly American culture — in this Age of Technology. He proposes that a turning point in society's polarization process has now been reached and argues that science and technology both are integral to mankind's culture and survival.

PEOPLE AND INFORMATION
Pergamon General Psychology Series, Volume 6
Edited by **Harold B. Pepinsky**, *The Ohio State University*

Featuring contributions from a carefully selected group of experts, this book provides vigorous discussion on the information revolution and its crucial personal and societal implications. Noted authorities, including George Klare, Karl Weick, Eldridge Adams, Arne Walle and James Robinson, raise important new questions about information processing and cover a broad spectrum of recent advances in both ideas and practices in this field.

THE PRACTICE OF MANAGERIAL PSYCHOLOGY
Pergamon Management and Business Series, Volume 1
By **Andrew J. DuBrin**, *Rochester Institute of Technology, New York*

Innovative conceptual schemes are presented to show the practitioner, student, and manager what the underlying factors are that determine if intervention will be helpful, meaningless, or harmful to organizations. The Managerial Psychology Matrix, for example, specifies the proper conditions for applying such techniques as sensitivity training, performance appraisal, psychological assessment, team development meetings, organizational analysis, and superior-subordinate counseling.

Managerial psychology, according to DuBrin's conception, is the application of psychological concepts, techniques, approaches, methods, and interventions toward increasing the effectiveness of managers and organizations. Clinical-industrial, personnel, and organizational psychology constitute the three fields upon which managerial psychology is based.

The author's experience as a consultant provides many of the illustrations and examples presented in this book. Practice, research, and theory provide the knowledge base for the innovations and new conceptualizations offered.

ENGINEERING: ITS ROLE AND FUNCTION IN HUMAN SOCIETY
Pergamon Unified Engineering Series, Volume 1
Edited by **William H. Davenport** and **Daniel Rosenthal**, *University of California, Los Angeles*

This book is primarily an anthology of readings bridging the gap between the engineer and the humanist. The thirty-nine contributors review the humanizing of the engineering curricula, the progress of man in his application of machines, the effects of humanizing the engineer and the need for engineering in the solution of society's problems.

SOCIOLOGY AND SOCIAL WORK
Perspectives and Problems
By Brian J. Heraud

From recent reviews . . .

"A British text, refreshingly free from jargon, that attempts to bring the two fields together by discussing what has kept them apart. The author describes the development of social work, its failure to become a profession, and its social function within the context of a sociological analysis of the structure of society. The final emphasis is on the possibilities and problems ahead for greater interaction between the two disciplines. Though written for the British arena, the book cites many American references and has applicability to the American scene."
—CHILD WELFARE

"This is an important, superbly written book. . . . It represents a major contribution to the sociology of social work."
—NEW SOCIETY

THE CHALLENGE OF CHANGE
Report on a Conference on Technological Change and Human Development at Jerusalem, 1969
By Sir Trevor Evans, *Foundation on Automation and Employment Ltd., England*

Presenting the main themes and problems facing all countries due to the inevitable expansion of technology, this volume is a report of a conference on Technological Change and Human Development held in Jerusalem in 1969. The author, noted former Industrial Correspondent of the British *Daily Express*, was one of the many distinguished employers, trade unionists, public servants, and academics at the conference which drew its representatives from the United States, the United Kingdom, and Israel. Among the subjects of special interest in this volume are the growing significance of international companies, the future of collective bargaining, the development of social and economic equality for ethnic minorities in developed countries, and the role of technology in raising economic standards in developing countries.

PERGAMON JOURNALS OF RELATED INTEREST . . .

AUTOMATICA
International Journal of ENGINEERING SCIENCE
Journal of the FRANKLIN INSTITUTE
INFORMATION STORAGE AND RETRIEVAL
LONG RANGE PLANNING
PATTERN RECOGNITION
SOCIAL SCIENCE AND MEDICINE
SOCIO-ECONOMIC PLANNING SCIENCES